Lecture Notes in Computer Science 15762

Founding Editors

Gerhard Goos
Juris Hartmanis

Editorial Board Members

Elisa Bertino, *Purdue University, West Lafayette, IN, USA*
Wen Gao, *Peking University, Beijing, China*
Bernhard Steffen, *TU Dortmund University, Dortmund, Germany*
Moti Yung, *Columbia University, New York, NY, USA*

The series Lecture Notes in Computer Science (LNCS), including its subseries Lecture Notes in Artificial Intelligence (LNAI) and Lecture Notes in Bioinformatics (LNBI), has established itself as a medium for the publication of new developments in computer science and information technology research, teaching, and education.

LNCS enjoys close cooperation with the computer science R & D community, the series counts many renowned academics among its volume editors and paper authors, and collaborates with prestigious societies. Its mission is to serve this international community by providing an invaluable service, mainly focused on the publication of conference and workshop proceedings and postproceedings. LNCS commenced publication in 1973.

Guido Tack
Editor

Integration of Constraint Programming, Artificial Intelligence, and Operations Research

22nd International Conference, CPAIOR 2025
Melbourne, VIC, Australia, November 10–13, 2025
Proceedings, Part I

Editor
Guido Tack
Monash University
Clayton, VIC, Australia

ISSN 0302-9743 ISSN 1611-3349 (electronic)
Lecture Notes in Computer Science
ISBN 978-3-031-95972-1 ISBN 978-3-031-95973-8 (eBook)
https://doi.org/10.1007/978-3-031-95973-8

© The Editor(s) (if applicable) and The Author(s), under exclusive license
to Springer Nature Switzerland AG 2025

This work is subject to copyright. All rights are solely and exclusively licensed by the Publisher, whether the whole or part of the material is concerned, specifically the rights of translation, reprinting, reuse of illustrations, recitation, broadcasting, reproduction on microfilms or in any other physical way, and transmission or information storage and retrieval, electronic adaptation, computer software, or by similar or dissimilar methodology now known or hereafter developed.
The use of general descriptive names, registered names, trademarks, service marks, etc. in this publication does not imply, even in the absence of a specific statement, that such names are exempt from the relevant protective laws and regulations and therefore free for general use.
The publisher, the authors and the editors are safe to assume that the advice and information in this book are believed to be true and accurate at the date of publication. Neither the publisher nor the authors or the editors give a warranty, expressed or implied, with respect to the material contained herein or for any errors or omissions that may have been made. The publisher remains neutral with regard to jurisdictional claims in published maps and institutional affiliations.

This Springer imprint is published by the registered company Springer Nature Switzerland AG
The registered company address is: Gewerbestrasse 11, 6330 Cham, Switzerland

If disposing of this product, please recycle the paper.

Preface

This book contains the proceedings of the 22nd International Conference on the Integration of Constraint Programming, Artificial Intelligence, and Operations Research (CPAIOR 2025). The conference was held in Melbourne, Australia, November 10–13, 2025. The conference organizers acknowledge that the conference was held on the unceded traditional lands of the Wurundjeri people of the Kulin Nation. We pay our respects to their Elders past, present and emerging. The conference was co-located with the 35th International Conference on Automated Planning and Scheduling (ICAPS) and the 22nd International Conference on Principles of Knowledge Representation and Reasoning (KR).

The conference received a total of 80 submissions of original unpublished research papers. After an initial screening, 68 papers were sent out to the Program Committee for single-blind peer review, with each paper receiving at least three reviews. The reviewing phase was followed by an author response period. The Program Committee then discussed each paper and made a recommendation for acceptance or rejection. At the end of this process, 32 papers were accepted for presentation at the conference and publication in these proceedings. In addition, a number of extended abstracts were accepted for presentation at the conference, but not included in these proceedings. The conference featured a masterclass and several joint invited talks that covered topics of interest at the intersection of Constraint Programming, Artificial Intelligence, Operations Research, Planning and Scheduling, and Knowledge Representation. It was preceded by a Summer School, which took place in San Remo, Victoria, Australia, November 3–7, 2025.

Of the accepted papers, the paper "Multitask Representation Learning for Mixed Integer Linear Programming" by Junyang Cai, Taoan Huang and Bistra Dilkina was selected for the Best Paper Award. The paper "A Dynamic Programming Approach for the Job Sequencing and Tool Switching Problem" by Emma Legrand, Vianney Coppé, Daniele Catanzaro and Pierre Schaus was selected for the Best Student Paper Award. The awards were selected by a sub-committee of seven members of the Program Committee.

I would like to thank our publicity chair, Jip Dekker, the sponsorship chair, Charles Gretton, and the masterclass chair, Buser Say, as well as the entire local organizing team for their contributions to making this conference a success. The conference would not have been possible to organise without the help of the local chairs of ICAPS, Nir Lipovetzky and Sebastian Sardina, whose hard work enabled us to run the two conferences as a single, joint event. Finally, I would like to thank our sponsors for their generous financial support.

May 2025   Guido Tack

Organization

Program and Conference Chair

Guido Tack   Monash University, Australia

Publicity Chair

Jip Dekker   Monash University, Australia

Masterclass Chair

Buser Say   Monash University, Australia

Sponsorship Chair

Charles Gretton   Australian National University, Australia

Program Committee

Chris Beck	University of Toronto, Canada
Nicolas Beldiceanu	IMT Atlantique, LS2N, France
Armin Biere	Universität Freiburg, Germany
Merve Bodur	University of Edinburgh, UK
Mats Carlsson	RISE Research Institutes of Sweden, Sweden
Margarida Carvalho	Université de Montréal, Canada
Margarita P. Castro	Pontificia Universidad Catolica de Chile, Chile
Andre Augusto Cire	University of Toronto, Canada
Simon de Givry	INRAE, France
Sophie Demassey	Mines Paris - PSL, France
Emir Demirović	Delft University of Technology, The Netherlands
María Andreína Francisco Rodríguez	Uppsala University, Sweden
Ambros Gleixner	HTW Berlin, Germany
Emmanuel Hebrard	LAAS-CNRS, France

John Hooker	Carnegie Mellon University, USA
Matti Järvisalo	University of Helsinki, Finland
Serdar Kadioglu	Brown University, USA
George Katsirelos	MIA Paris, INRAE, France
Zeynep Kiziltan	University of Bologna, Italy
Arnaud Lallouet	Huawei Technologies, France
Thi Thai Le	Zuse Institute Berlin, Germany
Pierre Le Bodic	Monash University, Australia
Jimmy H.M. Lee	The Chinese University of Hong Kong, China
Michele Lombardi	University of Bologna, Italy
Ciaran McCreesh	University of Glasgow, UK
Laurent Michel	University of Connecticut, USA
Nysret Musliu	Technische Universität Wien, Austria
Margaux Nattaf	Institut National Polytechnique de Grenoble, France
Barry O'Sullivan	University College Cork, Ireland
Sophie N. Parragh	Johannes Kepler University Linz, Austria
Justin Pearson	Uppsala University, Sweden
Laurent Perron	Google France, France
Gilles Pesant	École Polytechnique de Montréal, Canada
Claude-Guy Quimper	Université Laval, Canada
Jean-Charles Regin	Université Côte d'Azur, France
Louis-Martin Rousseau	École Polytechnique de Montréal, Canada
Domenico Salvagnin	University of Padua, Italy
Pierre Schaus	UCLouvain, Belgium
Thiago Serra	University of Iowa, USA
Paul Shaw	IBM, France
Mohamed Siala	INSA Toulouse & LAAS-CNRS, France
Helmut Simonis	University College Cork, Ireland
Christine Solnon	CITI, INSA Lyon/Inria, France
Kostas Stergiou	University of Western Macedonia, Greece
Peter J. Stuckey	Monash University, Australia
Michael Trick	Carnegie Mellon University, USA
Willem-Jan van Hoeve	Carnegie Mellon University, USA
Hélène Verhaeghe	UCLouvain, Belgium
Petr Vilím	OptalCP, Czech Republic
Mark Wallace	Monash University, Australia
Roland Yap	National University of Singapore, Singapore
Neil Yorke-Smith	Delft University of Technology, The Netherlands

Additional Reviewers

Valentin Antuori
Christian Artigues
Andrea Borghesi
Jip J. Dekker
Julien Ferry
Ernest Foussard
Maria Garcia de la Banda
Mohammed Ghannam
Luca Giuliani
Arthur Gontier

Richard Hua
Christoph Jabs
Edward Lam
Matthew J. McIlree
Xiao Peng
Angelo Quarta
Buser Say
Gaetano Signorelli
Mate Soos

Contents – Part I

Optimized Scheduling of Medical Appointment Sequences Using
Constraint Programming 1
  George Assaf, Sven Löffler, and Petra Hofstedt

An Integrated Optimisation Method for Aluminium Hot Rolling 17
  *Ioannis Avgerinos, Apostolos Besis, Ioannis Mourtos,
  Athanasios Psarros, Stavros Vatikiotis, and Georgios Zois*

Determining the Most Promising Selective Backbone Size for Partial
Knowledge Compilation 34
  Andrea Balogh, Guillaume Escamocher, and Barry O'Sullivan

Leveraging Quantum Computing for Accelerated Classical Algorithms
in Power Systems Optimization 52
  Rosemary Barrass, Harsha Nagarajan, and Carleton Coffrin

Hybridizing Machine Learning and Optimization for Planning Satellite
Observations 68
  Romain Barrault, Cédric Pralet, Gauthier Picard, and Eric Sawyer

Algorithm Configuration in Sequential Decision-Making 86
  *Luca Begnardi, Bart von Meijenfeldt, Yingqian Zhang,
  Willem van Jaarsveld, and Hendrik Baier*

Self-supervised Penalty-Based Learning for Robust Constrained
Optimization 103
  Wyame Benslimane and Paul Grigas

Revisiting Pseudo-Boolean Encodings from an Integer Perspective 113
  Hendrik Bierlee, Jip J. Dekker, and Peter J. Stuckey

Multi-task Representation Learning for Mixed Integer Linear Programming 134
  Junyang Cai, Taoan Huang, and Bistra Dilkina

Breaking the Symmetries of Indistinguishable Objects 152
  Özgür Akgün, Mun See Chang, Ian P. Gent, and Christopher Jefferson

Breaking Symmetries from a Set-Covering Perspective 169
  Michael Codish and Mikoláš Janota

Modeling and Solving the Generalized Test Laboratory Scheduling Problem ... 188
  Philipp Danzinger, Tobias Geibinger, Florian Mischek, and Nysret Musliu

Parallelising Lazy Clause Generation with Trail Sharing 205
  Toby O. Davies, Frédéric Didier, Laurent Perron, and Peter J. Stuckey

Learning Primal Heuristics for 0–1 Knapsack Interdiction Problems 222
  Luca Ferrarini, Stefano Gualandi, Letizia Moro, and Axel Parmentier

Bounded-Error Policy Optimization for Mixed Discrete-Continuous
MDPs via Constraint Generation in Nonlinear Programming 239
  Michael Gimelfarb, Ayal Taitler, and Scott Sanner

Minimising Source-Plate Swaps for Robotised Compound Dispensing
in Microplates 256
  *Ramiz Gindullin, María Andreína Francisco Rodríguez,
  Brinton Seashore-Ludlow, and Ola Spjuth*

Author Index 275

Contents – Part II

Combining Constraint Programming and Metaheuristics for Aircraft
Maintenance Routing with a Distribution Objective 1
    Ida Gjergji, Lucas Kletzander, Hendrik Bierlee, Nysret Musliu,
    and Peter J. Stuckey

Reducing Income Variability in Natural Resource Portfolios via Integer
Programming 18
    Laura Greenstreet, Qinru Shi, Marc Grimson, Franz W. Simon,
    Suresh A. Sethi, Carla P. Gomes, Andrea Lodi, and David B. Shmoys

Analyzing the Numerical Correctness of Branch-and-Bound Decisions
for Mixed-Integer Programming 35
    Alexander Hoen and Ambros Gleixner

LLMs for Cold-Start Cutting Plane Separator Configuration 51
    Connor Lawless, Yingxi Li, Anders Wikum, Madeleine Udell,
    and Ellen Vitercik

A Dynamic Programming Approach for the Job Sequencing and Tool
Switching Problem 70
    Emma Legrand, Vianney Coppé, Daniele Catanzaro, and Pierre Schaus

Time-Dependent Orienteering for High Altitude UAVs to Monitor
Greenhouse Gases: Mixed Integer Programming vs. Monte Carlo Tree
Search 86
    Richard Levinson, Vinay Ravindra, Jeremy Frank, and Meghan Saephan

A Column Generation Heuristic for Multi-depot Electric Bus Scheduling 103
    Yoann Sabatier Montanaro, Thomas Jacquet, Quentin Cappart,
    and Guy Desaulniers

On the Efficiency of Algebraic Simplex Algorithms for Solving MDPs 119
    Dibyangshu Mukherjee and Shivaram Kalyanakrishnan

Reinforcement Learning-Based Heuristics to Guide Domain-Independent
Dynamic Programming 137
    Minori Narita, Ryo Kuroiwa, and J. Christopher Beck

Acquiring and Selecting Implied Constraints with an Application
to the BinSeq and Partition Global Constraints 155
  Jovial Cheukam Ngouonou, Ramiz Gindullin, Claude-Guy Quimper,
  Nicloas Beldiceanu, and Rémi Douence

Integer and Constraint Programming for the Offline Nanosatellite Partition
Scheduling Problem 173
  Julien Rouzot, Mickaël Pereira, Christian Artigues, Romain Boyer,
  Frédéric Camps, Philippe Garnier, Emmanuel Hebrard, and Pierre Lopez

Accelerated Discovery of Set Cover Solutions via Graph Neural Networks 191
  Zohair Shafi, Benjamin A. Miller, Tina Eliassi-Rad,
  and Rajmonda S. Caceres

Constrained Machine Learning Through Hyperspherical Representation 209
  Gaetano Signorelli and Michele Lombardi

PySCIPOpt-ML: Embedding Trained Machine Learning Models
into Mixed-Integer Programs 218
  Mark Turner, Antonia Chmiela, Thorsten Koch, and Michael Winkler

Satellite Communication Resources Management in a Earth Observation
Federation of Constellations 235
  Henoïk Willot, Jean-Loup Farges, Gauthier Picard, and Philippe Pavero

Shaping Reward Signals in Reinforcement Learning Using Constraint
Programming 252
  Chao Yin, Quentin Cappart, and Gilles Pesant

Author Index 271

Optimized Scheduling of Medical Appointment Sequences Using Constraint Programming

George Assaf[(✉)], Sven Löffler, and Petra Hofstedt

Programming Languages and Compiler Construction Chair, Brandenburg University of Technology Cottbus-Senftenberg, Konrad-Wachsmann Allee 5, Cottbus, Germany
{george.assaf,sven.loeffler,hofstedt}@b-tu.de

Abstract. We propose a novel constraint-based model to efficiently tackle the medical appointment sequence scheduling problem (MASSP), inspired by a real-world problem at *Charité – Universitätsmedizin Berlin*. In many practical medical scenarios, scheduling a sequence of appointments, rather than a single appointment, has become increasingly essential for patients undergoing multi-stage treatments. The goal of the MASSP is to identify a set of medical resources with sufficient, consecutive, and available time slots in their calendars to create a sequence of appointments for effectively managing a treatment plan. The problem comprises various constraints, including the availability of both the intended patient and required medical resources, as well as the time and resource dependencies among the individual appointments that constitute the sequence. To address this, we formulate the problem as a constraint optimization problem ($\mathcal{COP}$) that not only captures the basic constraints of the MASSP but also optimizes resource assignment to ensure a fair workload distribution within the medical facility. The results of our experiments demonstrate that the model performs effectively under diverse conditions, which confirms the utility and robustness of the proposed model in optimizing resource allocation and ensuring equitable workload distribution.

Keywords: Medical appointment sequence scheduling problem · Constraint programming · Optimized scheduling · $\mathcal{COP}$

1 Introduction

As healthcare facilities expand and medical treatments grow more complex, there is an increasing need for sophisticated appointment scheduling strategies that can accommodate multiple visits for individual patients. Many patients with chronic or multifaceted health conditions require coordinated, sequential appointments across various departments, making efficient scheduling essential to enhance patient satisfaction, improve equity in access, reduce wait times, and control healthcare costs [1,2]. In this context, the medical appointment sequence scheduling problem (MASSP) is particularly critical for patients requiring frequent medical services due to, e.g. their complex health conditions. The goal of the MASSP is to find a set of suitable medical resources such as specialists with specific medical expertise or specialized medical equipment (according to given specific medical needs) and picking up time slots from the calendars of these resources under a set of constraints to schedule required appointments, each serving an individual medical purpose within a patient's treatment plan.

Scheduling a sequence of medical appointments requires coordinating a set of resources including physicians of different medical specializations, rooms, and other medical equipment. Each medical resource is linked to a calendar detailing both available and occupied time slots. Figure 1 presents a scheduled appointment sequence comprising three individual appointments. In this sequence, the following intuitive constraints must be satisfied:

- Each individual appointment requires a set of medical resources, all of which must be available simultaneously for the entire duration of the appointment.
- Each appointment within the sequence has a pre-determined duration, subject to a treatment plan.
- No two appointments within a sequence can occur simultaneously; instead, each appointment must be scheduled at a distinct time.
- It is prohibited to assign the same resource more than once when multiple instances of the same resource type are required. For example, if two cardiologists are needed, they must be distinct individuals.

In addition, the following constraints, which are driven by medical needs, must also be satisfied whenever required:

- Individual appointments must be scheduled in chronological order according to the intended medical plan, e.g. blood test must occur before CT Scan.
- A minimum time interval, measured in days or hours, must be maintained between appointments to, e.g. ensure sufficient time for the patient to prepare for the next stage of treatment. For example, there must be at least 5 days between *Appointment 1* and *Appointment 2*.
- A maximum time interval between appointments must also be adhered to. For instance, the gap between *Appointment 2* and *Appointment 3* must not exceed three days to perform *CT Scan* (in our example).
- The same resource may be required for multiple appointments to ensure the continuity of patient care. For example, both *Appointment 2* and *Appointment 3* may require the presence of the same specialist.

In modern healthcare, managing the scheduling of patient appointments is essential for enhancing patient experience and optimizing resource use. Our medical partner, *Charité – Universitätsmedizin Berlin*, Europe's largest university hospital, has highlighted that efficiently coordinating a sequence of appointments for a single patient not only significantly improves patient satisfaction by reducing wait times and minimizing back-and-forth visits but also enhances operational efficiency within the hospital. Current scheduling methods, which often

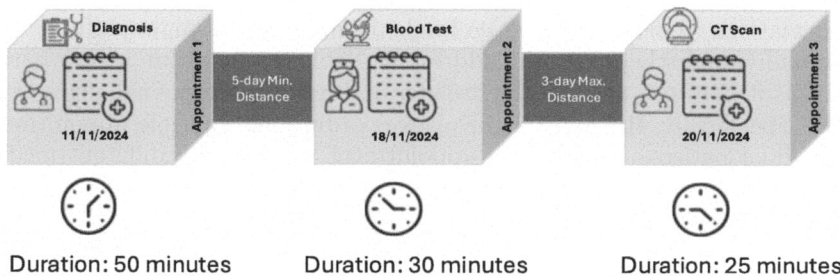

Fig. 1. A scheduled sequence of three medical appointments.

rely on spreadsheets or other basic tools, fall short in addressing the complexity of multi-appointment sequences, leading to suboptimal use of medical resources. This motivates the need for a sophisticated, constraint-based scheduling model capable of meeting the diverse requirements of the MASSP, as constraint programming ($\mathcal{CP}$) [3,6] is often seen as the *holy grail* of scheduling and optimization in complex domains like healthcare. To formalize the problem, we are going to utilize a constraint optimization problem ($\mathcal{COP}$) [5] as an approach to model as well as optimize the MASSP.

1.1 Related Work

$\mathcal{CP}$ has been widely utilized for the purpose of solving combinatorial problems including appointment scheduling and planning problems within various fields [4, 8].

Zhang et al. in [14] modeled the university timetabling problem using constraint programming to develop a fixed schedule for a teaching semester, assigning resources such as professors and laboratories to classes without conflicts or overlaps. While the university timetabling problem shares key similarities with the medical appointment sequence scheduling problem, such as allocating limited resources under strict constraints, the two problems differ significantly in several aspects. The nature of the tasks, the types of resources being managed, and their schedules are more varying and greatly depend on patients' conditions in MASSP. Specifically, the MASSP involves finding sequence-dependent appointments, often based heavily on patients' medical conditions.

Next, Said et al. in [10] developed a constraint model to solve the nurse scheduling problem, which involves assigning shifts to nurses under several constraints such as ensuring enough nurses per shift as well as considering individual nurse preferences. Compared to the MASSP, each nurse typically operates under a fixed schedule with predetermined shifts across a given period.

Further, Topaloglu et al. in [12] introduced a mixed-integer programming (MIP) model for scheduling residents' duty hours in the graduate medical education, focusing on adherence to residency program regulations, including on-call nights, days off, rest periods, and total work hours. This problem aims to minimize deviations from desired service levels during day and night shifts by configuring individual schedules. The addressed problem falls within a very narrow context, as it does not address the broader and more complex challenges of the MASSP. Specifically, the MASSP involves coordinating multiple appointments for individual patients, where each appointment may require different medical resources and may depend on the timing or completion of other appointments within the sequence.

Besides, Vermeulen et al. in [13] developed an adaptive model designed to dynamically allocate high-cost medical resources, such as CT scans, among different patient groups by considering factors like expected arrival times. The model enhances resource utilization by continually adjusting the resource schedule in real-time. Although this adaptive model focuses primarily on dynamically allocating capacity for a specific resource, it overlooks several aspects of scheduling a sequence of medical appointments.

Despite the substantial body of literature on scheduling problems, to the best of our knowledge, no existing study specifically addresses the scheduling of a sequence of medical appointments. This gap is particularly evident in the context of the MASSP, which is inherently complex due to the involvement of numerous medical resources, each with its own calendar spanning multiple days or even months. In the following, we formulate the problem as a $\mathcal{COP}$, encoding various medical requirements and incorporating an objective optimization function. The objective function ensures a balanced workload among medical staff when assigning appointments. It prevents overburdening specific staff members by considering their current workload, which is measured by the total time of scheduled appointments. This approach assigns resources to appointments more fairly, avoiding the repeated selection of the same resources while others remain underutilized.

2 Problem Description

In this section, we describe the decision variables required to construct the $\mathcal{COP}$ problem, which will be constrained to capture the diverse requirements of the MASSP. We begin by outlining the domains to which these decision variables will be assigned.

2.1 Problem Domains

The MASSP aims to schedule a sequence of appointments comprising medical resources that are associated with calendars sketching the scheduled appointments of the associated resources, e.g. a physicians. Each individual appointment occupies a specific number of time slots corresponding to its duration. These time slots must be available within the calendars of the resources such as specialists and medical equipment assigned to that appointment. Hence, we identify two domains: time slot domain and resource domain.

Time Slot Domain. Each resource calendar consists of a set of time slots of predefined length. Figure 2 presents a resource calendar spanning over ten days starting from *Monday 04/11/2024* to *Wednesday 13/11/2024*. The calendar is structured as a 2D matrix, whose rows and columns, represent time slots, and scheduling days, respectively. The number N of time slots per day is obtained using Eq. 1.

$$N = \frac{T_{\text{end}} - T_{\text{start}}}{D}, \quad (1)$$

where T_{end} and T_{start} denote latest and earliest times of the office hours (in Minutes) for the associated resource, respectively, and D^1 gives the duration of each time slot. For example, the resource calendar given in Fig. 2 comprises 24 time slots per day, spanning from 08:00 to 14:00, each with a duration of 15 min.

Each time slot is assigned a unique identifier to distinguish the day/date to which it belongs. A time slot identifier is calculated using Eq. 2.

$$T_{id} = \begin{cases} Day \times N + i & \text{if slot } i \text{ is free} \\ -1 & \text{otherwise} \end{cases} \quad (2)$$

Here, T_{id} represents the time slot identifier[2], Day denotes the day index (starting from 0), N is the total number of slots per day (see Eq. 1), and i is the row index of the time slot. It is worth noting that all slots indicating resource unavailability due to reasons such as previously scheduled appointments, vacations or public holidays are assigned negative values[3]. For example, the time slots covering the dates *09/11/2024* and *10/11/2024* are assigned -1, due to factors such as the weekend.

By using time slot identifiers, a specific date within a resource calendar can be obtained using Eq. 3.

$$d_t = \left\lfloor \frac{T_{id}}{N} \right\rfloor \quad (3)$$

Here, d_t represents the index of the date column corresponding to the slot identifier T_{id}. For example, if $T_{id} = 75$, the date is calculated as $\lfloor \frac{75}{24} \rfloor = 3$ (using zero-based indexing), which corresponds to the fourth column, giving the date *07/11/2024*.

Further useful information can be obtained from a slot identifier is the corresponding weekday, as shown in Eq. 4.

$$d_w = \left\lfloor \frac{T_{id}}{N} \right\rfloor \bmod 7 \quad (4)$$

Here, d_w represents the calendar day associated with the slot identifier T_{id}. Note that the *mod* operation is applied, given that a week consists of 7 days. For

---
[1] The time slot duration D remains constant across all resource calendars.
[2] The time slot identifiers are calculated for all resource calendars.
[3] When a booking is made, the identifiers for the corresponding time slots are assigned a value of -1.

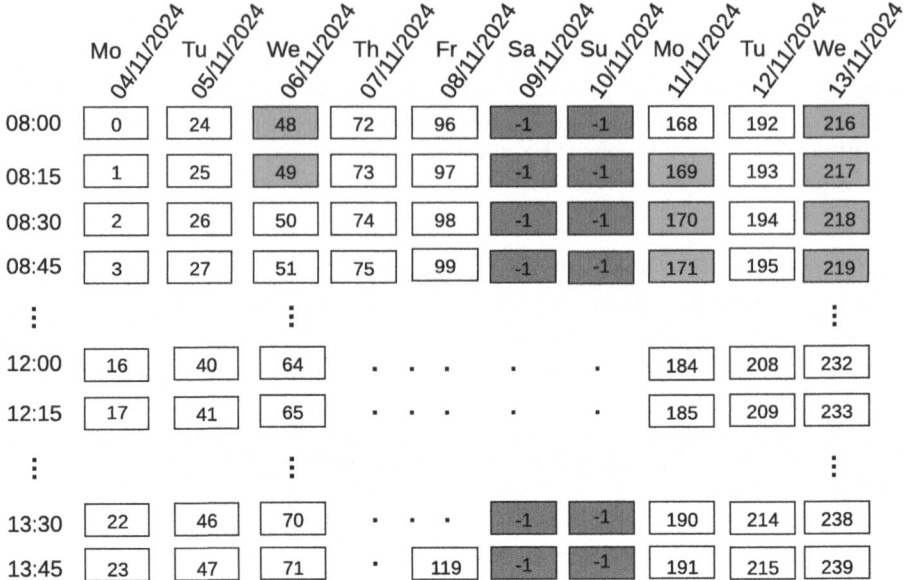

Fig. 2. A resource calendar sketching time slot domain of a resource within a medical facility. Time slots marked in red are invalid slots and cannot be used for scheduling. The green time slots reflect scheduled appointments for the resource. Note that the time slots are read by their starting time, e.g. 8:00 instead of 8:00–08:15. (Color figure online)

instance, for $T_{id} = 75$, the day is calculated as $\lfloor \frac{75}{24} \rfloor \mod 7 = 3 \mod 7 = 3$, which corresponds to *Thursday*. Crucially, to reveal the time corresponding to a given time slot identifier, we apply Eq. 5.

$$t_s = T_{id} \mod N \tag{5}$$

Here, t_s gives the time slot index corresponding to the slot identifier T_{id}. For instance, the time slot index of $T_{id} = 75$ is $75 \mod 24 = 3$ (zero-based indexing) yielding the fourth row index, which corresponds to 08:45.

To gain insight into the scheduled appointments for all resources, a global calendar is created from individual resource calendars, outlining the schedules of all resources within the medical facility. The start time slot of the global calendar is determined by the earliest time slot across all resource calendars, while the end time slot is set by the latest time slot among all involved resource calendars. To access a scheduled appointment for a resource within the global calendar, the index of the start slot within the global calendar is determined using Eq. 6.

$$S_g = \frac{S_l - G_s}{D} \tag{6}$$

Here, S_g denotes the start slot index of the appointment within the global calendar, S_l is the appointment start slot index within the local calendar, i.e. resource calendar, and G_s is the index of start time slot of the global calendar. For example, assuming $D = 15$ (minutes), if the start of the global calendar is at 7:30 (slot identifier 0 in the global calendar), and the start of a local calendar is at 8:00, then the identifier of the start slot in global calendar would be 2.

Resource Domain. Since appointments are assigned to specific resources as needed, each resource is identified by a unique identifier. Therefore, the values of this domain are the identifiers of the resources within medical facility. Medical resources are categorized into four main types: *medical staff*, including physicians with one or more specific specializations such as cardiologists, and neurologists, *medical equipment*, designed for specific purposes, e.g. MRI machines, *nurses*, and *rooms*, tailored to meet various medical needs, e.g. surgical operations. Based on this classification, resource identifiers are categorized into sub-domains, each comprises resource identifiers belonging to the same medical specialization or equipment type. For example, the sub-domain of cardiologists comprising n cardiologists is given as $D_c = (C_1, C_2, \ldots, C_n)$. Similarly, the sub-domain of medical equipment of type, e.g. *CT-Scan* is defined as $D_{CT} = (CT_1, CT_2, \ldots, CT_m)$. Likewise, sub-domains for other resources, such as *neurologists, nurses,* and *rooms*, can also be defined.

2.2 Decision Variables

Scheduling a potential appointment within a given sequence involves identifying suitable medical resources with available time slots (in their calendars) that cover the required duration and meet other constraints.

Let the tuple $\mathcal{A} = \langle \mathcal{R}, \mathcal{T} \rangle$ represent a medical appointment, where $\mathcal{R} = \{x_1, x_2, \ldots, x_r\}$ is the set of decision variables encoding the required resources (r resources) for the appointment, the sub-domain of each variable x_i is denoted by $\mathcal{D}(x_i)$, comprising all identifiers of facility resources that belong to the same medical specialization (for physicians) or the same medical type (for other resource types). The sub-domain $\mathcal{D}(x_i)$ comprising n resources of a certain type is formally defined as $\mathcal{D}(x_i) \in \{ID_1, ID_2, \ldots, ID_n\}$.

$\mathcal{T} = \{t_1, t_2, \ldots, t_d\}$ denotes a set of decision variables encoding time slot identifiers that are needed to cover the required duration. The number of time slots d can be intuitively determined by dividing the required appointment duration by the slot duration. The domain of the variable t_j, denoted by $\mathcal{D}(t_j)$, is the set of all available time slots (time slot identifiers) within the calendar of the resource x_i, given as $t_j \in \mathcal{D}(t_j)$ *for* $j = 0, 1, \ldots, d$, where $\mathcal{D}(t_j) = \{0, 1, \ldots, s\}$ represents the set of the identifiers of the available time slots ($t_j \neq -1$) within the calendars of all required resources in $\mathcal{R}$. Please note that s denotes the last slot identifier of a resource calendar.

For example, $\mathcal{A} = \langle \{C_2, CT_5\}, \{26, 27\} \rangle$ is an instantiated tuple which denotes an appointment requiring both a cardiologist identified by identifier C_2 and a CT-scan device identified by identifier CT_5 for a period of time covering two consecutive 15-min time slots, i.e. 30-min appointment starting from 08:30 to 9:00, see resource calender given in Fig. 2.

Now, let $\mathcal{S} = (\mathcal{A}_1, \mathcal{A}_2, \ldots, \mathcal{A}_l)$ represent a sequence of l medical appointments. Each appointment $\mathcal{A}_i = \langle \mathcal{R}_i, \mathcal{T}_i \rangle$ with $i = 1, 2, \ldots l$ is defined as follows:

- $\mathcal{R}_i = \{x_{i,1}, x_{i,2}, \ldots, x_{i,l}\}$ denotes the set of resources required by the i-th individual appointment.
- $\mathcal{T}_i = \{t_{i,1}, t_{i,2}, \ldots, t_{i,m}\}$ denotes the set of time slots needed by the i-th individual appointment.

For example, if $l = 3$, assuming the calendar shown in Fig. 2 represents the calendar of the cardiologist C_2. The sequence $\mathcal{S}$ could represent the following instantiated appointments, as follows:

- $\mathcal{A}_1 = \langle \{C_2, \text{CT}_5\}, \{48, 49\} \rangle$: An appointment with cardiologist C_2 and CT scanner CT_5 takes place on *Wednesday 06/11/2024* from 08:00 to 08:30.
- $\mathcal{A}_2 = \langle \{C_2, \text{MRI}_1\}, \{169, 170, 171\} \rangle$: An appointment with cardiologist C_2 and MRI machine MRI_1 takes place on *Monday 11/11/2024* from 08:15 to 09:00.
- $\mathcal{A}_3 = \langle \{C_2, Room_1\}, \{216, 217, 218, 219\} \rangle$: An appointment with cardiologist C_2 and room $Room_1$ takes place on *Wednesday 13/11/2024* from 08:00 to 9:00.

3 Problem Formulation

In this section, we outline how to constrain the decision variables introduced in the previous section to ensure compliance with the diverse requirements of the medical appointment sequence scheduling problem (MASSP). In this context, we identify two categories of constraints: *implicit* and *explicit*. The former are inherent to the MASSP definition, while the latter are explicitly stated and mathematically encoded in the model to ensure specific conditions are satisfied.

Table 1 presents the decision variables and parameters that are required to formulate the problem constraints.

Implicit Constraints: On the one hand, declaring the variables encoding the required resources $x_{i,k} \in \mathcal{R}_i$ for appointment $\mathcal{A}_i$, the constraint describing that right resources have to be chosen is inherently enforced because the resource variables are defined over sub-domains that are restricted to compatible medical types. For example, if the appointment $\mathcal{A}_i$ requires a cardiologist, the model must assign the corresponding variable $x_{i,k}$ an identifier belonging to sub-domain D_c which groups the identifiers of all cardiologists (see Sect. 2.1). Thus, the assignment of resources adheres to the necessary medical qualifications for each appointment. On the other hand, the time slot domain is restricted to available slots ($t_{i,j} \neq -1$), meaning time slots are not occupied within the calendars of all involved resources.

Table 1. Variables and parameters utilized in the MASSP constraint model.

Symbol	Description
l	A parameter describing the medical treatment sequence length, i.e. number of appointments
$\mathcal{S} = (\mathcal{A}_1, \mathcal{A}_2, \ldots, \mathcal{A}_l)$	A Sequence of l appointments, where $\mathcal{A}_i$ represents an individual appointment determined by the model
rn	A parameter determining the number of required resources in appointment $\mathcal{A}_i$
$\mathcal{R}_i = \{x_{i,1}, x_{i,2}, \ldots, x_{i,rn}\}$	A set of variables encoding the resources required for appointment $\mathcal{A}_i$
$H = \{1, \ldots, j_{rn}\}$	A set of resource indices in appointment $\mathcal{A}_j$ that require the occurrence of the same resource $x_{i,k} \in \mathcal{R}_i$ in $\mathcal{A}_i$
d_i	A parameter specifying the required duration of the appointment $\mathcal{A}_i$, which determines the number of required time slots, given by $i_m = \left\lceil \frac{d_i}{\text{slot duration}} \right\rceil$ within calendars of the chosen resources
$\mathcal{T}_i = \{t_{i,1}, t_{i,2}, \ldots, t_{i,i_m}\}$	A set of variables encoding the i_m time slots required for appointment $\mathcal{A}_i$
$\delta_{i,j}$	A parameter determining the minimum distance in days between appointment $\mathcal{A}_i$ and $\mathcal{A}_j$
$\mu_{i,j}$	A parameter determining the maximum distance in days between appointment $\mathcal{A}_i$ and $\mathcal{A}_j$
D_{ab}	A set of date indices on which the patient cannot attend an appointment
$O = [o_1, o_2, \ldots, o_l]$	A list of appointment indices specifying the required chronological order in which these appointments must occur
$W = \{w_1, w_2, \ldots, w_v\}$	The initial amount of scheduled time associated with each resource, where v represents the total number of resources involved in scheduling

Explicit Constraints: The constraints describing explicit requirements for the MASSP are given in Table 2.

Constraint (1) is an *AllDifferent* constraint describing that all resources $\mathcal{R}_i$ within an appointment $\mathcal{A}_i$ must be distinct, which means that a resource cannot be chosen twice in a single appointment. Alternatively, this constraint can also be expressed as an arithmetic constraint using the inequality operator. Constraint (2) ensures that a specific resource, such as a physician with a specific specialization or equipment with a specific type, referred to by $x_{i,k} \in \mathcal{R}_i$, is consistently allocated in other appointments $\mathcal{A}_j$ at given positions $h \in H$. This constraint addresses scenarios where continuity of care or specific resource utilization is essential. For example, if a patient requires follow-up appointments with the same cardiologist, this constraint mandates that the selected resource in the first appointment must be identical in a subsequent appointment (or appointments). Here, the function *Type* returns the domain of a given parameter. Constraint (3) describes that the time slots of an individual appointment must be consecutive, i.e. no gaps allowed. This constraint has to be respected for all calendars of resources belonging to each individual appointment. In con-

Table 2. List of MASSP constraints.

(1) AllDifferent $(x_{i,1}, x_{i,2}, \ldots, x_{i,rn})$ $\forall i \in \{1, \ldots, l\}$
(2) $\forall j, i \in \{1, 2, \ldots, l\}, i \neq j, \forall h \in H,$ if $\text{Type}(x_{i,k}) = \text{Type}(x_{j,h})$, then $x_{i,k} = x_{j,h}$
(3) $t_{i,j} = t_{i,j-1} + 1,$ $\forall j \in \{2, \ldots, i_m\},$ $t_{i,j} \in \mathcal{T}_i,$ $\forall x_{i,k} \in \mathcal{R}_i$
(4) AllDifferent $(\{t_{1,1}, t_{2,1}, \ldots, t_{l,1}\})$
(5) $t_{i,1} \geq (t_{j,1} + i_m + \Delta_{i,j}),$ $\forall i, j \in \{1, 2, \ldots, l\}, i \neq j$
(6) $t_{i,1} \leq (t_{j,1} + i_m + \alpha_{i,j}),$ $\forall i, j \in \{1, 2, \ldots, l\}, i \neq j$
(7) $\lfloor \frac{t_{i,1}}{N} \rfloor \neq d_{absence},$ $\forall i \in \{1, 2, \ldots, l\},$ $\forall d_{absence} \in D_{absence}$
(8) $t_{1,o} < t_{1,o+1},$ $\forall o \in O$
(9) $W_{r_i} = w_i + \sum_{a=1}^{l} b_{r,a} \cdot d_a,$ $\forall r_i \in \mathcal{R}_a$
(10) $W_{\max} = \max_{a \in \{1 \ldots l\}, k \in \{1 \ldots rn\}} W_{r_i}$
(11) $\min W_{\max}$

straint (4), non-overlapping appointments within the sequence S are ensured by enforcing that the start slots of the appointments (referred to by $t_{i,1}$) be distinct. Constraint (5) specifies a minimum-slot distance $\Delta_{i,j}$ (including appointment duration) between two appointments $\mathcal{A}_i$ and $\mathcal{A}_j$ for the purpose of ensuring there are at least δ_{ij} days between the appointments. The minimum-slot distance is equivalent to a given number of days ($\delta_{i,j}$) and is calculated as $\Delta_{i,j} = \delta_{i,j} \times N$, for example, 2 days in the calendar given in Fig. 2 are equivalent to 2×24 (slots) $= 48$ (slots). Intuitively, the distance in hours can also be considered by multiplying the number of hours by the number of slots forming one hour ($60 \div D$). Constraint (6) is similar to the constraint (5) but it ensures that at most $\mu_{i,j}$ days between the appointments $\mathcal{A}_i$ and $\mathcal{A}_j$. Here, $\alpha_{i,j}$ is the equivalent distance in slots. Constraint (7) eliminates the dates on which the patient cannot schedule an appointment. To this end, the start slots of the appointments have to be restricted to the slot identifiers that do not fall within the column (in the resource calendars) corresponding to a given absence date index $d_{absence_i} \in D_{ab}$. This is achieved by utilizing Eq. 3 to restrict the time slot variables (start time slot) to not get assigned the identifiers encoding an absence date. Similarly, a specific weekday can be excluded from scheduling due to the patient's unavailability, by utilizing Eq. 4. Constraint (8) ensures that the occurrence of individual appointments adheres to a specified order. This is achieved by enforcing that the first time slots of the appointments follow the given order O, thereby ensuring chronological consistency.

Constraint (9) calculates the *workload* W_{r_i} for a given resource in the set of resources. We introduce $b_{r,a}$: A binary variable indicating whether the resource r_i is assigned for an appointment $\mathcal{A}_a$. This means that the durations of all appointments involving the resource r_i will be accumulated in the auxiliary variable W_{r_i}. Constraint (10) determines the *maximum workload* (auxiliary variable $W_{\max}$) among all resources assigned to appointment $a \in \{1, \ldots, l\}$ for the purpose of identify the most heavily burdened resource. Equation (11) represents the *objective function*, aiming to minimize the maximum workload (accumulated scheduled time) across all resources. By minimizing $W_{\max}$, the workload distribution becomes more balanced, reducing the burden on the most heavily loaded resource.

4 Experimental Setup and Results

We utilized the *Choco* solver [9] to construct the model, as it provides a comprehensive set of constraint types and is equipped with a variety of search strategies. In this paper, we utilize the depth-first search (DFS), which is the default search strategy offered by the solver. Please note that we utilized the *Choco* built-in constraints such as *global constraints* [7] and the default constraint propagation algorithm.

In the following, we demonstrate the efficiency of the proposed constraint model. To comply with privacy regulations mandated by our medical partner (*Charité*), we utilize a fictitious dataset for evaluation. The dataset encompasses various types of medical resources, including physicians with diverse medical specializations, rooms, and CT-scan devices. Resource calendars are segmented into 15-min time slots and each calendar day follows office hours ranging from 8:00 to 16:00, resulting in a total of 32 time slots per day (according to 1). The experiments were conducted on a Ubuntu 22.04.4 LTS machine equipped with 16 GB of RAM and an 8-core AMD Ryzen 2.02GHz. Additionally, we utilized the *Choco* solver library, version 4.10.8 [11].

Experiment 1 - Performance in Relation to Problem Size. In this experiment, we highlight the model's performance with respect to the problem size measured in scheduling days, sequence length, and number of resources. Specifically, we capture the total time taken by the solver, referred to as the *resolution time* (RT), which encompasses finding the best solution, improving it, and proving its optimality (or fully exploring the search space). Table 3 outlines three problem instances (PI), each focusing on a single dimension of the MASSP. It is worth noting that these instances reflect realistic scenarios, except that we marginally increased the problem scales, e.g. number of scheduling days in PI_1, to test the model robustness as well as to observe if non-desirable performance issues might be raised. To imitate realistic calendar settings, we evaluated the model's performance under varying percentages of time slot availability of the involved resource calendars.

Figure 3 shows the resolution time based on scheduling days (PI_1) and sequence length (PI_2). The resolution time increases gradually with the problem scale under different percentages of available slots. The narrow confidence intervals further underscore the stability and predictability of the model's behavior, indicating minimal variability in resolution time across different runs. Notably, with a lower percentage of available slots in the resource calendars, the solver tends to find solutions more quickly compared to scenarios with a higher percentage of availability. This behavior can be attributed to the exclusion of occupied

Table 3. Configuration for the problem instances.

Parameter	PI$_1$	PI$_2$	PI$_3$
Scheduling days	varying: 200 to 2000	500	1000
Sequence length (in appointment)	5	varying: 5 to 40	5
Duration/appointment	60 min	60 min	30 min
Requested resources per appointment	5	5	5
Facility resources	100	100	varying: 50 to 500
Patient absence dates	5 dates	5 dates	3 dates
Patient care continuity	specialists in first and last appointment must be the same	specialists in first and last appointment must be the same	Not required
Min distance	2 days between each two appointments	2 h between sequential appointments	3 h between sequential appointments
Max distance	4 days between each two appointments	1 day between each two appointments	3 days between each two appointments

slots (i.e., slot identifiers with a value of -1), which reduces the size of the time slot domain and simplifies the problem. Figure 3b shows that increasing the sequence length significantly extends the resolution time compared to the number of scheduling days (Fig. 3a). To address the challenge of finding feasible solutions with longer sequences under the given dependencies between appointments over 500 days (according to PI$_2$), we intentionally increased the time slot availability percentage from 80% to 90%. This is reflected by close resolution times of availability variants. Figure 4 illustrates gradual increase in resolution time with the number of resources for different slot availability levels, highlighting a moderate and consistent trend over 1000 scheduling days (according to PI$_3$).

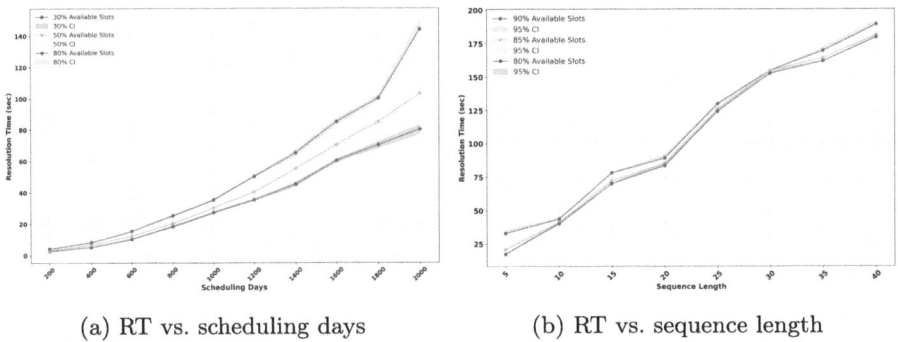

(a) RT vs. scheduling days (b) RT vs. sequence length

Fig. 3. Comparison of resolution time based on different parameters. Mean values are obtained over five individual runs.

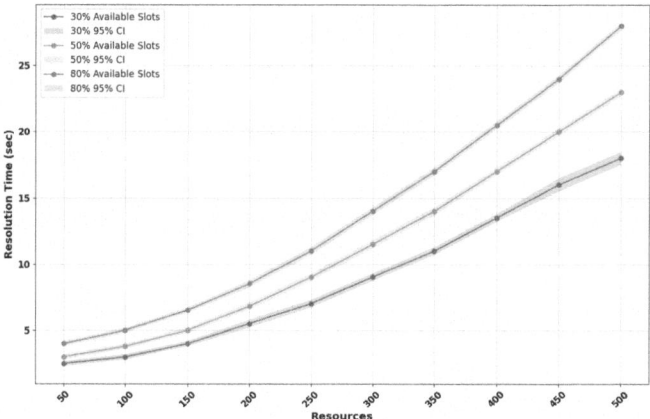

Fig. 4. RT vs. number of resources. Mean values are obtained over five individual runs.

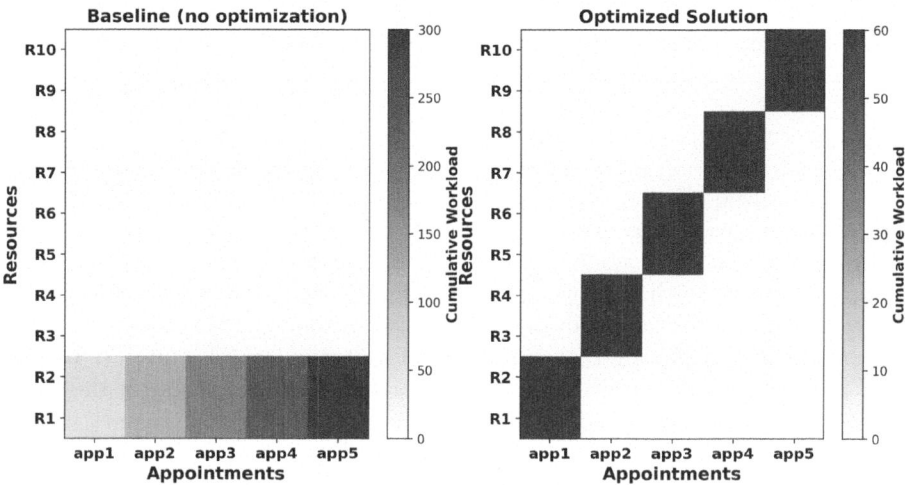

Fig. 5. Scheduling a sequence of five appointments with two resources per appointment. The duration of each individual appointment is 60 min.

Experiment 2 - Workload Distribution. To evaluate model optimization in distributing workloads among resources, we analyzed how the model assigns resources (specialists) to appointments within the sequence. In this experiment, we examined two scenarios. The first asks for scheduling a sequence comprising five 60-min appointments, each consisting of two resources (a cardiologist and a neurologist). The seconds extends the first sequence by assigning four resources per appointment. In this experimental, twenty available resources (with zero initial workload) sharing the same medical specializations. We also relaxed the other constraints. Figure 5 compares resource-to-appointment assignments without optimization (left subfigure) and with the inclusion of the *objective function*

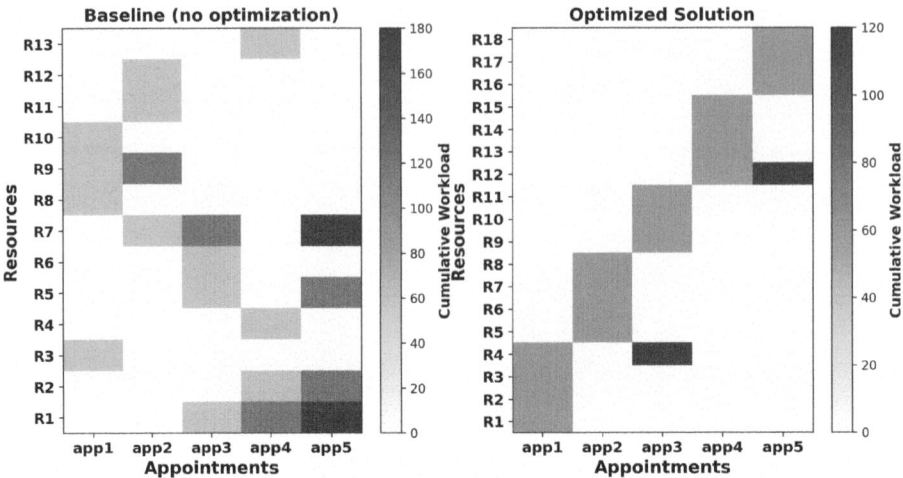

Fig. 6. Scheduling a sequence of five appointments with four resources per appointment. The duration of each individual appointment is 60 min.

$\min W_{\max}$ (right subfigure). Notably, the optimized solution effectively utilizes ten resources, ensuring a fair distribution across the appointments. In contrast, the non-optimized solution disproportionately assigns workload to only the first four resources, neglecting the others.

Figure 6 presents the assignment plan for the second scenario (four resources per appointment) with and without considering the *objective function*. Ultimately, the non-optimized solution assigned two resources three times, resulting in a cumulative workload (per resource) of 180 min for R1 and R7. In contrast, the optimized solution achieved a near-perfect distribution of the workload, which can be attributed to the heuristics employed by the solver.

5 Conclusion and Future Work

In this paper, we presented a novel optimization-based constraint model[4] to encode and solve the medical appointment sequence scheduling problem (MASSP). The significance of this problem stems from the potential large number of medical resources involved, each associated with an extensive calendar spanning numerous scheduling days, rendering traditional tools inadequate for timely solutions. Our model incorporates diverse medical requirements, such as resource and patient availability, as well as the need for resource continuity across individual appointments to ensure patient-care continuity. Importantly, our model includes an optimization function designed to balance the allocation of resources to individual appointments within the sequence.

---

[4] The sources of the model and the detailed experimental results are available upon request, subject to contractual agreements regulations.

The model demonstrated exceptional performance on complex problem instances compared to traditional scheduling approaches, consistently solving the scheduling problem efficiently. Furthermore, the model effectively assigned resources to the individual appointments comprising the sequence, showcasing the model's capability in optimizing resource allocation.

Last but not least, future work includes extending the model to incorporate patient preferences, such as selecting preferred dates and times for individual appointments and favoring certain specialists over others. Additionally, we intend to introduce further optimization objectives, such as profit maximization for the medical facility and scheduling all appointments within a sequence in the fewest possible days.

Our constraint model iterates over all optimal solutions it finds and proposes multiple solutions if they exist. However, as the number of potential solutions could be very large, we plan to incorporate a stop criterion to limit the number of solutions proposed. This threshold will help avoid overwhelming the end-user with too many options and ensure that the solutions presented are both manageable and meaningful.

Acknowledgments. The authors extend their gratitude to the anonymous reviewers for their valuable and insightful feedback. Additionally, we appreciate the fruitful collaboration with our medical partner, *Charité – Universitätsmedizin Berlin* (end user), and our industrial partner, *Doc Cirrus GmbH*, which provided the intermediate web-based application. This research was supported by the *German Ministry for Economic Affairs and Climate Action (BMWK)* as part of the *Zentrales Innovationsprogramm Mittelstand (ZIM – Central Innovation Programme for Small and Medium-sized Enterprises)* under grant number *FKZ: 16KN093238*.

Disclosure of Interests. The authors declare that no competing financial or personal interests that could have influenced the content of this work.

References

1. Abdalkareem, Z., Amir, A., Al-Betar, M., Ekhan, P., Hammouri, A.: Healthcare scheduling in optimization context: a review. Heal. Technol. **11**(3), 445–469 (2021). https://doi.org/10.1007/s12553-021-00547-5
2. Ala, A., Chen, F.: Appointment scheduling problem in complexity systems of the healthcare services: a comprehensive review. J. Healthc. Eng. **2022**(1), 5819813 (2022). https://doi.org/10.1155/2022/5819813
3. Apt, K.: Principles of Constraint Programming. Cambridge University Press (2003)
4. Baptiste, P., Laborie, P., Pape, C., Nuijten, W.: Constraint-based scheduling and planning. In: Rossi, F., van Beek, P., Walsh, T. (eds.) Handbook of Constraint Programming, Foundations of Artificial Intelligence, vol. 2, pp. 761–799. Elsevier (2006). https://doi.org/10.1016/S1574-6526(06)80026-X
5. Dechter, R.: Constraint networks. In: Dechter, R. (ed.) Constraint Processing, Chap. 2, pp. 25–49. The Morgan Kaufmann Series in Artificial Intelligence, Morgan Kaufmann, San Francisco (2003). https://doi.org/10.1016/B978-155860890-0/50003-7

6. Freuder, E.C., Mackworth, A.K.: Constraint satisfaction: an emerging paradigm. In: Rossi, F., van Beek, P., Walsh, T. (eds.) Handbook of Constraint Programming, Foundations of Artificial Intelligence, Chap. 2 , vol. 2, pp. 13–27. Elsevier (2006). https://doi.org/10.1016/S1574-6526(06)80006-4
7. Global Constraint Catalog: Global constraint catalog (2022). http://sofdem.github.io/gccat/
8. Hooker, J., van Hoeve, W.: Constraint programming and operations research. Constraints **23**(2), 172–195 (2018). https://doi.org/10.1007/s10601-017-9280-3
9. Prud'homme, C., Fages, J.: Choco-solver: a java library for constraint programming. J. Open Source Softw. **7**(78), 4708 (2022). https://doi.org/10.21105/joss.04708
10. Said, A., Mouhoub, M.: A constraint satisfaction problem (CSP) approach for the nurse scheduling problem. In: 2022 IEEE Symposium Series on Computational Intelligence (SSCI), pp. 790–795 (2022). https://doi.org/10.1109/SSCI51031.2022.10022250
11. Choco Team: Choco: an open-source java library for constraint programming, version 4.10.8. https://choco-solver.org/. Accessed 20 Nov 2024
12. Topaloglu, S., Ozkarahan, I.: A constraint programming-based solution approach for medical resident scheduling problems. Comput. Oper. Res. **38**(1), 246–255 (2011). https://doi.org/10.1016/j.cor.2010.04.018
13. Vermeulen, I.B., Bohte, S.M., Elkhuizen, S.G., Lameris, H., Bakker, P.J., Poutré, H.L.: Adaptive resource allocation for efficient patient scheduling. Artif. Intell. Med. **46**(1), 67–80 (2009). https://doi.org/10.1016/j.artmed.2008.07.019. Artificial Intelligence in Medicine AIME' 07
14. Zhang, L., Lau, S.: Constructing university timetable using constraint satisfaction programming approach. In: International Conference on Computational Intelligence for Modelling, Control and Automation and International Conference on Intelligent Agents, Web Technologies and Internet Commerce (CIMCA-IAWTIC 2006), vol. 2, pp. 55–60 (2005). https://doi.org/10.1109/CIMCA.2005.1631445

An Integrated Optimisation Method for Aluminium Hot Rolling

Ioannis Avgerinos[1](✉), Apostolos Besis[2], Ioannis Mourtos[1,3], Athanasios Psarros[2], Stavros Vatikiotis[1], and Georgios Zois[1,3]

[1] ELTRUN Research Lab, Department of Management Science and Technology, Athens University of Economics and Business, 104 34 Athens, Greece
{iavgerinos,mourtos,stvatikiotis,georzois}@aueb.gr
[2] Elval, Aluminium Rolling Division of ElvalHalcor SA, 320 11 Oinofyta, Greece
{abesis,apsaros}@elval.com
[3] Optiscale, 11472 Athens, Greece

Abstract. This paper addresses the scheduling of aluminium slabs in a hot rolling mill, which is a critical stage in aluminium rolling. Hot rolling defines a challenging variant of flowshop scheduling because of multiple side constraints imposed by quality specifications plus the need for synchronization with the preceding stage where slabs are preheated in furnaces. An additional complication arises from the objective of minimizing succession penalties in the rolling mill, in order to maximise roll quality. Current practices group slabs into batches and then schedule batches to reduce succession penalties and avoid idle times during the transition from preheating to hot rolling. Motivated by the operational requirements of a leading EU manufacturer of aluminium rolled products, we propose a novel approach that replaces the conventional batching approach by treating each aluminium slab individually. The aim is to expand the range of potential successions in the hot rolling mill and thus further reduce the sum of succession penalties. To formalise a set of elaborate (hard and soft) constraints, we introduce a Mixed Integer Linear Programming model for both the hot rolling and assignment of slabs to furnaces, and a Constraint Programming model for the precise schedule computation of each slab. These models are optimised sequentially to obtain a schedule (a 'production cycle') given a set of available slabs. Experiments on real inventory data demonstrate that the proposed approach significantly reduces the total succession penalties, while avoiding idle times between preheating and rolling. Therefore, our study could enhance production planning and scheduling for aluminium rolling.

Keywords: aluminium rolling industry · hot rolling · flowshop scheduling · integer programming · constraint programming

1 Introduction

Hot rolling is a crucial stage in the metal industry, playing a pivotal role in the production of steel, iron, and aluminium. The process operates under numerous constraints, broadly categorised as soft and hard. Soft constraints are evaluated through succession penalties, which are incurred for specific sequences of consecutive metal slabs. To improve product quality, the hot rolling process focuses on minimising total succession penalties while strictly complying with hard constraints. This requirement poses an optimisation challenge that should be integrated into the industry's Decision Support System [4]. Despite its inherent complexity, scheduling for hot rolling cannot be addressed in isolation, as it is contingent upon the preceding furnace preheating stage; Each slab must undergo preheating in one of several parallel-operating furnaces before it can proceed to the hot rolling stage. Motivated by the practical example of ELVAL[1], a leading EU manufacturer of aluminium rolled products, this paper addresses a flowshop scheduling problem that spans both stages and proposes a solution approach utilising exact methods for each.

Problem Description. The aluminium rolls manufacturer maintains an inventory of candidate aluminium slabs to meet a demand for orders arriving over time. Each slab is characterised by numerous parameters, such as alloy, product type, dimensions and weight, among others. For each order, a new slab with defined attributes is initially assigned to a furnace (Stage A), where it undergoes a preheating operation tailored to the specific product requirements, determined by the preheating conditions necessary for each slab to be processed by the rolling mill. Once preheating is complete, the slab must be promptly transferred to a rolling mill for the hot rolling stage (Stage B). The production flow, from inventory to order completion, is depicted in Fig. 1: Both Stages A and B impose numerous hard and soft constraints to ensure the production of high-quality slabs. Stage A involves preheating in parallel furnaces. Although any preheating operation can be conducted in any available furnace, each furnace has distinct capacity limitations. The furnace's capacity is segmented into individual slots, each accommodating a maximum specified length. Slabs that exceed this length cannot be assigned to certain furnaces due to these slot restrictions, while some furnaces are capable of holding multiple slabs per slot. The total weight of slabs undergoing concurrent preheating must not exceed the specified limit for each furnace. As already mentioned, each slab requires a preheating operation tailored to its product type, however, only compatible operations can

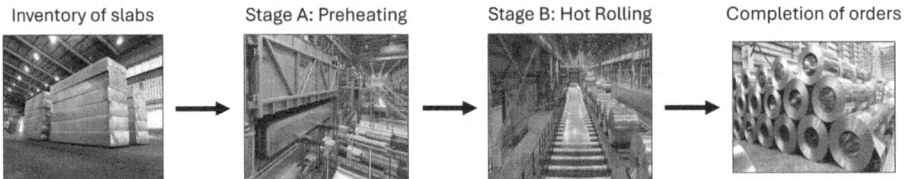

Fig. 1. Main stages of aluminium rolls production line

---

[1] https://www.elval.com/en/home.

be performed concurrently within the same furnace; otherwise, the furnace must be cleared before initiating a new preheating operation. The furnaces operate in a First-In, First-Out (FIFO) manner: once a slab is placed into the furnace, it can only be removed after all subsequent slots have been cleared.

Stage B takes place in a single hot rolling mill, which processes one slab at a time. The sequence of slabs affects the quality of the final product: hard constraints must always be met, while soft constraints add a succession penalty based on variations in the dimensions, product types, and alloys of consecutive slabs. The hot rolling mill utilises a consumable roll with a limited lifespan, determined by the cumulative wear incurred from processing the rolled slabs. This lifespan defines a *production cycle*. Each slab contributes a specific amount of wear to the roll and its rolling must adhere to the minimum and maximum roll wear limits; thus, a slab can only be rolled with respect to these wear limits.

Moreover, each slab preheats the roll's surface by a specified number of customised measurement units, referred to as *preheating units*. The roll's preheating level varies for each slab based on its unique characteristics. Additionally, each slab contributes a distinct number of preheating units to subsequent slabs during the hot rolling process, e.g., a narrower slab adds preheating units for subsequent slabs of similar width but has no effect on wider slabs. Notably, certain slabs do not contribute any preheating units to subsequent slabs, resulting in the roll's preheating level resetting to zero. To ensure proper hot rolling for each slab i, where at least T_i roll preheating units are required, the total preheating units provided by preceding slabs must be greater than or equal to T_i.

The preheating process in the furnaces spans several hours, while the hot rolling of each slab is completed within a few minutes. Additionally, the rolling mill can handle only one slab at a time, while each of the parallel furnaces can process multiple slabs simultaneously. Consequently, constructing a rolling sequence that satisfies all constraints must be accompanied by an efficient scheduling strategy for assigning slabs to furnaces. This ensures that each slab is adequately preheated and ready for hot rolling as soon as the previous slab is completed. The overall goal, therefore, is to simultaneously minimise the total succession penalties and the idle times between the furnace preheating and hot rolling stages.

Previous Work and Our Contribution. As previously noted, the hot rolling stage presents a significant scheduling challenge, prompting numerous studies to address it as an isolated problem. For instance, [2] proposed evolutionary algorithms to minimise the total succession penalty, also referred to as transition cost. Similarly, [3] integrated the minimisation of succession penalties with considerations for electricity costs. In another approach, [13] tackled the reduction of both succession penalties and penalties associated with unused inventory, incorporating elements of the Vehicle Routing Problem (VRP) into their mathematical model.

A review by [10] on scheduling in steel production highlights that most studies focus predominantly on the hot rolling mill stage. However, to maintain a seamless production line and minimise idle intervals, it is essential to integrate the scheduling of the hot rolling mill with the preceding furnace preheating stage. [11] proposed an algorithm to unify these two stages in the steel production industry. The algorithm presented by [7] addresses a problem closely related to our work, as its objective includes minimising total succession penalties and idle times between the completion of preheating and the start of hot rolling. [8] present a mathematical model and a recursive algorithm for the integrated problem in steel production industry. Another critical aspect of industrial settings is the uncertainty of demand, which evolves over a rolling horizon [5,12].

To summarise the previous work, it addresses critical hot rolling constraints [2,7], categorising them into hard and soft constraints, such as total roll wear limits and sequencing slabs based on their geometry. Additionally, it incorporates furnace preheating constraints [2,7,11,12], distinguishing between large slabs (occupying two slots) and smaller ones (occupying one slot), while accounting for furnace capacity in terms of slots and introducing slab eligibility criteria. The primary focus is on optimising quality, mill utilisation, and deadline-driven objectives [2,7,13], with hot rolling quality quantified through slab sequencing and minimising rolling mill idle time.

However, to the best of our knowledge, existing works lack consideration of: a) hot rolling constraints such as limits on roll preheating units and limits on the required roll wear for each slab, and b) furnace preheating constraints such as maximum allowable weight per furnace and compatibility of slab preheating operations within the same furnace. Moreover, a common limitation across these works is their reliance on a partitioning approach that treats the furnace preheating stage as a batch scheduling problem. In more detail, slabs are grouped into batches, each one associated to a furnace, thus containing slabs of compatible preheating operations. The batches are sequenced to the hot rolling mill to minimise idle times. Also, the order of slabs within the same batch is defined by a sequencing problem which minimises total succession penalties. A similar approach is currently implemented in the case of ELVAL.

Although the above method simplifies the problem, the resulting solutions are often significantly suboptimal. Hard constraints on compatible preheating operations heavily restrict the pool of candidate slabs for each batch, making it unlikely that consecutive slabs with low succession penalties will be grouped together. Furthermore, numerous hard constraints, such as the requirements for preheating roll units, are not taken into account. Constructing feasible schedules that account for these constraints is challenging via standard (meta-)heuristic methods, highlighting the need for mathematical programming based solution approaches. Although formulating a holistic mathematical model that captures the entire problem is impractical due to the overly complex constraints required for stage synchronisation, addressing different parts of the problem with the most suitable modeling approach has proven to be effective, as shown in Sect. 4.

To expand the pool of feasible solutions, we shift the focus from preheating batches to individual slabs, enabling consecutive slabs preheated in different furnaces to be considered eligible for scheduling in the hot rolling mill. This approach centers on constructing a production cycle in the rolling mill, aiming to improve the synchronisation between Stages A and B in the production line, which is identified as a bottleneck. First, we formulate a Mixed-Integer Linear Program (MILP) that encapsulates all hard and soft constraints of the hot rolling stage, with the objective of minimising total succession penalty. Additional constraints are then introduced to integrate Stage A (preheating) with Stage B: each scheduled slab is assigned to a specific furnace, ensuring compatibility of preheating operations with respect to furnace capacity constraints. The estimated completion time of each slab's furnace preheating is incorporated into the MILP's objective function to minimise idle time. The MILP produces set of slabs assigned to furnaces and determines the sequence of their hot rolling.

Subsequently, a Constraint Programming (CP) model is employed to refine the schedule by determining the precise start and end times for preheating and hot rolling of all slabs, based on the rolling mill sequence and furnace-related constraints. The primary goal is to assess the potential advantages of allowing consecutive slabs preheated in different furnaces, with a particular focus on enhancing product quality (i.e., minimising succession penalties) while preventing additional idle times that could disrupt industrial operations.

2 MILP for Hot Rolling and Furnace Assignment

To incorporate the hard and soft constraints of the problem into a mathematical model, we employ a Mixed-Integer Linear Program (MILP) that assigns slabs individually to specific ordered positions within the rolling mill. Such formulations are commonly described in the literature as *position-based* models [1,9] and share similarities with *event-based* formulations, where each event corresponds to the rolling of a slab [6]. Table 1 summarises the mathematical notation used in the MILP formulation for the hot rolling mill stage.

The set $\mathcal{J}$ represents candidate assignments for the rolling mill, where each assignment corresponds to a single slab. The order of hot rolling is determined by the assigned positions in the set $\mathcal{I}$. The exact number of positions depends on the lifespan of a roll and the average wear of candidate assignments in $\mathcal{J}$. Each slab $j \in \mathcal{J}$ has a restricted number of slabs that can succeed it and may add or eliminate preheating units on the roll. Therefore, we define subsets $\mathcal{J}_j$, $\mathcal{J}_j^+$ and $\mathcal{J}_j^-$ to represent each category, respectively.

The succession penalty for each pair of slabs j and k is denoted by c_{jk}. The roll preheating units of slab j for slab k are indicated by t_{jk}. If $t_{jk} = 0$, then j belongs to the subset $\mathcal{J}_k^-$; otherwise, j is included in $\mathcal{J}_k^+$. The minimum number of preheating units required for each slab j is denoted as T_j^*. Thus, j can be assigned to a position only if the number of preheating units in the previous position is greater than or equal to T_j^*. Each slab is subject to specific wear constraints that must be considered during its assignment. Specifically, if a slab

Table 1. Notation of the MILP formulation

Sets		
$\mathcal{I}$		Positions in the rolling mill
$\mathcal{J}$		(Groups of) Slabs
$\mathcal{J}_j$	$j \in \mathcal{J}$	Slabs which can precede j
$\mathcal{J}_j^+$	$j \in \mathcal{J}$	Slabs which add preheating units for slab j
$\mathcal{J}_j^-$	$j \in \mathcal{J}$	Slabs which eliminate preheating units for slab j
$\mathcal{F}$		Furnaces
$\mathcal{J}_j^\phi$	$j \in \mathcal{J}$	Slabs which can be preheated simultaneously with j
S_f	$f \in \mathcal{F}$	Slots of furnace f
Parameters		
$c_{j,k}$	$j, k \in \mathcal{J}$	Succession penalty if k succeeds j
$t_{j,k}$	$j, k \in \mathcal{J}$	Roll preheating units added by j to k
T_j^*	$j \in \mathcal{J}$	Minimum required preheating units of j
u_j	$j \in \mathcal{J}$	Maximum wear of rolls for j
l_j	$j \in \mathcal{J}$	Minimum wear of rolls for j
w_j	$j \in \mathcal{J}$	Wear added by j
p_j	$j \in \mathcal{J}$	Preheating time for slab j
r_j	$j \in \mathcal{J}$	Rolling time for slab j
v_j	$j \in \mathcal{J}$	Weight of slab j
V_f	$f \in \mathcal{F}$	Maximum weight of furnace f
q_f	$f \in \mathcal{F}$	Release time of furnace f
q_{mill}		Release time of rolling mill
Variables		
$x_{i,j}$	$i \in \mathcal{I}, j \in \mathcal{J}$	1 if j is assigned to position i, 0 otherwise
$T_{i,j}$	$i \in \mathcal{I}, j \in \mathcal{J}$	Roll preheating units for j on position i
C_i	$i \in \mathcal{I}$	Succession penalty of a slab assigned to position i
$b_{j,f}$	$j \in \mathcal{J}, f \in \mathcal{F}$	1 if j is preheated in furnace f, 0 otherwise
$n_{i,f}$	$i \in \mathcal{I}, f \in \mathcal{F}$	1 if position i is occupied by a slab in furnace f, 0 otherwise
D_f	$f \in \mathcal{F}$	Idle time of the rolling mill contributed by f

j contributes w_j wear units, the total wear prior to assigning slab j must be greater than or equal to l_j and less than or equal to u_j, where l_j, u_j represent the minimum and maximum allowable wear limits, respectively.

$\mathcal{F}$ denotes the set of parallel furnaces, each one having a number of slots S_f. We note that for furnaces which can fit more than one slab per slot, each slot counts as double. Set $\mathcal{J}_j^\phi$ includes all slabs with a compatible preheating operation with j, i.e. which can be preheated in the same furnace at the same time with j. Since the continuity of production cycles is critical for minimising idle times along the production line, it is essential to consider the release times of both the furnaces, q^f, and the rolling mill, q_{mill}, that is, the times at which previously scheduled operations are completed.

The following (MILP) formulates the hot rolling stage of aluminium slabs that exit the preheating in furnaces. Binary variables $x_{i,j}$ are set to 1 if slab j is assigned to position i. Continuous variables C_i and $T_{i,j}$ denote the succession penalty of position i and the total preheating units for slab j on position i. The objective function of the hot rolling stage is the minimisation of total succession penalties, a metric which is related to the quality of produced aluminium rolls. Each slab can be assigned to at most one position (1) and each position is occupied by exactly one slab (2). Constraints (3) ensure that a slab can be assigned to a position, only if the previous position is occupied by an eligible slab. The value of succession penalty is determined by pairs of consecutive slabs, as described in big-M constraints (4).

Constraints (5)–(9) determine the roll preheating units and ensure that the minimum requirements of assigned slabs are satisfied. Constraints (5) define the roll preheating units of each position, aggregating the units of the previous position and the units of the newly assigned slabs. Constraints (6) add the roll preheating units on the first position of the rolling mill. If any slab which eliminates preheating of roll is assigned to a position, then the total preheating units are set to 0, by constraints (7). Constraints (8) ensure that a slab can be assigned to a position, only if the total roll preheating units of the previous position are greater/equal than the minimum required. The first position of the rolling mill must be occupied by a slab which has no requirements of roll preheating units (i.e., $T_j^* = 0$), as ensured by (9). Constraints (10) and (11) ensure that the aggregated wear before assigning a new slab j should be within limits l_j and u_j. Constraints (12) combine information from furnace preheating, Stage A, and are described in the next subsection.

(MILP)

$$\min \sum_{i \in \mathcal{I}} C_i$$

$$\sum_{i \in \mathcal{I}} x_{i,j} \leq 1 \qquad \forall j \in \mathcal{J} \qquad (1)$$

$$\sum_{j \in \mathcal{J}} x_{i,j} = 1 \qquad \forall i \in \mathcal{I} \qquad (2)$$

$$x_{i,j} \leq \sum_{k \in \mathcal{J}_j} x_{i-1,k} \qquad \forall j \in \mathcal{J}, i \in \mathcal{I} : i > 0 \qquad (3)$$

$$\sum_{k \in \mathcal{J}_j} c_{k,j} \cdot x_{i-1,k} - C_i \leq M \cdot (1 - x_{i,j}) \qquad \forall j \in \mathcal{J}, i \in \mathcal{I} : i > 0 \qquad (4)$$

$$T_{i,j} \leq T_{i-1,j} + \sum_{k \in \mathcal{J}_j^+} t_{k,j} \cdot x_{i-1,k} \qquad \forall j \in \mathcal{J}, i \in \mathcal{I} : i > 0 \qquad (5)$$

$$T_{0,j} \leq \sum_{k \in \mathcal{J}_j^+} t_{k,j} \cdot x_{0,k} \qquad \forall j \in \mathcal{J} \qquad (6)$$

$$T_{i,j} \leq M \cdot (1 - \sum_{k \in \mathcal{J}_j^-} x_{i-1,k}) \qquad \forall j \in \mathcal{J}, i \in \mathcal{I} : i > 0 \qquad (7)$$

$$T_{i-1,j} \geq T_j^* \cdot x_{i,j} \qquad \forall j \in \mathcal{J}, i \in \mathcal{I} : i > 0 \qquad (8)$$

$$T_j^* \cdot x_{0,j} \leq 0 \qquad \forall j \in \mathcal{J} \qquad (9)$$

$$\sum_{l \in \mathcal{I} : l < i} \sum_{k \in \mathcal{J}} w_k \cdot x_{l,k} \leq \sum_{j \in \mathcal{J}} u_j \cdot x_{i,j} \qquad \forall i \in \mathcal{I}, j \in \mathcal{J} \qquad (10)$$

$$\sum_{l \in \mathcal{I} : l < i} \sum_{k \in \mathcal{J}} w_k \cdot x_{l,k} \geq \sum_{j \in \mathcal{J}} l_j \cdot x_{i,j} \qquad \forall i \in \mathcal{I}, j \in \mathcal{J} \qquad (11)$$

[Constraints for the assignment of slabs to furnaces] (12)

$$x_{i,j} \in \{0,1\} \qquad \forall i \in \mathcal{I}, j \in \mathcal{J}$$

$$C_i \geq 0 \qquad \forall i \in \mathcal{I}$$
$$T_{i,j} \geq 0 \qquad \forall i \in \mathcal{I}, j \in \mathcal{J}$$

We note that our industrial partner may prioritise certain slabs or product types, requiring immediate inclusion in the current production cycle, thus it is possible that cover constraints for certain subgroups of slabs may be added.

2.1 Connecting Stages A and B in the Same Model

As discussed in Sect. 1, focusing only on minimising total succession penalties can disrupt coordination between Stages A and B, potentially leaving the rolling mill idle as it waits for the next slab to be preheated. To address this, we incorporate the minimisation of total idle time by defining a set of mathematical expressions for Constraints (12) within the MILP.

$$\sum_{i \in \mathcal{I}} x_{i,j} = \sum_{f \in \mathcal{F}} b_{j,f} \qquad \forall j \in \mathcal{J} \qquad (13)$$

$$\sum_{f \in \mathcal{F}} n_{i,f} = 1 \qquad \forall i \in \mathcal{I} \qquad (14)$$

$$\sum_{j \in \mathcal{J}} p_j \cdot x_{i,j} - \sum_{k \in \mathcal{J}} \sum_{l \in \mathcal{I}: l < i} r_k \cdot x_{l,k} - D_f \leq M \cdot (1 - n_{i,f}) \qquad \forall i \in \mathcal{I}, f \in \mathcal{F} \qquad (15)$$

$$\sum_{j \in \mathcal{J}} b_{j,f} \leq |S_f| \qquad \forall f \in \mathcal{F} \qquad (16)$$

$$\sum_{j \in \mathcal{J}} v_j \cdot b_{j,f} \leq V_f \qquad \forall f \in \mathcal{F} \qquad (17)$$

$$b_{j,f} + b_{k,f} \leq 1 \qquad \forall f \in \mathcal{F}, j \in \mathcal{J}, k \notin \mathcal{J}_j^\phi \qquad (18)$$

$$1 + n_{i,f} \geq x_{i,j} + b_{j,f} \qquad \forall i \in \mathcal{I}, j \in \mathcal{J}, f \in \mathcal{F} \qquad (19)$$

$$b_{j,f} \in \{0,1\} \qquad \forall j \in \mathcal{J}, f \in \mathcal{F}$$
$$n_{i,f} \in \{0,1\} \qquad \forall i \in \mathcal{I}, f \in \mathcal{F}$$
$$D_f \geq 0 \qquad \forall f \in \mathcal{F}$$

Variables b_{jf} are set to 1 if slab j is assigned to furnace f, or 0 otherwise. Variables n_{if} are set to 1 if position i of the rolling mill handles a slab preheated in furnace f, or 0 otherwise. Continuous variables D_f indicate the idle time of the rolling mill, attributed to furnace f. As part of preprocessing, variables b_{jf} can be set to 0 if j is ineligible for furnace f, meaning that the length of j cannot fit in the slots of f.

Constraints (13) ensure that if a slab is scheduled in the rolling mill, then it is also assigned to one furnace. Each position is associated with a furnace (14). Big-M constraints (15) determine the idle time of the rolling mill which is attributed to furnace f. Each slab requires p_j minutes before being inserted in the assigned position i of the rolling mill. The time interval between the start of the preheating and the start of the hot rolling of j should be occupied by the hot rolling of all slabs in the previous positions of the mill, thus eliminating idle times. Constraints (16) and (17) ensure that the capacity constraints of furnaces will be satisfied. If a pair of slabs have incompatible preheating operations, then they cannot be assigned to the same furnace (17). Finally, by constraints (18), if a slab j is scheduled in position i of the mill and assigned to furnace f, then position i is dedicated to furnace f. Finally, to incorporate the minimisation of idle times, the objective function is replaced with:

$$\min \alpha \cdot \sum_{i \in \mathcal{I}} C_i + \beta \cdot \sum_{f \in \mathcal{F}} D_f$$

Succession penalties are measured in thousands of penalty units, while idle times are measured in minutes. Therefore, an appropriate convex combination of coefficients α and β must be chosen to balance these objectives effectively, ensuring that $\alpha + \beta = 1$. For the use-case of ELVAL, an approximate distribution between the coefficients is $\alpha \approx 0.01$ and $\beta \approx 0.99$.

We note that constraints (17) limit the number of slabs per production cycle to the maximum capacity of all available furnaces. In some cases, the furnaces may have a lower capacity than the rolls' lifespan could potentially handle, allowing for additional slabs to be scheduled within the same cycle. Also, it is possible that the MILP solution might not fully utilise the capacity of certain furnaces. In both situations, any empty slots can be filled by locally inserting slabs from the inventory, ensuring their preheating operations align with that of the assigned slabs. Additionally, a position at the rolling mill can be appended, adhering to the stage's quality constraints.

After the addition of Constraints (12), each solution of the MILP generates a sequence $R = [j^0, j^1, ..., j^{|\mathcal{I}|}]$, where j^i is the slab assigned to position i, i.e. $x_{i,j} = 1$. Additionally, each furnace f is allocated a dedicated group of slabs $\bar{J}_f = \{j \in \mathcal{J} | b_{j,f} = 1\}$. Due to the FIFO operation of the furnaces, the sequence of slabs in R also determines the order in which slabs are processed in each furnace's dedicated group.

3 Constraint Programming for Precise Scheduling

As discussed in Sect. 1, the construction of a production cycle begins with generating a sequence at the rolling mill, as this stage is the bottleneck in the flowshop scheduling problem. Assigning scheduled slabs to furnaces allows for the precise calculation of the start and end times of their furnace preheating operation, which precedes hot rolling.

The strict precedence constraints imposed by the FIFO operation of the furnaces can be effectively managed through standard Constraint Programming (CP) modeling. Table 2 presents the mathematical annotation of the CP model. Based on the order of hot rolling in R and the assignment of slabs to furnaces, $\bar{J}_f$, we construct the following CP formulation.

(CP)

$$\min \sum_{j^i \in R} \texttt{length\_of}(idle_{j^i})$$

$$\texttt{no\_overlap}(mill) \qquad (20)$$

$$\texttt{start\_at\_end}(roll_{j^i}, roll_{j^{i-1}}) \qquad \forall j \in \mathcal{J} \text{ (under certain conditions)} \qquad (21)$$

$$\texttt{previous}(mill, roll_{j^{i-1}}, roll_{j^i}) \qquad \forall j^i \in R : i > 0 \qquad (22)$$

Table 2. Notation of the CP formulation

Sets		
R		Sequence of slabs in the hot rolling mill
$\bar{J}_f$	$f \in \mathcal{F}$	Slabs assigned to furnace f
Parameters		
p_j	$j \in \mathcal{J}$	Furnace preheating time for slab j
r_j	$j \in \mathcal{J}$	Hot tolling time for slab j
q_f	$f \in \mathcal{F}$	Release time of furnace f
q_{mill}		Release time of rolling mill
λ_f	$\forall f \in \mathcal{F}$	Length of each slot of furnace f
Variables		
$roll_{j^i}$	$j^i \in R$	Rolling of the ith slab of R
$mill$		Sequence of scheduled slabs
$idle_{j^i}$	$\forall j^i \in R$	Idle time of the rolling mill for the ith slab of R
prh_{j^i}	$\forall j^i \in R$	Preheating of the ith slab of R
$slot_{f,s}$	$\forall f \in \mathcal{F}, s \in S_f$	Occupation of slot s in furnace f
$position_{j,f,s}$	$\forall f \in \mathcal{F}, s \in S_f, j^i \in \bar{J}_f$	Occupation of slot s in furnace f by slab j

$$\texttt{start\_of}(idle_{j^0}) = q_{mill} \tag{23}$$

$$\texttt{start\_at\_end}(idle_{j^i}, roll_{j^{i\text{-}1}}) \quad \forall j^i \in R : i > 0 \tag{24}$$

$$\texttt{start\_at\_end}(roll_{j^i}, idle_{j^i}) \quad \forall j^i \in R : i > 0 \tag{25}$$

$$\texttt{start\_at\_end}(roll_{j^i}, prh_{j^i}) \quad \forall j^i \in R \tag{26}$$

$$\texttt{start\_of}(prh_{j^k}) \leq \texttt{start\_of}(prh_{j^i}) \quad \forall f \in \mathcal{F}, j^i \in \bar{J}_f, j^k \in \bar{J}_f : k < i \tag{27}$$

$$\sum_{s \in S_f} \texttt{presence\_of}(position_{j,f,s}) = 1 \quad \forall f \in \mathcal{F}, j \in \bar{J}_f \tag{28}$$

$$\texttt{synchronize}(position_{j,f,s}, prh_j) \quad \forall f \in \mathcal{F}, j \in \bar{J}_f, s \in S_f \tag{29}$$

$$\sum_{j \in \bar{J}_f} \texttt{presence\_of}(position_{j,f,s}) \leq \lambda_f \quad \forall f \in \mathcal{F}, s \in S_f \tag{30}$$

$$\texttt{span}(slot_{f,s}, [position_{j,f,s} | j \in \bar{J}_f]) \quad \forall f \in \mathcal{F}, s \in S_f \tag{31}$$

$roll_{j^i} : \texttt{interval}, size = r_{j^i}$ $\quad \forall j^i \in R$

$mill : \texttt{sequence}, [roll_{j^i} | j^i \in R]$

$idle_{j^i} : \texttt{interval}$ $\quad \forall j^i \in R$

$prh_{j^i} : \texttt{interval}, size \geq p_{j^i}$ $\quad \forall j^i \in R$

$slot_{f,s} : \texttt{interval}, \texttt{optional}$ $\quad \forall f \in \mathcal{F}, s \in S_f$

$position_{j^i,f,s} : \texttt{interval}, size \geq p_{j^i}, \texttt{optional}$ $\quad \forall j^i \in R$

We define interval variables for the two processes each slab must undergo: preheating, prh, and hot rolling, $roll$. To minimise the total idle time of the rolling mill, we also construct a set of interval variables, $idle$, which indicate the idle time associated with each slab, i.e., the time between the completion of hot rolling for the previous slab and the beginning of the next one. To allocate each slab to slots within the furnaces, we introduce optional interval variables, *position*, which specify the time each slab occupies a slot. Additionally, to precisely schedule each slot within a furnace, we define optional interval variables, *slot*, as some slots may remain empty during the scheduled production cycle. All interval variables involving furnaces have a minimum start time set to the respective release time q_f.

Constraint (20) ensures that the rolling mill processes at most one slab at any time. Constraints (21) stipulate that once a slab has completed processing, hot rolling for the next slab can begin. It is important to note that delays between consecutive slabs at the rolling mill may impair product quality: since many rolled products require a minimum level of roll preheating units, T_j^*, an idle mill effectively resets the preheating level. If a scheduled slab does not have such roll preheating requirements, the corresponding constraint (21) can be omitted. The order of processed slabs, as determined by R, must be respected (22).

The idle time of a slab begins once the previous slab has completed processing (24). For the first slab in sequence R, idle time starts at the release time of the rolling mill (23). Idle time ends when hot rolling of the slab begins (25). Additionally, hot rolling must commence immediately following the end of the furnace preheating operation (26). Due to the FIFO operation of the furnaces, preheating for each subsequent slab within the same furnace must start later (27). By constraint (28), each slab must be assigned to one slot within the designated furnace. The slot occupation time is synchronised with preheating time by constraints (29). It is noted that the solver-specific constraint `synchronize` involves an interval variable and a list of interval variables, and ensures that all present variables in the list have the same start and end times as the first interval variables. As mentioned earlier, some furnaces contain slots with larger maximum lengths, enabling the placement of more than one slab. The total number of placed slabs must not exceed this capacity, as indicated by constraint (30). Finally, each slot remains occupied from the earliest placement of a slab to the latest removal, as enforced by (31). Constraint `span` involves an interval variable and a list of interval variables: it imposes that the interval variable starts at the minimum starting time and ends at the maximum end time of all present variables of the list. It should be noted that although the furnaces have a maximum weight limit for inserted slabs, this constraint is already satisfied during the assignment of slabs to furnaces, as specified by (17).

4 Experimental Evaluation

The proposed algorithm leverages the available inventory to construct consecutive production cycles. A sequential application of the MILP and CP components generates a production cycle, updating the release times of the furnaces and the rolling mill for subsequent cycles. Hereinafter, our approach is referred to as *slab-based*, emphasizing the transition from batch preheating to an individual slab-focused process, enabling consecutive slabs to be preheated in different furnaces while simultaneously determining their hot rolling sequence.

The slab-based approach is compared to a two-stage batch scheduling method, which reflects the current practice at ELVAL, while also aligning with the predominant approaches in the literature. In the first stage, each furnace is allocated to a specific preheating operation based on the availability of a sufficient number of compatible slabs to fully utilise its capacity. Each furnace is dedicated to a distinct batch of slabs, whose hot rolling process will remain uninterrupted by slabs from other furnaces. Although no slabs are directly assigned at this stage, the preheating time for each batch is determined based on the dedicated preheating operation, while the estimated rolling time is calculated using the average rolling times of compatible slabs. Therefore, the furnaces are scheduled in a sequence that minimises the expected idle times of the hot rolling mill. In the second stage, a variant of the Traveling Salesman Problem (TSP) is solved sequentially for each furnace, following the order of batches determined in the first stage. For each TSP, a subset of slabs with compatible preheating operations is selected, and their preheating sequence is scheduled to minimise succession penalties. Since the hot rolling of each batch is continuous, the preheating sequence directly determines the rolling sequence for the assigned slabs. This process accounts for penalties incurred between the last slab of the preceding batch and the first slab of the current batch, as well as between the last slab of the current batch and the first slab of the subsequent batch.

Experimental Setting. All experiments were conducted on a server equipped with 32 AMD Ryzen Threadripper PRO 5955WX 16-core processors and 32 GB of RAM, running Ubuntu 22.04.5 LTS. To solve the (MILP), we employed the *Gurobi 11.0.3* optimiser via the open-source optimisation library *Pyomo 6.8.0* in *Python 3.10.12*. To solve the (CP), we used *CPLEX 22.11* CP Optimizer via the DOCplex module.

We apply the two approaches to a set of five real inventory instances provided by ELVAL. We also consider six pusher-type furnaces working in parallel for the preheating of slabs, and a single hot rolling mill. The number of candidate slabs in each inventory instance ranges from 200 to 400. Note that, for larger instances (300–400 slabs) the two methods are employed to generate schedules spanning three production cycles, while for smaller ones (200–300 slabs) the schedule is limited to two production cycles. A time limit of 3600 s is imposed on solving the (MILP). Given the large scale and the complexity of constraints, the MILP model fails to reliably indicate the gap from optimality, as the dual bounds remain close to zero after the time limit is reached in most experiments.

For each inventory, the results of the two approaches are compared by evaluating the total succession penalties. Table 3 presents the results for the five inventories per production cycle. Column 'Total time' indicates the total duration of each production cycle, and column 'Idle time' presents the total duration of idle intervals within the cycle. All time values are measured in minutes. Column 'Succession penalties' indicates the total succession penalties between all consecutive slabs. In the columns 'Idle Time' and 'Succession Penalties', which

An Integrated Optimisation Method for Aluminium Hot Rolling   29

Table 3. Comparison between Batching and Slab-based approaches

Inventory	Cycle	Total time		Idle time		Succession penalties	
		Batching	Slab-based	Batching	Slab-based	Batching	Slab-based
1	1	1059	1027	**7**	14	30168	**3851**
	2	1115	1136	98	**11**	19888	**6533**
	3	1020	1000	**3**	20	24579	**10008**
2	1	1015	1141	**0**	15	32177	**3773**
	2	1219	1282	**7**	122	19606	**16425**
	3	1196	884	48	**9**	34083	**9008**
3	1	988	948	**14**	18	35196	**2672**
	2	1226	1227	46	**10**	34689	**2310**
4	1	1145	896	26	**15**	19924	**3333**
	2	1095	1194	**12**	33	16841	**7992**
5	1	1020	699	18	**17**	23237	**6908**
	2	1265	1613	**136**	180	45835	**15310**

correspond to the objectives of the mathematical models, the dominant values are highlighted in bold. Regarding the 'Total Time' column, the differences in results between the two methods are mainly due to the selection of different slabs, which lead to varying total rolling times, as the value of r_j is specific to each slab j. Since total time is not part of the objective function, the values in the respective columns are unrelated to the quality of the solutions.

The comparison of succession penalties reveals a notable disparity between the two approaches. Across all production cycles, the slab-based approach consistently outperforms the batching method, often achieving remarkable improvements. On average, adopting the holistic approach reduces total succession penalties per cycle by approximately 70.8% compared to the batching approach. The reduction ranges from 16.2% in cycle 2 of Inventory 2 to an impressive 93.3% in cycle 2 of Inventory 3.

Importantly, this improvement is achieved without compromising production continuity, as evidenced by the idle times per cycle. To evaluate this, we compare the two approaches by analyzing mill utilisation per cycle, which represents the percentage of total time the hot rolling mill remains operational. Mill utilisation is calculated by considering the total production cycle duration (total time) and idle times: mill utilisation = $\frac{\text{total time} - \text{idle time}}{\text{total time}}$%.

Table 4 summarises the mill utilisation per production cycle and per inventory instance for both approaches. Mill utilisation consistently remain above 97% in the majority of production cycles; 10 out of 12 cycles for the slab-based and 9 out of 12 for the batching approach, respectively.

The batching approach achieves slightly higher mill utilisation in 8 out of 12 production cycles and 3 out of 5 inventories. However, the difference in overall average utilisation across all inventories is negligible. Notably, production cycles

Table 4. Comparison of mill utilisation between the batching and slab-based approaches

Inventory	1			2			3		4		5	
Cycle	1	2	3	1	2	3	1	2	1	2	1	2
Batching	**99.34**	91.21	**99.71**	**100.00**	**99.43**	95.99	**98.58**	96.25	97.73	**98.90**	**98.24**	89.25
		96.62			**98.40**		97.29		**98.30**		93.26	
Slab-based	98.64	**99.03**	98.00	98.69	90.48	**98.98**	98.10	**99.19**	**98.33**	97.24	97.57	88.84
		98.58			95.59		**98.71**		97.70		91.48	

with mill utilisation exceeding 95% are effectively uninterrupted. This is because idle intervals caused by unsynchronised preheating in furnaces and hot rolling in the mill are typically offset by operational delays inherent to industry practices - factors not explicitly considered in the optimisation models.

Insights on Minimised Succession Penalties. To illustrate the significant reduction in succession penalties achieved with the slab-based approach, we construct Gantt charts for the proposed schedules, using Inventory 3 as a case study. Among the parameters influencing succession penalties, the difference in width between consecutive slabs is the most illustrative. Therefore, Inventory 3 has been selected as it best highlights these width variations.

Each Gantt chart is divided into two sections. The upper section represents the preheating stage across the six parallel furnaces, with preheated slabs depicted as coloured blocks. Each colour corresponds to a specific furnace. The lower section, separated by a dashed red line, displays the sequence of slabs in the hot rolling mill. Here, the height of each coloured block represents the width of the rolled slab. While the succession penalty between two consecutive slabs depends on various parameters, the most visually prominent factor is the difference in their widths. Consequently, this parameter has been chosen as a representative aspect of slab transitions. Finally, black blocks in the Gantt chart indicate the end of a production cycle, followed by the replacement of serviceable rolls, a standard procedure that requires 60 min.

As illustrated in Fig. 2, which presents the schedules of the batching and slab-based approaches, allowing the succession of slabs preheated in different furnaces results in smoother transitions in the hot rolling mill. In the batching approach, solving a sequencing problem for each batch ensures smooth transitions between consecutive slabs within the same batch but fails to minimise penalties when connecting slabs between batches. In contrast, the slab-based approach leverages a broader pool of feasible successions, enabling the selection of transitions with smaller succession penalties throughout the entire production cycle.

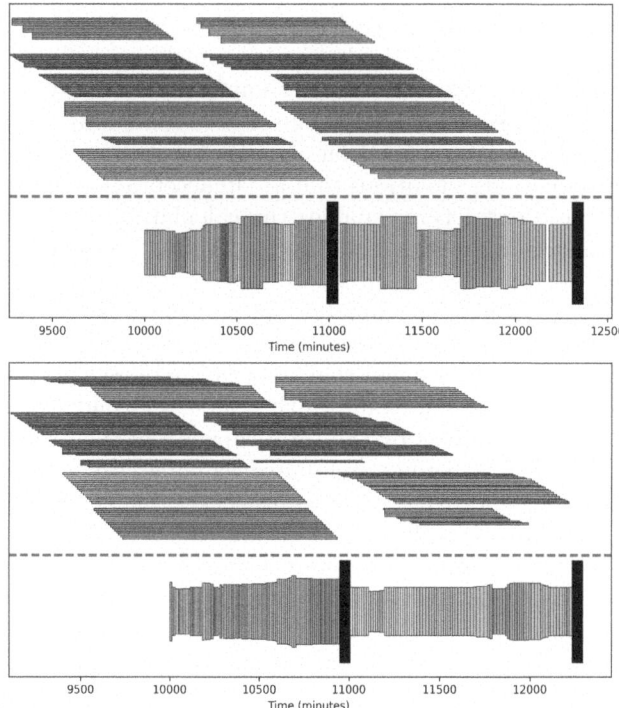

Fig. 2. Production cycles of Inventory 3: Batching (above) and Slab-based (below) approach

5 Conclusions

This study addresses the integrated optimisation problem of scheduling aluminium slabs in hot rolling mills in the aluminium rolling industry. Current practices often rely on simplifications that limit the pool of feasible solutions, creating challenges in scheduling inventory-dependent production cycles. Motivated by a real-world use case from ELVAL, a leading EU aluminium manufacturer, we propose a novel partitioning approach that uses mathematical models to accurately represent the soft and hard constraints of the problem. Applying this approach to a set of inventory instances demonstrates a significant reduction in succession penalties, leading to higher-quality products and preventing idle times caused by inadequate synchronisation between the preheating of slabs and the hot rolling process.

The results highlight the often-underappreciated contribution of mathematical models solved with exact methods in addressing practical problems. While metaheuristic frameworks are commonly recommended for large-scale problems, they may fail to guarantee feasible solutions due to numerous case-specific constraints. This can lead to the relaxation of critical restrictions related to safety and machines preservation - violations that are unacceptable. In contrast, explicitly formulating these restrictions enables the calculation of feasible schedules, ensuring high-quality solutions that adhere to all strict constraints.

Although the problem described is case-specific, the methodological framework can easily be generalised for broader application in the metal industry, as the flow of operations is similar across aluminium, steel, and copper production. Additionally, aspects of the problem that were not explored in this study could be addressed in future work. For example, this paper's formulation assumes that each production cycle is limited by the total capacity of the available furnaces. However, a production cycle could potentially accommodate more slabs, suggesting an extension of our approach to allow furnaces to manage multiple preheating operations. More importantly, linking production scheduling with inventory planning could further optimise production lines. Given the strong dependence of the solution on available inventory, an extended method could provide valuable insights into inventory requirements, supporting long-term planning and strategic decision-making in industry operations.

Acknowledgment. This research has been supported by the European Union through the Horizon Europe projects Tec4MaaSEs and MODAPTO, grant numbers 101138517 and 101091996, respectively.

References

1. Bektaş, T., Hamzadayı, A., Ruiz, R.: Benders decomposition for the mixed no-idle permutation Flowshop scheduling problem. J. Sched. **23**(4), 513–523 (2020). https://doi.org/10.1007/s10951-020-00637-8
2. Chen, Y.W., Lu, Y.Z., Ge, M., Yang, G.K., Pan, C.C.: Development of hybrid evolutionary algorithms for production scheduling of hot strip mill. Comput. Oper. Res. **39**, 339–349 (2012)
3. Chen, L., Cao, L.L., Wen, Y.M., Chen, H., Jiang, S.L.: A knowledge-based NSGA-II algorithm for multi-objective hot rolling production scheduling under flexible time-of-use electricity pricing. J. Manuf. Syst. **69**, 255–270 (2023)
4. Cowling, P.: A flexible decision support system for steel hot rolling mill scheduling. Comput. Ind. Eng. **45**, 307–321 (2003)
5. Huang, Y., Yang, Q., Liu, J., Li, X., Zhang, J.: Sustainable scheduling of the production in the aluminum furnace hot rolling section with uncertain demand. Sustainability **13**, 7708 (2021)
6. Koné, O., Artigues, C., Lopez, P., Mongeau, M.: Event-based MILP models for resource-constrained project scheduling problems. Comput. Oper. Res. **38**(1), 3–13 (2011)
7. de Ladurantaye, D., Gendreau, M., Potvin, J.Y.: Scheduling a hot rolling mill. J. Oper. Res. Soc. **58**(3), 288–300 (2007)

8. Li, F., Zhang, Y., Wei, H., Lai, X.: Integrated problem of soaking pit heating and hot rolling scheduling in steel plants. Comput. Oper. Res. **108**, 238–246 (2019)
9. Naderi, B., Ruiz, R.: The distributed permutation flowshop scheduling problem. Comput. Oper. Res. **37**(4), 754–768 (2010)
10. Özgür, A., Uygun, Y.: A review of planning and scheduling methods for hot rolling mills in steel production. Comput. Ind. Eng. **151**, 106606 (2021)
11. Schiefer, E.: An algorithm for the hot rolling mill scheduling problem in a high-grade steel production. In: 20th Annual Conference of the Production and Operations Management Society, POM 2009, 011-0358 (2009)
12. Stauffer, L., Liebling, T.M.: Rolling horizon scheduling in a rolling-mill. Ann. Oper. Res. **69**, 323–349 (1997)
13. Zhang, T., Chaovalitwongse, W.A., Zhang, Y.J., Pardalos, P.M.: The hot-rolling batch scheduling scheduling method based on the prize collecting vehicle routing problem. J. Ind. Manage. Optim. **5**(4), 749–765 (2009)

Determining the Most Promising Selective Backbone Size for Partial Knowledge Compilation

Andrea Balogh[✉], Guillaume Escamocher, and Barry O'Sullivan

Insight Research Ireland Centre for Data Analytics, School of Computer Science
and Information Technology, University College Cork, Cork, Ireland
{andrea.balogh,guillaume.escamocher}@insight-centre.org,
b.osullivan@cs.ucc.ie

Abstract. Knowledge compilation allows for quick retrieval of a wide variety of information about an instance of a constraint satisfaction problem or Boolean formula. Unfortunately, the compiled representation can be so large that it cannot be used in practical settings, especially when there are restrictions on the amount of available memory. It has been recently observed that assigning some carefully chosen variables can significantly reduce the size of the representation, with only a small loss of data about the instance. However determining how many, and which, variables should be assigned to create a compact compiled instance requires making a number of expensive computations, including completely compiling the full initial instance, which might not be achievable for large instances. Our work in this paper involves removing large compilations from the process of constructing selective backbones. First we look at how statistical properties of literals can be used as fast heuristics to pick which variable to assign next. Then we show how the behaviour of selective backbones for small instances can help determine the size at which selective backbones for large instances are expected to offer the best compromise between the quantity of knowledge lost and the amount of space saved. We also extend previous work to take into account weighted preferences on models. This is beneficial since various probabilistic models can be expressed in such terms.

1 Introduction

Decision Diagrams have been around for a long time, originally introduced for circuit design and formal verification [6,17]. The earliest and most widely used language is the Ordered Binary Decision Diagram (OBDDs), which has been used in circuit analysis and synthesis [7]. Deterministic Decomposable Negational Normal Form (d-DNNF) became a popular representation since it is a superset of OBDDs, and probabilistic inference and Maximum a-Posteriori (MAP) can be answered in polynomial time in the size of the representation. A variety of problems in machine learning, expert systems, social network analysis, bioinformatics and information theory can be formulated as probabilistic maximum a-posteriori inference.

Knowledge Compilation is the field that considers different representations and their methods to compile. The Knowledge Compilation Map [12] describes a list of representations and when to use each of them, according to what query and transformations are required by the user. These representations are particularly useful for settings where one can afford to spend time on creating an initial compilation of all the solutions to a problem and later use this to answer queries. Such scenarios include diagnosis, configuration etc.

There has been significant effort in speeding up compilation methods but less related to time and memory limitations. In many applications, such as onboard diagnosis or reasoning on embedded processors, large problems might still fail to compile leaving the user without any practical representation of the problem. Other cases might lead to a representation that is too large to be used in practice. In these cases partial compilation could provide the user with a subset of solutions represented in the target language [18]. Since many subsets of solutions exist, representing the most preferred, most likely, or most important, ones for the user is important.

The novelty in this paper is threefold. First, we apply previous heuristics to weighted problems, where weights can represent a preference or probability. Weighted model counting has many applications, especially in probabilistic inference problems [10,11,13,21]. Second, we introduce new heuristics that are computationally much faster than in [2]. Third, we define an approach to determine a number of iterations (the size of the selective backbone) for a heuristic without evaluating the partial compilation at each step. This, in particular, allows us to build selective backbones for very large instances, which was not feasible at all with only the tools provided by the previous paper. The best partial compilation of a weighted CNF is one whose weighted model count is as close as possible to the original and its representation size is minimized. We evaluate[1] eight heuristics: two computationally expensive ones, five simple heuristics that do not rely on anything more than basic properties of the input CNF, and a hybrid heuristic that combines one of these simple heuristics with a limited number of calls to a weighted model counter. The proposed scoring functions used to define the subset of the solutions to be part of the partial compilation do not depend on the compilation tool or the target representation used. Our method should have a similar behaviour using a different target representation.

Our approach transforms the original instance into a more compact one by assigning some variables. We call this set of variables the selective backbone, which is the relaxation of the proper backbone such that assigning the variables we want to maximise the number of solutions kept. To know the precise number of variables to assign to get an adequate level of compactness, one could compute the weighted model count and the size of the representation every time that a variable is assigned, and stop when the ratio of the former divided by the latter

---

[1] https://github.com/baloghAndi/PartialKC_WMC.

has improved enough compared to its original value. However this would require making many compilations, including one of the possibly large original instance, wasting the potential of the simpler heuristics, whose whole purpose is to avoid exactly that. We show in this paper that we can instead use results on small instances to have a good indication for much bigger instances of which number of assigned variables is most likely to yield high compactness, removing the need for any large compilation.

The rest of the paper is organised as follows. Section 2 presents the most important related work. Section 3 introduces the notation and definitions used throughout the paper. Section 4 describes the heuristics and scoring functions used to create the Selective Backbone for weighted CNFs. Here we also describe a method we used for identifying the most promising size of a Selective Backbone. Section 5 presents the results obtained and Sect. 6 concludes the paper.

2 Related Work

Decision Diagrams (DDs) represent a set of solutions. The size of the representation will influence its efficiency for different queries and transformations. It has been shown that the size of DDs heavily depends on the variable ordering used for them, which optimization problem is known to be NP-hard. Several approaches have looked into obtaining a better ordering [9]. In our work we do not look at different variable orderings.

Another way of reducing the size of the representation is to relax or restrict the decision diagram [3]. By relaxing or reducing the decision diagram the problem becomes less constrained, allowing more solutions. This way the decision diagram decreases in size since some of the branches can be merged. Another way to decrease the size of the decision diagram is by restricting it and thus eliminating some solutions. In this case parts of the decision diagram that represents the eliminated solutions can also be eliminated, leading to a smaller representation. Both relaxation and restriction have a wide range of usage. Explanations have also been explored as a means to achieve approximate weighted model counting [20].

We define partial compilation as a compiled representation that omits some solutions from the initial set of solutions. Partial compilations have been an area of interest for multiple reasons [18]. For example, they can act as bounds in an optimization problem. A partial compilation can be looked at as a restriction on the original decision diagram, this way used as an upper bound. Decision Diagrams have been used as upper and lower bounds for optimization problems. They have shown significant advantages compared to the widely used linear relaxations, but the size of DDs heavily depends on the variable ordering used for them, which optimization problem is known to be NP-hard. To tackle this problem [9] look at Reinforcement Learning in order to learn the best bounds. We do not explore different variable orderings. Our approach for partial compilation is by fixing part of the problem and eliminating some solutions.

Approximate MDDs have also been explored as generators for lower bounds. A top-down compilation method has been developed for this purpose that merges nodes when the width of the representation gets larger than a limit [4]. Both MDDs and BDDs have been used as domain stores for Constraint Programming systems.

Knowledge Compilation has also been explored as a method for uniform sampling [22]. A given CNF formula is compiled into d-DNNF form, and then performing only two passes over the d-DNNF representation they generate as many identically and independently distributed samples as specified by the user. But this approach needs the initial compilation as well.

Weighted projected model counting is an extension of weighted model counting such that it considers a set of variables Y to be forgotten. The tool d4Max [1], extends d4 to deal with such problems as well. Our approach differs from this since we do not forget variables, but fix their values and eliminate solutions that are not consistent with the assignment.

An algorithm has been developed for obtaining the top k models, given a d-DNNF circuit [5]. Our approach is different from this as we aim to obtain the top models without compiling the initial problem, only the subproblem that represents these solutions. A similar incremental approach has been explored by introducing a collapsed compilation approach for approximate inference [14].

3 Definitions

We now formally define the concepts that will come into play throughout the paper. A *literal* is either a Boolean variable or its negation. A *clause* is a disjunction of literals. If a clause only contains one literal, then we call it a *unit clause*. A *CNF instance* consists of n Boolean variables and m clauses. A *solution*, or *model* of an instance is a set of n literals, one for each variable in the instance, such that every clause contains a literal that is present in the model. The set of variables that take the same value in all models to an instance is called the *backbone* of the instance [24]. This is not to be confused with *backdoors*, which are sets of variables that can be assigned in such a way that the resulting problem becomes easy to solve. The *model count* of an instance is the number of models to this instance. The model counting problem is #P-Complete, even when no clause is allowed to have more than two literals [23].

In this paper we focus on *weighted model counting* (WMC) instances, which are CNF instances where every literal l is given a weight $w(l)$. The *weight of a model* is the product of the weights of the literals present in the model. The *weighted model count* of an instance is the sum of the weights of the models to the instance. The WMC problem is #P-Complete, like its unweighted version (because the latter is a special case of the former where all literals weigh 1).

Our work aims to remove a few models from a WMC instance so that the representation of the instance becomes much more compact, while maintaining the weighted model count as high as possible. The set S of models to keep is characterised by the set of literals that appear in all models from S. This set of literals is called a *selective backbone* [2]. Our approach will thus be to find a selective backbone B such that removing all models that are not supersets of B

significantly decreases the size of the instance representation, and the sum of the weights of the models that are an extension of B is as close as possible to the original weighted model count. Note that the set of variables assigned by this selective backbone B is a subset of the backbone of the instance obtained after removing all models that do not include B.

Decision diagrams are a graphical representation of a set of solutions to a given problem. They are represented as a directed acyclic graph (DAG), where nodes represent AND or OR operations and edges denote variable assignments. There is a variety of specialized decision diagrams, each defined by properties that make the representation best suited for some queries and problem types. In this paper, we chose to look at the d4 compiler [16], which is a top-down compiler that has ranked highly in the Weighted Model Counting competitions in the past years. D4 compiles to a Decision-DNNF which is an extension of d-DNNF. This is a rooted DAG, where each node is either a decomposable AND gate or a deterministic OR gate. Decomposable AND gates are defined so that the sets of variables of each subtree of an AND gate are pairwise disjoint. Deterministic OR gates means that subtrees of the OR gate are pairwise inconsistent. Decision-DNNF are defined the same as d-DNNF except the OR gates are replaced by decision gates. A decision gate N on decision variable X has the form $(\neg X \wedge Y) \vee (X \wedge Z)$, meaning if X then Z else Y. The size of the representation is defined by the number of edges of the graph.

4 Building a Selective Backbone

4.1 Overview

In our experiments we compared eight different ways to build a selective backbone for a weighted model counting instance. Two of them are based on the heuristics that achieved the best performance in the unweighted version of the problem [2]. They both make a high number of calls to an exact model counter. Five heuristics that we examined do not require any expensive computation, so they can be used even when the input instance is too large for the previous two methods. The final heuristic makes some calls to an exact model counter, but these calls are a lot fewer than the calls made by the two most expensive heuristics.

The eight heuristics that we studied can be sorted into two categories: dynamic heuristics and static heuristics. Heuristics from the former type iteratively choose a literal to add to the selective backbone, then assign and propagate this literal before repeating the process on the resulting instance, until the selective backbone is of the desired size. Literals that form a unit clause in the instance are given priority because, since these literals appear in all models for the instance, we know that picking one of these literals cannot decrease the weighted model count. If no such literal is present in the instance, we pick the literal that has the highest score according to some given metric f, breaking ties by picking the literal with the highest weight. In our experiments we looked at four different scoring functions, all described in the next subsection.

Once a decision has been made on which literal to assign, we remove from the instance clauses that are satisfied by the chosen literal, and remove its opposite from the remaining clauses. We also introduce a unit clause composed of the literal that was picked, to record that the corresponding variable has been assigned and is not unconstrained. Finally, we remove both the chosen literal and its opposite from consideration for future iterations. This ensures that no variable is assigned twice.

Because dynamic heuristics modify the instance every time they add a literal to the selective backbone, they need to update the score of each literal before each pick, and also to keep track of clauses that have become unit clauses after the opposites of previously chosen literals have been pruned out. On the one hand, the information uncovered in the early steps of dynamic approaches can be made available for later steps, improving the quality of the selective backbones that are built. On the other hand, if the scoring function is expensive to compute, then determining its value for each remaining eligible literal at every step could be a burden on the runtime of dynamic methods. In contrast, static heuristics compute the scoring function only once for each literal, and at each iteration pick the remaining literal that achieved the highest score in this initial ranking.

The procedure for choosing the first literal of the selective backbone is the same for static and dynamic heuristics: compute the value of the scoring function for every literal, then pick the literal with the highest score. The difference is that static heuristics do not make any expensive operation beyond this first step, they just follow the ranked list of literals. While this prevents them to acquire updated information about the instance, it also means that from the second pick onwards they are much faster than dynamic heuristics.

4.2 Scoring Functions

Some of the best performing heuristics for creating selective backbones for unweighted model counting [2] assigned a score to each literal that was equal to the number of models for the instance containing this literal. Adapting this measure for unweighted model counting gives us *actual_WMC*, the first of our four scoring functions:

Definition 1. *Given an instance I and a literal l from I, let I' be the instance obtained after adding the unit clause l to I. We say that actual_WMC(l) is the weighted model count of I'.*

This function returns the sum of the weights of the models that contain the given literal. Calling it is equivalent to computing the weighted model count of an instance that is about the same size as the original instance, so it can be costly when the input instance is large, especially for dynamic heuristics that run it for each eligible literal at every step. For massive instances, it might therefore be wiser to use simpler scoring functions, that only look at the basic structure of the CNF instance and do not require any kind of model counting or compilation.

We now define the *relative weight* of a literal. Given a literal l, we call $W(l)$ the weight of l, and $W_{rel}(l) = \frac{W(l)}{W(l)+W(\bar{l})}$ the weight of l relative to its opposite

$\bar{l}$. Because $W_{rel}(l) + W_{rel}(\bar{l}) = 1$ for every literal l, we can treat the cases where the sum of the weight of a literal and the weight of its opposite is not the same for all variables.

The second of our four scoring functions, *relative_weight*, is the relative weight itself:

Definition 2. *Given a literal l, we say that relative_weight(l) is equal to $W_{rel}(l)$.*

For the third scoring function, we compute an estimate of the number of models that contain a literal l. In general, a rough estimate $Est_{inst}(I)$ of the model count of a CNF instance I with m clauses $C_1, C_2, \ldots, C_m$ is:

$$Est_{inst}(I) = 2^n \prod_{i=1}^{m} (1 - (\frac{1}{2})^{k_i})$$

where k_i is the number of literals of C_i. Indeed, the number of assignments on the k_i variables that appear in the ith clause is 2^{k_i}, and exactly one of them is forbidden by the clause, so we can expect the clause to multiply the number of models by $\frac{2^{k_i}-1}{2^{k_i}} = 1 - (\frac{1}{2})^{k_i}$. This number is of course only a crude approximation, but it is easy to compute.

In the same manner we can compute $Est_{lit}(l)$, an estimate of the number of models containing literal l:

$$Est_{lit}(l) = \prod_{i=1}^{m(l)} (1 - (\frac{1}{2})^{k_i - 1})$$

where $m(l)$ is the number of clauses containing $\bar{l}$, and k_i is the number of literals in the ith such clause. There are three differences between the formula for Est_{inst} and the formula for Est_{lit}:

1. The factor 2^n does not appear in the formula for Est_{lit}. This is because within an instance the value of n does not change so the term 2^n becomes a large constant that we can ignore.
2. Not all clauses are considered in Est_{lit}. This is because if neither l nor $\bar{l}$ appears in a clause C, then C will have the same impact on the expected number of models containing l as on the expected number of models containing $\bar{l}$. Furthermore, if l appears in a clause, then this clause will be satisfied by all models containing l, so we can ignore it.
3. The exponent in Est_{lit} is $k_i - 1$ instead of k_i. This is because we already know that the literal $\bar{l}$ that appears in all considered clauses is not present in any model containing l, so the number of possible assignments to the variables of one of these clauses is only $2^{k_i - 1}$.

We can now introduce the scoring function *estimated_WMC*:

Definition 3. *Given a literal l, we say that*

$$estimated\_WMC(l) = W_{rel}(l) \times \frac{Est_{lit}(l)}{Est_{lit}(l) + Est_{lit}(\bar{l})}$$

Our last scoring function, *random*, actually simulates returning a random score by picking a completely random literal. It is intended to serve as a baseline. Note that dynamic heuristics using this function still pick unit clauses when available, so it is possible for them to build selective backbones of decent sizes before reaching a conflict. We however found out in our early experiments that the static heuristic with random literal ranking often reaches a conflict in the first few steps, and therefore most of the selective backbones that it obtains lead to a weighted model count of 0 and an empty instance representation. Thus, we will only present the results for the dynamic version of the function *random*.

4.3 Hybrid Heuristic

The eighth and last heuristic, *hybrid_WMC*, is dynamic. To pick the next literal to add to the selective backbone, it first computes *estimated_WMC* (Definition 3) for all remaining literals. Let e be the highest score obtained, and let $E = \{l \mid estimated\_WMC(l) \geq \frac{90e}{100}\}$. The heuristic then computes *actual_WMC* (Definition 1) for all literals in E, and picks the literal of E that got the highest score with *actual_WMC*.

This hybrid heuristic is faster than computing *actual_wmc* for every single literal and, as we will show in the next Section, it obtains higher ratios than *estimated_wmc*.

4.4 A Good Size for the Selective Backbone

Our goal is to produce partial compilations that are as compact as possible with regard to the compilation of the original instance. A formal measure for this relative compactness is what we call the *adjusted ratio* (AR), that we define as follows, with *rep_size* being the size of the compiled representation:

$$\text{adjusted ratio} = \frac{\text{WMC/initial WMC}}{\text{rep\_size/initial rep\_size}}$$

An adjusted ratio of 1 is equivalent to the compactness of the initial instance, whereas the higher the adjusted ratio the more compact the partial compilation.

We aim to build a selective backbone for a large instance I with a high adjusted ratio, without any intermediate compilation of I. This means that we do not compute the adjusted ratio at each iteration to check whether it is good enough to stop, instead we decide on the number of variables to assign before starting building the selective backbone.

We determine the selective backbone size that is most likely to lead to a high adjusted ratio in the following manner. Let $\mathcal{I}$ be a set of instances. We partition

$\mathcal{I}$ into two sets $\mathcal{I}_{small} = \{I_1, I_2, \ldots, I_{n_1}\}$ and $\mathcal{I}_{large} = \{J_1, J_2, \ldots, J_{n_2}\}$. For a given heuristic h, we run h on all n_1 instances from $\mathcal{I}_{small}$. For each $i \in [1, n_1]$, for each percentage p (sampled depending on the number of variables of instances in $\mathcal{I}_{small}$), we compute the adjusted ratio $AR_{i,p}$ obtained after assigning $p\%$ of the variables of I_i with h. Computing the adjusted ratio requires compilation, but because the instances in $\mathcal{I}_{small}$ are small, this should not take too much time. Let p_{max} be the percentage of variables assigned where the highest median adjusted ratio is observed.

For each $j \in [1, n_2]$, we run h to compute a selective backbone B_j of size $p_{max}\%$ of the number of variables in J_j. Using h we run $|B_j|$ number of iterations to obtain the required partial model.

This prediction approach is very simple but effective since instances seem to have a similar behaviour when looking at the progression of the adjusted ratio. Ideally, instances from $\mathcal{I}_{small}$ and $\mathcal{I}_{large}$ should be part of the same dataset, representing similar problems on smaller and larger scales. If this is the case, then the trend of the adjusted ratio of the larger instances should be similar to the smaller instances trend. This way the location of the best adjusted ratio from the small instances should be relevant for the large instances as well.

5 Results

We looked at the same dataset as in [2]: a set of generated instances denoted as AB containing 10 instances with 15 variables and another 10 with 30 variables; a subset of the ISCAS suite containing 14 instances; and 134 instances from the planning suite. The number of models for these instances ranges from 4e0 to 6.2e85, with a median of 8.1e6. The number of variables ranges between 26–252 for ISCAS and 5–577 for planning, the number of clauses ranges between 66–639 for ISCAS and 10–1901 for planning. These benchmarks have previously been used by numerous other papers about knowledge compilation and model counting [8,15,19].

We generated a weight w between 0 and 3 for each positive literal following the uniform distribution and assigned its negation the weight of $3 - w$. The choice of 3 is an arbitrary scaling factor since the weights 0–1 were producing small weighted model count values leading to underflow problems. We used preprocessing on all instances before running the heuristics in order to eliminate duplicate/unit clauses, as well as duplicate/opposing literals within clauses.

We evaluated eight approaches to obtain a compact compilation with respect to weighted model counting. We looked at the two types of heuristics, static and dynamic, each with the scoring functions: *actual_WMC*, *estimated_WMC* and *relative_weight*. As a baseline, we use the *random* scoring function with the dynamic formulation. The heuristic *hybrid_WMC* is only taken into account as a dynamic heuristic. We refer to *estimated_WMC* and *relative_weight*, both static and dynamic, as well as *random* dynamic, as "simple heuristics".

Table 1. Execution time of each heuristic in seconds without extra compilation at each iteration. Dashed line implies timeout of an hour was reached.

	Min	Max	Average	Median
actual_WMC Stat	1.246	395.674	66.577	43.581
actual_WMC Dyn	5.857	-	1906.339	1895.844
hybrid_WMC	0.001	639.823	65.457	32.174
estimated_WMC Stat	0.002	0.471	0.078	0.046
estimated_WMC Dyn	0.003	27.239	1.563	0.515
relative_weight Stat	0.001	0.079	0.014	0.010
relative_weight Dyn	0.003	2.441	0.145	0.049
random Dyn	0.849	3.017	0.145	0.040

We used d4 on the command line with its default parameters to compile the instances and perform weighted and unweighted model counting.[2] All experiments were performed on a machine with Intel(R) Xeon(R) CPU E5620 @2.40 GH running Ubuntu 22.04.2 LTS. Each instance was run for 1 h with each heuristic.

5.1 Efficiency and Relative Ratio

As defined earlier each heuristic implements an iterative approach, where each variable gets assigned the best value according to the scoring function used. Satisfiability of the instance before making the assignment at every iteration is only ensured for the dynamic heuristic with *actual_WMC* scoring function and *hybrid_WMC*. Every other combination of heuristic and scoring function could lead to making an assignment that would lead to a conflict, making the instance unsatisfiable. In case a heuristic reaches a model count of zero or times out within one hour, for the rest of the iterations that it did not complete zero is taken as a value for its model count and weighted model count. For all eight heuristics, we compiled the instances and calculated the weighted model count at each iteration for post-analysis only.

Runtime for each heuristic can be found in Table 1. Note that runtime for *actual_WMC* includes an extra step for performing model counting at each iteration. In each iteration, this heuristic spends time compiling each possible assignment which includes writing out CNF files so that d4 can be called from the command line. The extra step of one model counting operation per iteration is not as significant as the $2 \times n$ compilations to obtain the best next assignment. We can see there are huge differences in runtimes between the heuristics. Dynamic heuristics tend to run longer than their static version since they update the CNF at each iteration and also spend some time calculating the new scores. On average it takes very little time to run the simple heuristics,

---
[2] https://github.com/crillab/d4.

while *actual_WMC* dynamic is by far the slowest. In terms of average runtime *hybrid_WMC* behaves similarly to *actual_WMC* static, but its median is lower.

Table 2. Variables assigned before conflict and/or timeout. Last column denotes the number of instances that assigned all variables out of the dataset.

	Percentage of variables assigned				No. finished
	Min	Max	Average	Median	
actual_WMC Stat	61.21	100	96.00	100	94/168
actual_WMC Dyn	2.71	100	73.78	100	112/168
hybrid_WMC Dyn	11.7	100	94.24	100	146/168
estimated_WMC Stat	2.46	100	29.02	18.56	4/168
estimated_WMC Dyn	4.93	100	87.12	100	127/168
relative_weight Stat	0.69	66.67	10.55	3.97	0/168
relative_weight Dyn	0.69	100	72.58	100	94/168
random Dyn	0.55	100	74.31	100	98/168

Only *actual_WMC* dynamic times out, for 56 out of 168 instances. For other heuristics for some instances, a conflict is reached and consequently not all variables are assigned. As seen in Table 2, the heuristics that seem the best at successfully assigning variables are *actual_WMC* static and *hybrid_WMC* dynamic. The former has the higher minimum and average number of assigned variables, always assigning more than 60% of the variables, while the latter also has a high average and is the heuristic that completes the most instances. The dynamic heuristics tend to complete more instances, which is understandable since they propagate the assignments after each iteration, thus a more informed decision of the actual state of the CNF is made at each step. That is the reason the *random* dynamic heuristic seemingly performs well, completing assignments for a high number of instances, but we will later see that the quality of those partial compilations is not good.

The previous tables shine a light on some behaviours of the heuristics but not on the quality of the partial compilations. In order to evaluate the performance of the heuristics we look at two metrics: the efficiency and the adjusted ratio. The aim of the heuristics is to obtain the best selective backbone that leads to the most compact compilation. We define compactness in terms of the ratio between WMC and the size of the representation, maximizing the WMC and minimizing the representation size. Since the size of the representation is canonical with respect to the variable ordering used, and we are not optimizing the variable ordering, in general when the representation gets smaller it is due to fewer models being represented. In [2] it was shown that there can be a partial compilation that decreases more the size of the representation than the number of models it does not represent. In this paper, we aim to prove that there exists partial compilations of weighted CNF problems such that the tradeoff between smaller WMC and size of representation is not directly proportional. To show this we plotted the percentage reduction of WMC with respect to the initial WMC against the percentage of reduction of size with respect to the initial compiled representation size. We refer to this as efficiency. Figure 1 represents the median of all instance efficiencies. The grey 45° line denotes the efficiency when the loss of WMC and compiled representation size is directly proportional. We want to be above this line, as close as possible to the top right corner, in order to obtain a compact compilation. Both timeout and inconsistency are treated the same way, weighted model count and size is taken as 0. Because instances have vastly different number of variables, we show 50 data points for each heuristic, with the ith data point being the median of the values achieved after $2i\%$ of the variables have been assigned for each instance. The same will be done in Figs. 2 and 3.

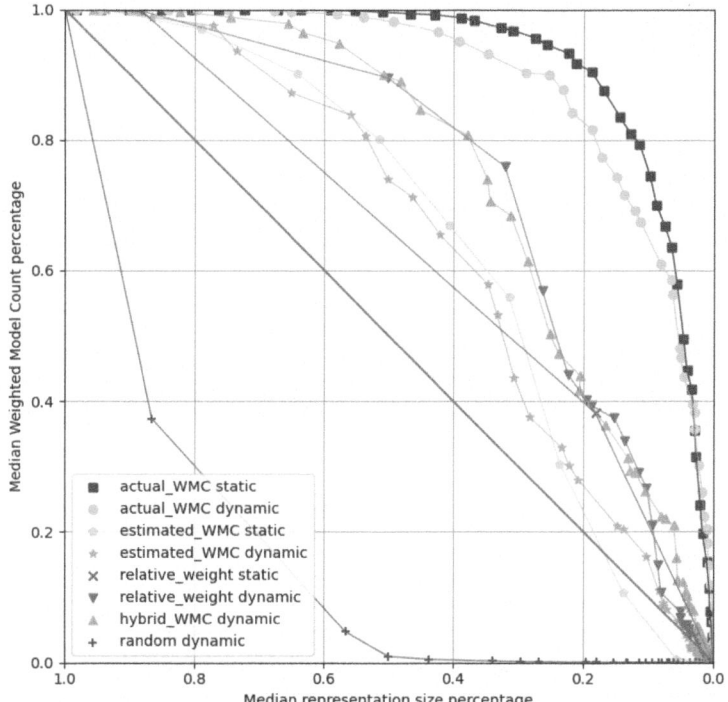

Fig. 1. Median of efficiency over all instances.

In Fig. 1 *random* dynamic is below the 45° line, meaning it does not achieve worthy partial compilations. The *actual_WMC* dynamic and static heuristics perform the best, with static occasionally outperforming dynamic, because in some cases the dynamic times out, bringing its median value down. The next best heuristics are the dynamic *relative_weight* and *hybrid_WMC*, with the latter one better since more observations occur closer to the high WMC percentage. The rest of the heuristics, *relative_weight* and *estimated_WMC* both static and dynamic seem to behave very similarly, with no clear best.

The efficiency plots are good for visualizing the evolution of the compactness of the compilations but are not suitable to show how instances behave with the same sized selective backbone, so in Fig. 2 we present the median of the adjusted ratio for all percentages of selective backbone size. Here too *actual_WMC* dynamic and static reach the best performance, with static outperforming dynamic due to the latter's timeout. Among the other heuristics, the best one is *hybrid_WMC*, which keeps a somewhat steady adjusted ratio of 2.

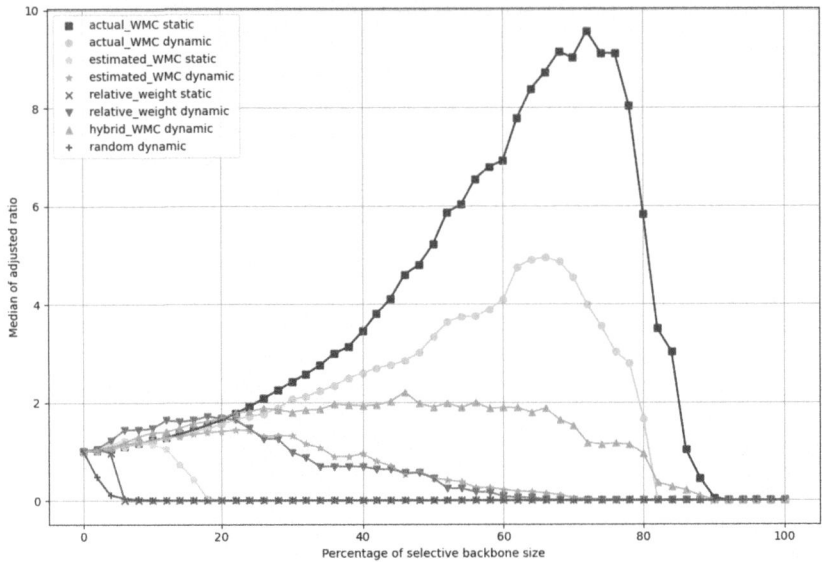

Fig. 2. Median of the adjusted ratio over all instances.

Table 3 presents the number of times each heuristic was the one that reached the highest adjusted ratio of an instance, excluding instances where more than one heuristic reached the highest adjusted ratio. The heuristics that most often get the highest adjusted ratio are unsurprisingly the two most computationally expensive ones, but *hybrid_WMC* also manages to reach a higher adjusted ratio than any other heuristic for quite a few instances, while it is very rare that a simple heuristic is the one with the best result.

Table 3. Number of best adjusted ratio per heuristic

	Number of times best AR	Percent of times best AR
actual_WMC Stat	63	45.99%
actual_WMC Dyn	55	40.15%
hybrid_WMC	11	8.03%
all five simple heuristics combined	8	5.84%

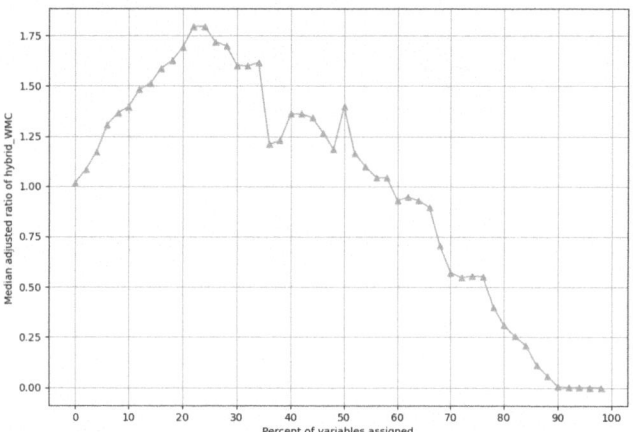

Fig. 3. Median Adjusted ratio of *hybrid_WMC* over the ISCAS and planning instances with 50–300 variables

5.2 A Good Size for the Selective Backbone

In this subsection, we test whether the approach described in the last subsection of Sect. 4 can successfully pick a selective backbone size with a high adjusted ratio for large instances, without compiling any of these large instances.

In this set of experiments we look at the heuristic *hybrid_WMC*, but any heuristic could be used. $\mathcal{I}_{small}$ dataset contains a subset of the 168 instances we previously looked at. We eliminate 45 instances that have less than 50 variables since they are too small. The obtained dataset contains 123 instances with 50–300 variables. Given the $\mathcal{I}_{small}$ dataset we observe that the highest median adjusted ratio is 1.79, obtained at $p_{max} = 22\%$ of variables of the specific instance. At this percentage, 68.3% of the instances from $\mathcal{I}_{small}$ have an adjusted ratio higher than 1. Figure 3 represents the median adjusted ratio of the $\mathcal{I}_{small}$ at each p. We notice this has a slightly different behaviour than the *hybrid_WMC* from Fig. 2. This is due to the fact that Fig. 3 represents a different dataset, where instances with fewer than 50 variables are eliminated.

After identifying p_{max} on the smaller instances we look at evaluating the partial compilation of the larger instances given B_j. $\mathcal{I}_{large}$ contains 171 instances with 300–900 variables. We eliminated 7 since they had no initial compilation and there is one instance that times out before obtaining any selective backbone.

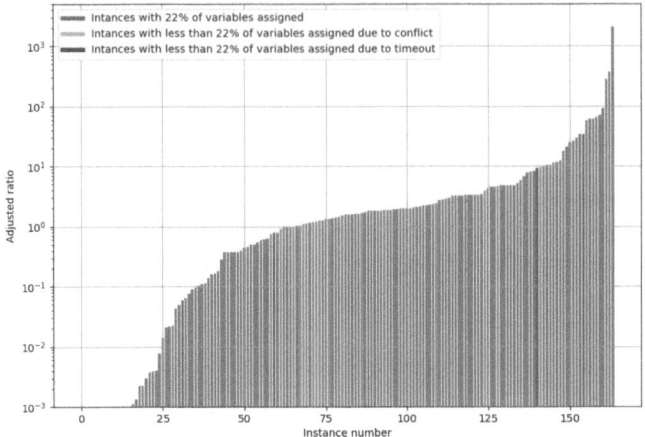

Fig. 4. Adjusted ratio at 22% Selective Backbone

We define a 6 h timeout but some instances obtain the selective backbone within 0.8 s, with a median time of 376.23 s. There are 25 instances that do not reach 22% variables assigned or reach a conflicting assignment. Figure 4 represents the adjusted ratio in increasing order for instances in $\mathcal{I}_{large}$ given the partial compilation defined by the selective backbone of size 22% of variables. Each green bar represents the adjusted ratio of an instance with 22% of variables assigned, orange represents the compilation of an instance before a conflicting assignment is reached and blue represents the partial compilation for instances where a timeout occurred before assigning 22% of the variables.

The median of the adjusted ratio of the compilations from $\mathcal{I}_{large}$ is 1.56 and 60% of the instances are compact, with an adjusted ratio larger than 1. These values are close to the ones observed for $\mathcal{I}_{small}$, indicating that the behavior of selective backbones on small instances can indeed provide an idea of the behavior of selective backbones on larger instances.

6 Conclusion

We have compared eight methods to select models to eliminate from a weighted CNF instance such that the representation size of the resulting instance is as small as possible, while the weighted model count remains as high as possible. Two of these methods are based on existing approaches for the unweighted version of the problem and are computationally expensive. The other six methods are novel and are less taxing on resources.

Our paper contains two main results. First, we have shown that fast approaches to build compact partial compilations do achieve decent results for some selective backbone sizes. Second, we have shown that it is possible to determine in an efficient manner at which selective backbone size these highly compact partial compilations can be expected to be found. This proves that a compact representation of a weighted CNF instance can be obtained with minimal information loss without any kind of expensive computation.

In the future, more sophisticated machine learning approaches could be explored to predict the size of the selective backbone and explore the impact of the diversity of the instances. Further experimentation on even larger instances, where compilation times out for the initial instance but compiles for a predicted size of selective backbone could further strengthen the approach.

Acknowledgments. This paper is based upon research supported by the Science Foundation Ireland under Grants 12/RC/2289-P2 and 16/RC/3918 which are co-funded under the European Regional Development Fund.

Disclosure of Interests. The authors have no competing interests to declare that are relevant to the content of this article.

References

1. Audemard, G., Lagniez, J., Miceli, M.: A new exact solver for (weighted) max#sat. In: Meel, K.S., Strichman, O. (eds.) 25th International Conference on Theory and Applications of Satisfiability Testing, SAT 2022, 2–5 August 2022, Haifa, Israel. LIPIcs, vol. 236, pp. 28:1–28:20. Schloss Dagstuhl - Leibniz-Zentrum für Informatik (2022). https://doi.org/10.4230/LIPIcs.SAT.2022.28
2. Balogh, A., Escamocher, G., O'Sullivan, B.: Partial compilation of SAT using selective backbones. In: Gal, K., Nowé, A., Nalepa, G.J., Fairstein, R., Radulescu, R. (eds.) ECAI 2023 - 26th European Conference on Artificial Intelligence, 30 September– 4 October 2023, Kraków, Poland - Including 12th Conference on Prestigious Applications of Intelligent Systems (PAIS 2023). Frontiers in Artificial Intelligence and Applications, vol. 372, pp. 174–181. IOS Press (2023). https://doi.org/10.3233/FAIA230268
3. Balogh, A., O'Sullivan, B.: Breaking symmetry for knowledge compilation. In: Hong, J., Lanperne, M., Park, J.W., Cerný, T., Shahriar, H. (eds.) Proceedings of the 38th ACM/SIGAPP Symposium on Applied Computing, SAC 2023, Tallinn, Estonia, 27–31 March 2023, pp. 991–994. ACM (2023). https://doi.org/10.1145/3555776.3577863
4. Bergman, D., van Hoeve, W.-J., Hooker, J.N.: Manipulating MDD relaxations for combinatorial optimization. In: Achterberg, T., Beck, J.C. (eds.) CPAIOR 2011. LNCS, vol. 6697, pp. 20–35. Springer, Heidelberg (2011). https://doi.org/10.1007/978-3-642-21311-3_5
5. Bourhis, P., Duchien, L., Dusart, J., Lonca, E., Marquis, P., Quinton, C.: Pseudo polynomial-time top-k algorithms for d-DNNF circuits. CoRR abs/2202.05938 (2022). https://arxiv.org/abs/2202.05938
6. Bryant, R.E.: Graph-based algorithms for Boolean function manipulation. IEEE Trans. Comput. **C-35**, 677–691 (1986). https://api.semanticscholar.org/CorpusID:1911887
7. Bryant, R.E.: Symbolic boolean manipulation with ordered binary-decision diagrams. ACM Comput. Surv. **24**, 293–318 (1992). https://api.semanticscholar.org/CorpusID:1933530

8. Capelli, F., Lagniez, J., Marquis, P.: Certifying top-down decision-DNNF compilers. In: Thirty-Fifth AAAI Conference on Artificial Intelligence, AAAI 2021, Thirty-Third Conference on Innovative Applications of Artificial Intelligence, IAAI 2021, The Eleventh Symposium on Educational Advances in Artificial Intelligence, EAAI 2021, Virtual Event, 2–9 February 2021, pp. 6244–6253. AAAI Press (2021). https://doi.org/10.1609/aaai.v35i7.16776
9. Cappart, Q., Bergman, D., Rousseau, L., Prémont-Schwarz, I., Parjadis, A.: Improving variable orderings of approximate decision diagrams using reinforcement learning. INFORMS J. Comput. **34**(5), 2552–2570 (2022). https://doi.org/10.1287/ijoc.2022.1194
10. Chavira, M., Darwiche, A.: On probabilistic inference by weighted model counting. Artif. Intell. **172**(6–7), 772–799 (2008). https://doi.org/10.1016/j.artint.2007.11.002
11. Chavira, M., Darwiche, A., Jaeger, M.: Compiling relational Bayesian networks for exact inference. Int. J. Approx. Reason. **42**(1–2), 4–20 (2006). https://doi.org/10.1016/j.ijar.2005.10.001
12. Darwiche, A., Marquis, P.: A knowledge compilation map. J. Artif. Intell. Res. (2002). https://doi.org/10.1613/jair.989
13. Fierens, D., et al.: Inference and learning in probabilistic logic programs using weighted Boolean formulas. Theory Pract. Log. Program. **15**(3), 358–401 (2015). https://doi.org/10.1017/S1471068414000076
14. Friedman, T., den Broeck, G.: Approximate knowledge compilation by online collapsed importance sampling. In: Bengio, S., Wallach, H., Larochelle, H., Grauman, K., Cesa-Bianchi, N., Garnett, R. (eds.) Advances in Neural Information Processing Systems, vol. 31, Curran Associates, Inc. (2018)
15. Lagniez, J., Lonca, E., Marquis, P.: Definability for model counting. Artif. Intell. **281**, 103229 (2020). https://doi.org/10.1016/j.artint.2019.103229
16. Lagniez, J., Marquis, P.: An improved decision-DNNF compiler. In: Sierra, C. (ed.) Proceedings of the Twenty-Sixth International Joint Conference on Artificial Intelligence, IJCAI 2017, Melbourne, Australia, 19–25 August 2017, pp. 667–673. ijcai.org (2017). https://doi.org/10.24963/ijcai.2017/93
17. Lee, C.Y.: Representation of switching circuits by binary-decision programs. Bell Syst. Tech. J. **38**, 985–999 (1959). https://api.semanticscholar.org/CorpusID:123528408
18. O'Sullivan, B., Provan, G.M.: Approximate compilation for embedded model-based reasoning. In: Proceedings, The Twenty-First National Conference on Artificial Intelligence and the Eighteenth Innovative Applications of Artificial Intelligence Conference, 16–20 July 2006, Boston, Massachusetts, USA, pp. 894–899. AAAI Press (2006). http://www.aaai.org/Library/AAAI/2006/aaai06-141.php
19. Oztok, U., Darwiche, A.: A top-down compiler for sentential decision diagrams. In: Yang, Q., Wooldridge, M.J. (eds.) Proceedings of the Twenty-Fourth International Joint Conference on Artificial Intelligence, IJCAI 2015, Buenos Aires, Argentina, 25–31 July 2015, pp. 3141–3148. AAAI Press (2015). http://ijcai.org/Abstract/15/443
20. Renkens, J., Kimmig, A., den Broeck, G.V., Raedt, L.D.: Explanation-based approximate weighted model counting for probabilistic logics. In: Statistical Relational Artificial Intelligence, Papers from the 2014 AAAI Workshop, Québec City, Québec, Canada, 27 July 2014. AAAI Technical Report, vol. WS-14-13. AAAI (2014). http://www.aaai.org/ocs/index.php/WS/AAAIW14/paper/view/8803

21. Sang, T., Beame, P., Kautz, H.A.: Performing Bayesian inference by weighted model counting. In: Veloso, M.M., Kambhampati, S. (eds.) Proceedings, The Twentieth National Conference on Artificial Intelligence and the Seventeenth Innovative Applications of Artificial Intelligence Conference, 9–13 July 2005, Pittsburgh, Pennsylvania, USA, pp. 475–482. AAAI Press/The MIT Press (2005). http://www.aaai.org/Library/AAAI/2005/aaai05-075.php
22. Sharma, S., Gupta, R., Roy, S., Meel, K.S.: Knowledge compilation meets uniform sampling. In: Barthe, G., Sutcliffe, G., Veanes, M. (eds.) LPAR-22. 22nd International Conference on Logic for Programming, Artificial Intelligence and Reasoning, Awassa, Ethiopia, 16–21 November 2018. EPiC Series in Computing, vol. 57, pp. 620–636. EasyChair (2018). https://doi.org/10.29007/h4p9
23. Valiant, L.G.: The complexity of enumeration and reliability problems. SIAM J. Comput. **8**(3), 410–421 (1979). https://doi.org/10.1137/0208032
24. Williams, R., Gomes, C.P., Selman, B.: Backdoors to typical case complexity. In: Gottlob, G., Walsh, T. (eds.) IJCAI-03, Proceedings of the Eighteenth International Joint Conference on Artificial Intelligence, Acapulco, Mexico, 9–15 August 2003, pp. 1173–1178. Morgan Kaufmann (2003). http://ijcai.org/Proceedings/03/Papers/168.pdf

Leveraging Quantum Computing for Accelerated Classical Algorithms in Power Systems Optimization

Rosemary Barrass[1](✉), Harsha Nagarajan[2], and Carleton Coffrin[2]

[1] Industrial & Systems Engineering, Georgia Institute of Technology, Atlanta, GA, USA
rbarrass3@gatech.edu
[2] Los Alamos National Laboratory, Los Alamos, NM, USA
{harsha,cjc}@lanl.gov

Abstract. The recent advent of commercially available quantum annealing hardware (QAH) has expanded opportunities for research into quantum annealing-based algorithms. In the domain of power systems, this advancement has driven increased interest in applying such algorithms to mixed-integer problems (MIP) like Unit Commitment (UC). UC focuses on minimizing power generator operating costs while adhering to physical system constraints. Grid operators solve UC instances daily to meet power demand and ensure safe grid operations. This work presents a novel hybrid algorithm that leverages quantum and classical computing to solve UC more efficiently. We introduce a novel Benders-cut generation technique for UC, thereby enhancing cut quality, reducing expensive quantum-classical hardware interactions, and lowering qubit requirements. Additionally, we incorporate a k-local neighborhood search technique as a recovery step to ensure a higher quality solution than current QAH alone can achieve. The proposed algorithm, QC4UC, is evaluated on a modified instance of the IEEE RTS-96 test system. Results from both a simulated annealer and real QAH are compared, demonstrating the effectiveness of this algorithm in reducing qubit requirements and producing near-optimal solutions on noisy QAH.

Keywords: Quantum Computing · Unit Commitment · Benders' Decomposition · Hybrid Quantum-Classical Algorithm

1 Introduction

We stand on the edge of a technological revolution, where the rise of quantum computing promises to redefine our capabilities across a myriad of fields, unlocking new pathways for solving a wide range of complex optimization problems. Notably, quantum computers have demonstrated great promise in tackling combinatorial optimization problems, offering a glimpse into the potential for quantum computing to reshape complex decision-making processes. One area poised for significant transformation is power systems optimization, where mixed-integer optimization problems like unit commitment (UC) must be solved multiple times every day to maintain safe operations of massive power grids. UC is a fundamental optimization problem in power grid operations that determines which generating units should be online at any given time to meet anticipated electricity demand over a specified planning horizon, typically ranging from hours to days. The goal of UC is to minimize overall energy production costs while satisfying engineering constraints, such as generation limits, minimum up and down times of power plants, and operational reliability requirements. Effective UC decision-making is crucial for maintaining the stability and efficiency of power systems, as it ensures that there is a sufficient supply of electricity to meet consumer needs while minimizing wastage of resources. For this reason, researchers have started to explore the potential of quantum and hybrid quantum-classical (HQC) algorithms for speeding up the solution of UC and other critical power systems optimization problems.

Research on quantum and HQC algorithms for power systems optimization remains in its early stages, hindered by the limited availability of QPUs and the inherent constraints of NISQ hardware. Demonstrating quantum advantage in practical applications remains challenging due to three fundamental issues. First, QPUs suffer from noise and decoherence, leading to unreliable computations [2,18]. Second, their small scale restricts computational capacity, delaying the realization of quantum advantage. While IBM and Google have recently surpassed the 1000-qubit threshold, the effective number of programmable qubits remains significantly lower due to the overhead of error correction [19]. Third, quantum circuit generation and embedding introduce substantial time complexity, with D-Wave's own embedding algorithm exhibiting exponential scaling in practical implementations [3]. These challenges underscore the pressing need for algorithmic strategies that can adapt to noisy outputs, limited qubit resources, and high computational overhead.

Despite the limitations of NISQ-era QPUs, evidence suggests that certain power system problems are well-suited for testing on small, noisy quantum hardware [8]. HQC algorithms have shown promise as an intermediary approach for solving MIPs with existing quantum technology. Hybrid quantum-classical algorithms have been developed for a range of mixed-integer power system applications, including unit commitment [4,6,13,17,20,21], micro-grid scheduling [9,16], multi-energy system optimization [14], and optimal transmission switching [5]. Among HQC-based unit commitment approaches, research is largely divided between quantum approximation optimization algorithms (QAOA) and quantum annealing (QA). Koretsky et al. applied QAOA within a hybrid framework, partitioning UC into an inner binary optimization problem solved by a QPU and an outer linear optimization problem handled classically [13]. This formulation, common in HQC algorithms, reduces qubit requirements for large-scale problems, enabling demonstrations of quantum speed-up on current hardware. In Koretsky et al.'s study, this speed-up is observed; however, due to hardware limitations, the quantum circuit is simulated using simulated annealing (SA) rather than executed on a physical QPU.

Paterakis introduced an HQC algorithm for solving the UC problem, leveraging a QPU to optimize cut selection within a Benders' decomposition framework [1,17]. The algorithm demonstrated the benefits of quantum-assisted optimal cut selection on an 8-bus system. However, the constraints of NISQ hardware led to excessive embedding overhead, diminishing its viability compared to purely classical approaches. Notably, Paterakis' approach struggled with scalability, failing to improve performance on a larger 30-bus system–despite the expectation that quantum advantage should manifest as problem size grows. This study underscores the inherent challenges of current noisy QPUs, which remain ill-suited for handling large-scale combinatorial optimization in their present form.

These algorithms highlight the necessity for designing HQC methods that account for NISQ-era constraints, including noise-induced errors in large problems, qubit embedding challenges, and excessive iteration overhead from quantum-classical data transfer. This paper presents QC4UC, a novel HQC algorithm tailored for the UC problem, explicitly addressing the limitations of small-scale noisy quantum hardware. By harnessing quantum computing's unique capabilities alongside classical optimization, QC4UC enhances solution accuracy while reducing qubit requirements compared to previous hybrid approaches.

2 Methodology

This paper presents a novel hybrid algorithm for operation on NISQ-era quantum computers which implements a recovery step for smoothing out noise and clever cut generation to reduce both qubit requirements and quantum-classical interactions. The proposed algorithm is comprised of two sub-algorithms:

- A classical Benders' decomposition algorithm with a QUBO main problem (MP) and linear programming (LP) sub-problems (SP). The QUBO MP is solved using either SA or QA, gathering n samples to at each iteration to be used in n SP. In this paper, the QUBO is solved using both SA and QA for comparison. The n SP are solved on a classical computer to find the largest violating QUBO solution. The SP with the largest violating solution is then used to generate a rounded cut, which is added to the QUBO. This algorithm is repeated until the load and generation penalties are small, with some upper bound set by the user. If, during the execution of this algorithm, some i iterations do not show large improvements in the lower bound, a general variable neighborhood search (GVNS) is employed [15]. This step effectively "shakes" the solution if a local minimum has been found that is difficult to escape with annealing techniques. This solution shaking acts as a way to "reset" the current solution without requiring step size tuning in annealing algorithm (Fig. 1).
- A k-local neighborhood search [7] is employed after the completion of the Benders' decomposition algorithm to act as a feasibility recovery step. The user picks some k integer value, which should be small. This k is the upper bound for a maximum Hamming distance cut, which is added to the original full MIP formulation. This cut greatly reduces the search space for the MIP

and acts as a recovery step to bring poor QA solutions to the feasible region of the original problem.

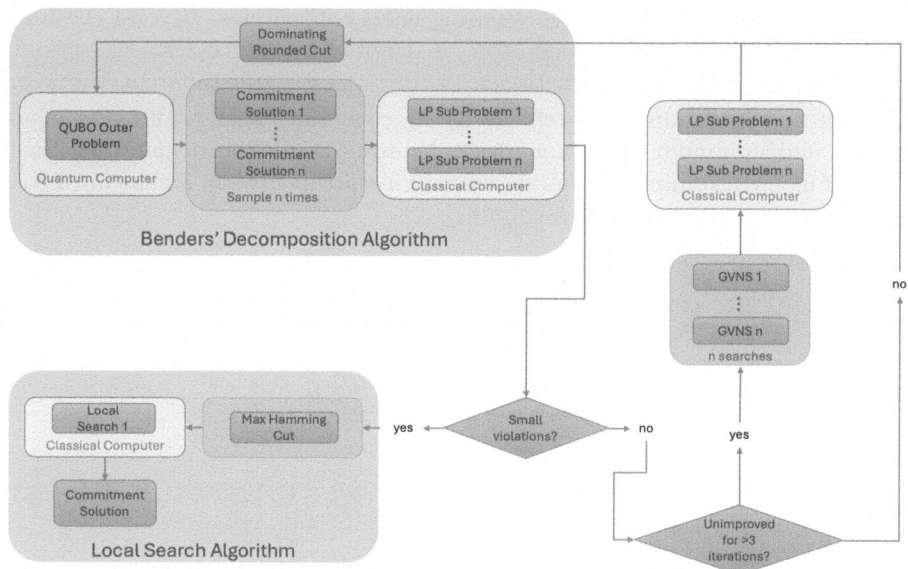

Fig. 1. The proposed iterative algorithm with rounded cuts, general variable neighborhood search step, and k-local neighborhood search recovery step.

This new algorithm is applied to a realistic UC formulation, which reflects the pertinent power systems and engineering constraints. The formulation includes DC power flow constraints and thermal line limits, the latter of which is not included in prior HQC literature. These constraints, however, are standard in power systems and represent critical operating constraints in real-world systems. The inclusion of thermal line constraints introduces additional complexity to the MIP, which our proposed QC4UC algorithm can successfully solve for larger instances than previous work. We compare our QC4UC algorithm with a classical version of QC4UC which uses SA to solve the QUBO MP.

The full UC formulation is described in Sect. 2.1. The Benders' decomposition algorithm with GVNS and dynamic binary encoding is described in Sect. 2.2. The k-local neighborhood search recovery step is described in Sect. 2.4. Computational results are presented in Sect. 3.

2.1 Problem Formulation

The fundamental UC formulation, which serves as the basis for our approach, is given below. In this formulation, $\mathcal{T} = \{1, 2, \ldots, T\}$ represents the set of time steps t from 1 to T, and $\mathcal{G} = \{1, 2, \ldots, G\}$ denotes the set of generators g from

1 to G. All variables are bolded to differentiate them from constants. Each generator g at time t is characterized by three binary variables: $\mathbf{u}_{g,t}$ (on/off), $\mathbf{v}_{g,t}$ (startup), and $\mathbf{w}_{g,t}$ (shutdown). A generator must remain online for at least T_g^{minup} time steps after startup and offline for at least T_g^{mindn} after shutdown. Marginal generator dispatch—representing energy dispatched beyond the minimum required for an online generator—is denoted by the continuous variable $\mathbf{p}_{g,t}$. This value is constrained between 0 and $\overline{p}_{g,t} - \underline{p}_{g,t}$, the difference between maximum and minimum dispatch levels. At each time step t, the sum of all marginal and minimum generator dispatch must satisfy the demand d_t. Each generator incurs a minimum running cost c_1, a startup cost c_2, a shutdown cost c_3, and quadratic dispatch costs c_4 and c_5.

$$\min_{\mathbf{u},\mathbf{p}} \quad c_1^T \mathbf{u} + c_2^T \mathbf{v} + c_3^T \mathbf{w} + (c_4)^T (\mathbf{p} \otimes \mathbf{p}) + (c_5)^T \mathbf{p} \qquad \text{(Basic UC)}$$

$$\text{s.t.} \quad \mathbf{u}_{g,t} - \mathbf{u}_{g,t-1} = \mathbf{v}_{g,t} - \mathbf{w}_{g,t}, \qquad \forall g, t \in \mathcal{G}, \mathcal{T} \qquad (1)$$

$$\sum_{\tau=t-T_g^{\text{minup}}+1}^{t} \mathbf{v}_{g,\tau} \leq \mathbf{u}_{g,t}, \qquad \forall g, t \in \mathcal{G}, \{T_g^{\text{minup}}, \ldots, \mathcal{T}\} \qquad (2)$$

$$\sum_{\tau=t-T_g^{\text{mindn}}+1}^{t} \mathbf{w}_{g,\tau} \leq 1 - \mathbf{u}_{g,t}, \qquad \forall g, t \in \mathcal{G}, \{T_g^{\text{mindn}}, \ldots, \mathcal{T}\} \qquad (3)$$

$$\sum_{g \in \mathcal{G}} \mathbf{p}_{g,t} + \sum_{g \in \mathcal{G}} \left(\underline{p}_{g,t} \cdot \mathbf{u}_{g,t} \right) = d_t, \qquad \forall t \in \mathcal{T} \qquad (4)$$

$$\mathbf{u}_{g,t}, \mathbf{v}_{g,t}, \mathbf{w}_{g,t} \in \{0,1\}, \qquad \forall g, t \in \mathcal{G}, \mathcal{T} \qquad (5)$$

$$0 \leq \mathbf{p}_{g,t} \leq \overline{p}_{g,t} - \underline{p}_{g,t}, \qquad \forall g, t \in \mathcal{G}, \mathcal{T} \qquad (6)$$

In this paper, costs are linearized using a piecewise approximation. The generator dispatch variables for each generator g are partitioned into L segments between the minimum and maximum dispatch levels. The fraction of the total energy dispatch allocated to segment l for generator g at time t is denoted by $\boldsymbol{\alpha}_{g,t}^l$. The cost coefficient for segment l in the piecewise linear approximation of the generation cost is given by $c6_{g,t}^l$. Thus, the total marginal energy dispatch for generator g at time t is the sum of these segments:

$$\mathbf{p}_{g,t} = \sum_{l \in \mathcal{L}_g} (p_{g,t}^l - \underline{p}_{g,t}) \cdot \boldsymbol{\alpha}_{g,t}^l, \qquad \forall g, t \in \mathcal{G}, \mathcal{T} \qquad (7)$$

All of these fractions must add up to 1 when the generator is online and 0 when the generator is offline:

$$\sum_{l \in \mathcal{L}_g} \boldsymbol{\alpha}_{g,t}^l = \mathbf{u}_{g,t}, \qquad \forall g, t \in \mathcal{G}, \mathcal{T} \qquad (8)$$

In addition to these linearizing constraints, practical engineering constraints are incorporated to enhance solution quality. We define $B_{n,j}$ as the susceptance of the power line connecting bus n to bus j, and $\boldsymbol{\theta}_{n,t}$ as the phase angle at bus n

at time t. The set $\mathcal{N} = \{1, \ldots, N\}$ represents all buses in the network. To ensure that the thermal limits $F_{n,j}^{\max}$ of each line (n,j) are not exceeded, we impose the following constraint:

$$B_{n,j}|(\boldsymbol{\theta}_{n,t} - \boldsymbol{\theta}_{j,t})| \leq F_{n,j}^{\max}, \quad \forall n \in \mathcal{N} \tag{9}$$

The following ramping constraints are introduced to limit the change in power dispatch that each generator can achieve between consecutive time steps. R_g^{startup} and R_g^{up} define the maximum allowable increases in energy dispatch during startup and between successive time steps, respectively. Similarly, R_g^{shutdown} and R_g^{down} specify the maximum allowable decreases during shutdown and between consecutive time steps. The variable $\mathbf{r}_{g,t}$ represents the spinning reserve provided by generator g at time t.

$$\mathbf{p}_{g,t} + \mathbf{r}_{g,t} \leq (\overline{p}_{g,t} - \underline{p}_{g,t}) \cdot \mathbf{u}_{g,t}$$
$$- \max\{\overline{p}_{g,t} - R_g^{\text{startup}}, 0\} \cdot \mathbf{v}_{g,t}, \quad \forall g, t \in \mathcal{G}, \mathcal{T} \tag{10}$$

$$\mathbf{p}_{g,t} + \mathbf{r}_{g,t} \leq (\overline{p}_{g,t} - \underline{p}_{g,t}) \cdot \mathbf{u}_{g,t}$$
$$- \max\{\overline{p}_{g,t} - R_g^{\text{shutdown}}, 0\} \cdot \mathbf{w}_{g,t+1}, \quad \forall g, t \in \mathcal{G}, \mathcal{T}\backslash\{T\} \tag{11}$$

$$\mathbf{p}_{g,t} + \mathbf{r}_{g,t} - \mathbf{p}_{g,t-1} \leq R_g^{\text{up}}, \quad \forall g, t \in \mathcal{G}, \mathcal{T}\backslash\{1\} \tag{12}$$

$$\mathbf{p}_{g,t-1} - \mathbf{p}_{g,t} \leq R_g^{\text{down}}, \quad \forall g, t \in \mathcal{G}, \mathcal{T}\backslash\{1\} \tag{13}$$

Including constraints (7) through (13) to (Basic UC) gives us the final MILP UC formulation. $\underline{c_6}_{g,t}$ is added to represent the minimum generation dispatch cost for generator g at time t ($\underline{c_6}_{g,t}^1 = \underline{c_6}_{g,t}$).

$$\min_{\mathbf{u},\mathbf{p}} \; c_1^T \mathbf{u} + c_2^T \mathbf{v} + c_3^T \mathbf{w} + \sum_{l \in \mathcal{L}_g}(c_{6g,t}^l - \underline{c_6}_{g,t})\boldsymbol{\alpha}_{g,t}^l \quad \text{(Full UC)}$$

s.t. $(1), (2), (3), (6), (7), (8), (9), (10), (11), (12), (13)$

$$\sum_{g \in \mathcal{G}} \mathbf{p}_{g,t} + \sum_{g \in \mathcal{G}} \left(\underline{p}_{g,t} \cdot \mathbf{u}_{g,t}\right) - d_{n,t} =$$
$$\sum_{(n,j) \in \mathcal{L}} B_{n,j}(\boldsymbol{\theta}_{n,t} - \boldsymbol{\theta}_{j,t}), \quad \forall n, t \in \mathcal{N}, \mathcal{T} \tag{14}$$

$$\mathbf{u}_{g,t}, \mathbf{v}_{g,t}, \mathbf{w}_{g,t} \in \{0,1\}, \quad \forall g, t \in \mathcal{G}, \mathcal{T} \tag{15}$$

$$\mathbf{r}_{g,t} \geq 0, \quad \forall g, t \in \mathcal{G}, \mathcal{T} \tag{16}$$

$$\boldsymbol{\alpha}_{g,t}^l \geq 0, \quad \forall l, g, t \in \mathcal{L}_g, \mathcal{G}, \mathcal{T} \tag{17}$$

$$\boldsymbol{\theta}_{\text{ref},t} = 0, \quad \forall t \in \mathcal{T} \tag{18}$$

2.2 Benders' Decomposition Algorithm

This section presents the proposed Benders' decomposition algorithm (QC4UC), incorporating a novel combination of a dynamic precision encoding technique, rounded cut generation, and a general variable neighborhood search. The MILP formulation (Full UC) is decomposed into a binary optimization MP, solved using QAH, and a continuous SP, handled by a commercial branch-and-bound solver on a CPU. η represents the MP's estimation of the SP objective value. The MP formulation is as follows:

$$\min_{\mathbf{u},\mathbf{p}} c_1^T \mathbf{u} + c_2^T \mathbf{v} + c_3^T \mathbf{w} + \eta \qquad \text{(MP)}$$

$$\text{s.t.} \quad (1), (2), (3), (15)$$

The MP generates solutions u^*, v^*, w^* which represent the commitment, startup, and shutdown decisions. In order to ensure all MP solutions are feasible in the subproblem, δ_t^+, δ_t^- are added to the SP as penalty terms for relaxing the power balance, where δ_t^+ represents overproduction of energy and δ_t^- represents underproduction of energy. The SP take the form:

$$Q(u^*) = \min_{\boldsymbol{\alpha},\mathbf{p}} \sum_{l \in \mathcal{L}_g} (c_{6_{g,t}}^l - \underline{c_{6}}_{g,t}) \alpha_{g,t}^l + c_t^{\text{pen}} (\delta_t^+ - \delta_t^-) \qquad \text{(SP)}$$

$$\text{s.t.} \quad (6), (7), (8), (9), (12), (13), (16), (17), (18)$$

$$\sum_{g \in \mathcal{G}} \mathbf{p}_{g,t} + \sum_{g \in \mathcal{G}} \left(\underline{p}_{g,t} \cdot u_{g,t}^* \right) - d_{n,t} =$$

$$\sum_{(n,j) \in \mathcal{L}} B_{n,j} (\boldsymbol{\theta}_{n,t} - \boldsymbol{\theta}_{j,t}) + (\delta_t^+ + \delta_t^-), \qquad \forall n, t \in \mathcal{N}, \mathcal{T} \quad (19)$$

$$(10) \text{ with fixed values } u_{g,t}^*, v_{g,t}^* \qquad (20)$$

$$(11) \text{ with fixed values } u_{g,t}^*, w_{g,t}^* \qquad (21)$$

$$\delta_t^+ \geq 0, \ \delta_t^- \leq 0, \qquad \forall t \in \mathcal{T} \quad (22)$$

These penalty terms have an associated penalty cost c_t^{pen} which is larger than the highest generator dispatch cost in the system. λ_i^*'s are introduced as the dual variables for the constraints of the SP. Note that λ_1^* is always zero, as the first SP constraint has a right-hand side (RHS) value of 0. The optimality cut generated by each SP takes the form

$$(\mathbf{u}_{g,t})^T \lambda_2^* \qquad \text{(OptCut)}$$

$$+ \left(d_{n,t} - \sum_{g \in \mathcal{G}_n} \left(\underline{p}_{g,t} \cdot \mathbf{u}_{g,t} \right) \right)^T \lambda_3^*$$

$$+ \left((\overline{p}_{g,t} - \underline{p}_{g,t}) \cdot \mathbf{u}_{g,t} - \max\{\overline{p}_{g,t} - R_g^{\text{startup}}, 0\} \cdot \mathbf{v}_{g,t} \right)^T \lambda_4^*$$

$$+ \left((\overline{p}_{g,t} - \underline{p}_{g,t}) \cdot \mathbf{u}_{g,t} - \max\{\overline{p}_{g,t} - R_g^{\text{shutdown}}, 0\} \cdot \mathbf{w}_{g,t+1} \right)^T \lambda_5^*$$

$$+ (R_g^{\text{up}})^T \lambda_6^* + (R_g^{\text{down}})^T \lambda_7^* + (F_{n,j}^{\text{max}})^T \lambda_8^*$$

$$\leq \eta$$

In order to solve this problem with QAH, the MP and optimality cuts are converted to a QUBO form. To do this, we introduce P_j, $j = 1, \ldots, 4$ as penalty values that are tuned empirically to penalize distance from the feasible region. $\mathbf{s}_1, \mathbf{s}_2, \mathbf{s}_3$ are added as slack variables to relax constraints in the MP. The expression $\text{OptCut}_{\text{LHS}}(\lambda_2^*, \lambda_2^*, \ldots, \lambda_8^*)$ is introduced to represent left-hand side (LHS) of the optimality cut with fixed dual variable solutions $\lambda_i^*, i = 2, \ldots, 8$ and unfixed binary variables $\mathbf{u}, \mathbf{v}$, and $\mathbf{w}$. The QUBO form of the MP thus becomes

$$\min_{\mathbf{u},\mathbf{v},\mathbf{w}} c_1^T \mathbf{u} + c_2^T \mathbf{v} + c_3^T \mathbf{w} + \eta \tag{MP-QUBO}$$

$$+ P_1 \sum_{g \in \mathcal{G}} \sum_{t \in \mathcal{T}} (\mathbf{u}_{g,t} - \mathbf{u}_{g,t-1} - \mathbf{v}_{g,t} + \mathbf{w}_{g,t})^2 \tag{23}$$

$$+ P_2 \sum_{g \in \mathcal{G}} \sum_{t \in \{T_g^{\text{minup}}, \ldots, T\}} \left(\sum_{\tau = t - T_g^{\text{minup}} + 1}^{t} \mathbf{v}_{g,\tau} + \mathbf{s}_1 - \mathbf{u}_{g,t} \right)^2 \tag{24}$$

$$+ P_3 \sum_{g \in \mathcal{G}} \sum_{t \in \{T_g^{\text{mindn}}, \ldots, T\}} \left(\sum_{\tau = t - T_g^{\text{mindn}} + 1}^{t} \mathbf{w}_{g,\tau} + \mathbf{s}_2 - 1 + \mathbf{u}_{g,t} \right)^2 \tag{25}$$

$$+ P_4 \left(\text{OptCut}_{\text{LHS}}(\lambda_2^*, \lambda_2^*, \ldots, \lambda_8^*) - \eta + \mathbf{s}_3 \right)^2 \tag{26}$$

Note that the slack variables $\mathbf{s}_1, \mathbf{s}_2$ are binary by nature of the inequalities that they are used in. We can show this by analyzing the minimum up time constraint:

$$\sum_{\tau = t - T_g^{\text{minup}} + 1}^{t} \mathbf{v}_{g,\tau} \leq \mathbf{u}_{g,t}$$

Since that $\mathbf{v}_{g,t}$ and $\mathbf{u}_{g,t}$ are binary for all $g, t \in \mathcal{G}, \mathcal{T}$, $\sum_{\tau=t-T_g^{\text{minup}}+1}^{t} \mathbf{v}_{g,\tau} - \mathbf{u}_{g,t}$ is bounded between -1 and 0. With slack variable $\mathbf{s}_1$ to create an equality, we have,

$$\sum_{\tau = t - T_g^{\text{minup}} + 1}^{t} \mathbf{v}_{g,\tau} - \mathbf{u}_{g,t} + \mathbf{s}_1 = 0,$$

implying that $\mathbf{s}_1$ can only be a discrete value between 0 and 1. A similar argument holds for the minimum down time constraint to confirm its binary nature.

In contrast, $\mathbf{s}_3$ must be discretized for the QUBO formulation. To address the constraints of NISQ-era quantum hardware, which offers a very limited number of qubits, we implement a novel dynamic precision encoding step and rounded cuts, essential for minimizing the qubits required to binary encode $\mathbf{s}_3$ and η in each iteration of Benders' decomposition.

Dynamic Precision Encoding. The slack variable $\mathbf{s}_3$ can be bounded, which allows for a novel dynamic qubit allocation regime for the cut slack variable at each iteration. To do this, consider the original optimality cut (OptCut). In the MP, the decision variables in this cut will be $\mathbf{u}, \mathbf{v}, \mathbf{w}$ and η. For a minimization problem, we can assume that η will be chosen such that the left hand side will be equal to η. Thus, a worst-case lower bound on the left-hand side can be used to determine an approximate precision necessary for η. So, at each iteration of our algorithm, we solve a small optimization problem of the form:

$$\max_{\mathbf{u},\mathbf{v},\mathbf{w}} \quad \mathrm{OptCut}_{\mathrm{LHS}}(\lambda_2^*, \lambda_2^*, \ldots, \lambda_8^*) \qquad \text{(MaxEta)}$$

$$\mathbf{u}_{g,t} - \mathbf{u}_{g,t-1} = \mathbf{v}_{g,t} - \mathbf{w}_{g,t}, \qquad \forall g, t \in \mathcal{G}, \mathcal{T}$$

$$\sum_{\tau=t-T_g^{\mathrm{minup}}+1}^{t} \mathbf{v}_{g,\tau} \leq \mathbf{u}_{g,t}, \qquad \forall g, t \in \mathcal{G}, \{T_g^{\mathrm{minup}}, \ldots, \mathcal{T}\}$$

$$\sum_{\tau=t-T_g^{\mathrm{mindn}}+1}^{t} \mathbf{w}_{g,\tau} \leq 1 - \mathbf{u}_{g,t}, \qquad \forall g, t \in \mathcal{G}, \{T_g^{\mathrm{mindn}}, \ldots, \mathcal{T}\}$$

$$\mathbf{u}_{g,t}, \mathbf{v}_{g,t}, \mathbf{w}_{g,t} \in \{0, 1\}, \qquad \forall g, t \in \mathcal{G}, \mathcal{T}$$

which is bounded. The solution of this IP is then used to inform the precision parameter p to be used in the slack variable encoding for each cut.

Rounded Optimality Cuts. In addition to dynamic precision encoding, we introduce a novel rounding technique for the L-shaped cuts of Benders' decomposition to further reduce qubit requirements while ensuring validity as a lower bound to the subproblem's convex recourse function. Consider the optimality cuts (OptCut), simplified to the form:

$$a_1^T \mathbf{x} + a_2^T \mathbf{y} + a_3^T \mathbf{z} + b \leq \eta,$$

where η is a decision variable that is minimized in the MP's objective function to provide a valid lower approximation to the true recourse function of the SP. Here, $a_1, a_2, a_3 \in \mathbb{R}^n$, $b \in \mathbb{R}$, and $\mathbf{x}, \mathbf{y}, \mathbf{z} \in \{0,1\}^n$. To embed this inequality into the QUBO model, we introduce a slack variable $\mathbf{s}_3 \geq 0$: $a_1^T \mathbf{x} + a_2^T \mathbf{y} + a_3^T \mathbf{z} + b + \mathbf{s}_3 = \eta$. Since a_1, a_2, a_3, b are real numbers, the slack variable $\mathbf{s}_3$ is also continuous. To encode $\mathbf{s}_3$ with p decimal places of precision, at least

$$N = \lceil \log_2(\eta^*(10^p) + 1) \rceil$$

binary variables are required, where $\mathbf{s}_3 \in [0, \eta^*]$ and η^* is the objective value of (MaxEta). To reduce the binary variable count, we apply a *conservative rounding* to the coefficients. This ensures that the resulting integer-based cut is never more restrictive than the original fractional Benders' cut, thereby preserving a valid lower bound to the convex recourse as η is minimized. Specifically, we define the rounded coefficients $\tilde{a}_1, \tilde{a}_2, \tilde{a}_3 \in \mathbb{Z}^n$ and $\tilde{b} \in \mathbb{Z}$, yielding the new constraint:

$$\tilde{a}_1^T \mathbf{x} + \tilde{a}_2^T \mathbf{y} + \tilde{a}_3^T \mathbf{z} + \tilde{b} + \mathbf{s}_3 = \eta.$$

To maintain validity, we round each coefficient as follows:

- Positive coefficients ($[a_i]_j > 0$): round down, $[\tilde{a}_i]_j = \lfloor [a_i]_j \rfloor$.
- Negative coefficients ($[a_i]_j < 0$): round up, $[\tilde{a}_i]_j = \lceil [a_i]_j \rceil$.
- Constant term ($b > 0$): round down, $\tilde{b} = \lfloor b \rfloor$.

With rounded integer coefficients, the slack variable s_3 can also be discretized to an integer domain, taking $\eta^* + 1$ possible integer values, requiring only:

$$M = \lceil \log_2(\eta^* + 1) \rceil$$

binary variables. For large η^*, the resulting difference in the number of binary variables before and after rounding, $N - M$, simplifies approximately to:

$$\lceil p \log_2 10 \rceil \approx \lceil 3.322 p \rceil .$$

Thus, rounding the coefficients to the nearest integers and using an integer-valued slack variable significantly reduces the binary variables needed to represent s_3, thereby simplifying the QUBO formulation and preserving a valid lower bound, while balancing the trade-off between precision and solution quality.

2.3 General Variable Neighborhood Search

QA and SA do not guarantee convergence, meaning the Benders' algorithm may iterate indefinitely without reaching the optimal solution, especially given the reduced precision of the rounded cuts. This often occurs when the annealer gets trapped in a local minimum that the current step size does not allow "escape" from. One way to address this is through heuristic methods, such as variable flipping or tabu search [10], to force solutions out of difficult-to-escape local minima. In classical computing, these methods iterate from a single solution, making random modifications to generate new ones. Since they do not guarantee convergence to the global optimum, the final solution from a General Variable Neighborhood Search (GVNS) can vary. This provides an opportunity to leverage multiple solution samples from the quantum or simulated annealer for improved results. The proposed algorithm iterates Benders' cuts until no further solution improvements are observed. Here, we define some small distance between objective values at each iteration. If the algorithm produces a solution with an objective value within this distance from the last iteration, then this iteration is considered to be non-improving. After i consecutive non-improving iterations, it is assumed that the annealer is stuck and the GVNS is employed. Parallel GVNS will then be performed using n solutions sampled from the quantum or simulated annealer, with n defined by the user. Each solution is used to solve the SP, generating a new cut to add to the MP for further iterations of the Benders' decomposition algorithm.

The GVNS includes two stages: a local improvement stage and a shaking stage. In the local improvement stage, k_1 commitment variables are randomly flipped from 0 to 1 or vice versa until a solution that improves the objective value is found. In the shaking stage, k_2 commitment variables are randomly flipped without considering the objective value. These stages are applied iteratively across the n QUBO samples.

2.4 k-Local Neighborhood Search Algorithm

For the k-local neighborhood search algorithm, the full MILP is limited by a maximum Hamming distance constraint which significantly reduces the size of the search space to a ball around the final Benders' decomposition solution. The maximum Hamming distance constraint takes the form

$$\sum_{g,t:\mathbf{u}_{g,t}^*=0} \mathbf{u}_{g,t} + \sum_{g,t:\mathbf{u}_{g,t}^*=1} 1 - \mathbf{u}_{g,t} = k + \mathbf{s}_4 \qquad (27)$$

where $\mathbf{s}_4$ is some integer valued slack variable. This constraint is added to (Full UC) and solved on a classical computer.

3 Computational Results

The proposed algorithm, tested on a significantly larger test case than any previous work, demonstrates effective qubit reduction and achieves near-optimal solutions on noisy QAH. This is evidenced by comparative results from both SA and real QAH. In this section, QC4UC is tested on a modified IEEE RTS-96 bus system, which contains 8 thermal generators, 27 renewable generators, and 24 buses. For the MP, we compare the D-Wave SA package "dwave-neal" version 0.6.0 [11] on a classical computer and the D-Wave QA package "dwave-system" version 1.26.0 [12] on QAH. The SA package was developed to emulate a noiseless version of D-Wave's current QAH and acts as a way to compare the current NISQ-era QAH to what possible perfect hardware could produce. The classical computer used to solve the SPs and the SA tests has an Apple M3 Max 16-core CPU. The QA computations are performed on the Advantage_system4.1 QPU, which comprises 5612 operational qubits arranged in the Pegasus architecture with a connectivity range of $[-18, 15]$ per qubit, operating at a cryogenic temperature of 15.4 mK. The qubit thermal width is 0.198 Ising units, and the quantum critical point is observed at 1.391 GHz, ensuring robust performance during annealing. The SPs and recovery step are solved on the classical computer using Gurobi 11.0.2 with default settings. A single solution of (Full UC) is produced using Gurobi 11.0.2 with default settings on a classical computer to evaluate the solution quality of QC4UC. It is important to note here that QC4UC is not compared to the Gurobi solution to demonstrate any kind of improvement over classical methods. Instead, this Gurobi solution is used to demonstrate that QC4UC solutions are good.

3.1 Solution Quality on Quantum Annealing Hardware

The proposed QC4UC algorithm with QA is solved on the same test case 100 times to assess its consistency and mitigate the impact of noise. Among the 100 runs, the algorithm produced a solution within 1% of the Gurobi objective value 51 times. Of these, 24 solutions were within 0.01% of the Gurobi optimal value, as shown in Fig. 2. This is despite the fact QAH is only guaranteed to converge in probability to a solution by the law of large numbers. These results demonstrate the efficacy of the algorithm on real NISQ-era quantum hardware. They also highlight the recovery step's capability to produce high-quality solutions despite significant noise and the potential challenges posed by rounded cuts, such as sub-optimality or cutting into the feasible region of the UC problem. Figure 2 demonstrates that some of the very large optimality gaps are caused by penalty costs from the load shed penalties. The penalty costs for load shed are set to $5e5$ \$/MW to act as a big M for keeping load shed to 0, when possible. Of the 85 tests with no penalties, 81 of the tests are within 5% of the Gurobi objective value. This level of solution quality has yet to be achieved on a problem of this size in the literature. Overall, this reflects the strength of QC4UC in deriving near-optimal solutions for the UC problem using noisy, stochastic hardware.

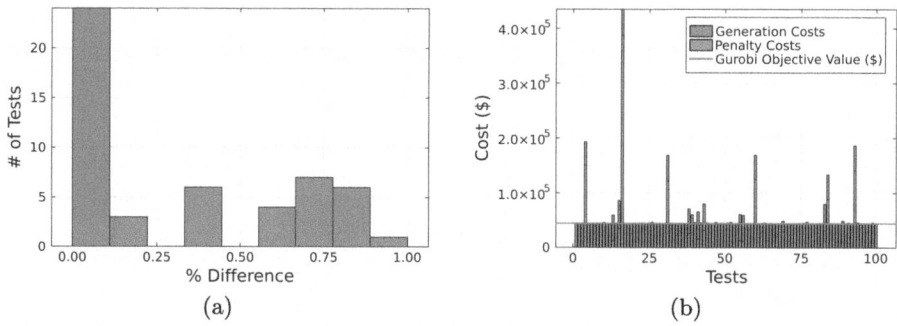

Fig. 2. (a) Histogram of the 51 out of 100 QC4UC with QA tests achieving within 1% of the Gurobi's optimal objective value. (b) Bar plot of penalty and generation costs of all 100 tests of QC4UC with QA.

Across the same number of independent runs, the QC4UC algorithm with simulated annealing (SA) achieved the optimal objective value for all runs, despite SA's lack of guaranteed convergence properties. The same rounded cuts and decomposition methods were applied for these tests as those conducted for QA. This demonstrates that the proposed algorithm will perform well on noiseless annealing methods. The consistency of results with SA also highlights the potential for future noiseless QAH.

3.2 Embedding Algorithm Solve Time and Algorithm Run Time

Figure 3 plots the embedding algorithm solve time, which is the total amount of time that it takes for the QA sampler to find an embedding for the problem using the methodology described in Cai et al.'s work [3]. The results show that the embedding algorithm solve time increases for each iteration of the algorithm. As the problem size of the QUBO MP increases with each cut, the computational complexity of the embedding search problem increases. These results highlight

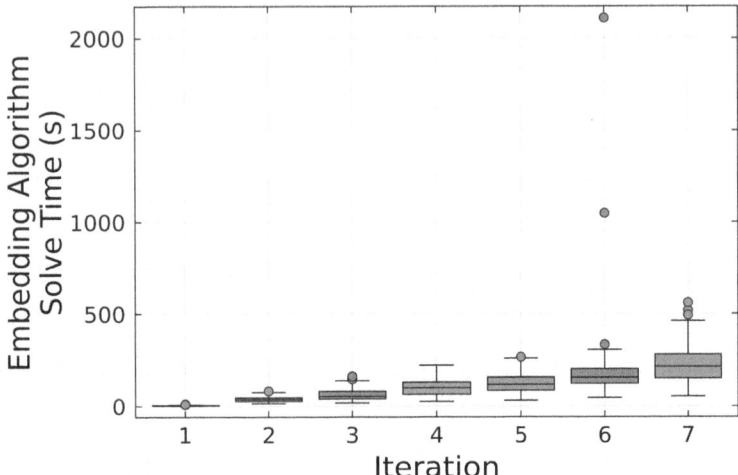

Fig. 3. Embedding algorithm solve time for QC4UC with QA for 100 independent tests.

the importance of qubit reduction techniques, such as the proposed rounded cuts and dynamic encoding regime implemented in QC4UC.

Outside of the embedding algorithm solve time, the total algorithm run time for each iteration of QC4UC with QA demonstrates the linear scaling advantage of QAH, as shown in Fig. 4. The total algorithm run time includes the time to solve the MP, time to solve the SP, and, when using QA, the actual embedding time. As the problem size grows with each iteration, QC4UC with QA scales better than QC4UC with SA. It is seen clearly that, despite the stochasticity of the QA solver, QC4UC with QA can solve the larger problems faster on average than QC4UC with SA. This demonstrates the necessity to test quantum algorithms on real quantum hardware, as classical computers cannot physically emulate the time scaling advantages of QA. It is important to note that the first iteration of QC4UC with QA has a much longer run time than the rest of the iterations. This is a result of the embedding of the QUBO into the QAH. In the first iteration, the full problem must be embedded, whereas the subsequent iterations must only add the new cuts to the embedding.

3.3 Dynamic Precision Encoding

From Fig. 4, we can see that the dynamic encoding of η allows us to reduce the qubit requirement over a static η encoding regime. In these tests, QC4UC solved within 7 to 8 iterations and varied between 16 and 21 qubits per iteration for encoding the estimated secondary cost. Without rounded cuts, 60 to 83 qubits were used per iteration to estimate the secondary cost. This total qubit savings represents a significant portion of the available qubits for currently available QAH. As QAH continues to increase in size, this dynamic encoding step will

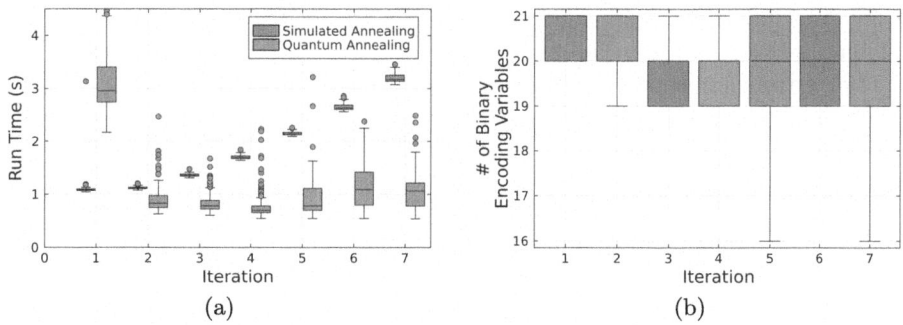

Fig. 4. (a) Total QC4UC run time per iteration, including MP solve time, SP solve time, and problem build time for QC4UC with SA and QC4UC with QA, excluding embedding algorithm solve time for QC4UC with QA. (b) Number of binary encoding variables used to approximate η at each iteration of QC4UC with QA.

allow the solution of larger and more realistic systems sooner than traditional static encoding regimes.

These results showcase the ability of QC4UC to enhance the solution accuracy of QA while minimizing qubit requirements. This is in contrast to results in prior literature, which demonstrate algorithmic performance using SA rather than QA or used small test cases. Our results have demonstrated remarkable ability to achieve high-quality results for large-scale problems on current NISQ-era hardware. QC4UC is able to reduce qubit requirements for cuts by about one third per iteration and achieve near-optimal solutions in about 50% of cases. Neither of these results have been seen before in the literature for a large-scale, complex UC instance like the one presented here.

4 Conclusion

The novel QC4UC algorithm introduces a rounded Benders' optimization cut formulation that significantly reduces binary precision encoding requirements compared to previous HQC algorithms. This improvement allows QC4UC to significantly lower qubit requirements, enabling the solution of larger UC instances on real NISQ hardware than any previous work, to the best of our knowledge. The QA-enabled algorithm achieves objective values within 1% of the Gurobi optimal solution in 51% of tests, showcasing its effectiveness in handling moderately sized power systems optimization problems. A key feature of this algorithm is the novel k-local search recovery step, which addresses noisy or near-feasible solutions. It refines suboptimal or infeasible solutions caused by the inherent QA noise, enhancing overall solution quality. However, a trade-off between solution quality and qubit requirements is observed, as expected in HQC approaches on NISQ hardware. This is addressed using a dynamic binary encoding step, enabling QC4UC to reduce qubit requirements where possible. This dynamic binary encoding approach also applies to broader MIPs using Benders' decomposition, enabling research on larger problems with real NISQ hardware.

The current method significantly improves scalability; however, further research could focus on several areas. First, reducing qubit requirements while maintaining tighter bounds on solution quality by refining the rounding technique or exploring alternative encoding strategies. Second, embedding algorithm solve time can be substantially reduced by dynamic cut management techniques. Finally, incorporating more complex and realistic UC formulations, including non-convex power flow physics and combined-cycle generators.

Acknowledgments. The authors acknowledge funding from the U.S. Department of Energy (DOE) under the Advanced Grid Modeling (AGM) program for the project "Rigorous Assessment and Development of Quantum and Hybrid Quantum-Classical Computing Approaches for Solving the Unit Commitment Problem." They also acknowledge the use of resources provided by the LANL's Institutional Computing Program, supported by the DOE National Nuclear Security Administration under Contract No. 89233218CNA000001. The authors would like to acknowledge Dr. Rabab Haider of University of Michigan Ann Arbor for her helpful feedback.

Disclosure of Interests. The authors have no competing interests to declare that are relevant to the content of this article.

References

1. Benders, J.: Partitioning procedures for solving mixed-variables programming problems. Numerische Mathematik **4**, 238–252 (1962). http://eudml.org/doc/131533
2. Brooks, M.: Quantum computing is taking on its biggest challenge: noise. MIT Technol. Rev. (2024). https://www.technologyreview.com/2024/01/04/1084783/quantum-computing-noise-google-ibm-microsoft/
3. Cai, J., Macready, W.G., Roy, A.: A practical heuristic for finding graph minors (2014). https://arxiv.org/abs/1406.2741
4. Chang, C.Y., Jones, E., Yao, Y., Graf, P., Jain, R.: On hybrid quantum and classical computing algorithms for mixed-integer programming (2022). https://arxiv.org/abs/2010.07852
5. Ellinas, P., Chevalier, S., Chatzivasileiadis, S.: A hybrid quantum-classical algorithm for mixed-integer optimization in power systems (2024). https://arxiv.org/abs/2404.10693
6. Feng, F., Zhang, P., Bragin, M.A., Zhou, Y.: Novel resolution of unit commitment problems through quantum surrogate Lagrangian relaxation. IEEE Trans. Power Syst. **38**(3), 2460–2471 (2022)
7. Fischetti, M., Lodi, A.: Local branching. Math. Prog. **98**, 23–47 (2003). https://doi.org/10.1007/s10107-003-0395-5
8. Ganeshamurthy, P.A., Ghosh, K., O'Meara, C., Cortiana, G., Schiefelbein-Lach, J.: Bridging the gap to next generation power system planning and operation with quantum computation (2024). https://arxiv.org/abs/2408.02432
9. Gao, F., Huang, D., Zhao, Z., Dai, W., Yang, M., Shuang, F.: Hybrid quantum-classical general benders decomposition algorithm for unit commitment with multiple networked microgrids (2022). https://arxiv.org/abs/2210.06678

10. Glover, F.: Tabu search–part i. ORSA J. Comput. **1**(3), 190–206 (1989)
11. dwave-neal: An implementation of a simulated annealing sampler (2017). https://docs.ocean.dwavesys.com/projects/neal/en/latest/index.html#. Python package version 0.6.0
12. Dwave-system: a basic API for easily incorporating the D-Wave system as a sampler in the D-Wave Ocean software stack, directly or through Leap's cloud-based hybrid solvers (2018). https://docs.ocean.dwavesys.com/projects/system/en/stable/#. Python package version 1.26.0
13. Koretsky, S., et al.: Adapting quantum approximation optimization algorithm (QAOA) for unit commitment (2021). https://arxiv.org/abs/2110.12624
14. Leenders, L., Sollich, M., Reinert, C., Bardow, A.: Integrating quantum and classical computing for multi-energy system optimization using benders decomposition. Comput. Chem. Eng. **188**, 108763 (2024). https://doi.org/10.1016/j.compchemeng.2024.108763. https://www.sciencedirect.com/science/article/pii/S0098135424001819
15. Mladenović, N., Hansen, P.: Variable neighborhood search. Comput. Oper. Res. **24**(11), 1097–1100 (1997) https://doi.org/10.1016/S0305-0548(97)00031-2. https://www.sciencedirect.com/science/article/pii/S0305054897000312
16. Nikmehr, N., Zhang, P., Bragin, M.A.: Quantum distributed unit commitment: an application in microgrids. IEEE Trans. Power Syst. **37**(5), 3592–3603 (2022). https://doi.org/10.1109/TPWRS.2022.3141794
17. Paterakis, N.G.: Hybrid quantum-classical multi-cut benders approach with a power system application. Comput. Chem. Eng. **172**, 108161 (2023). https://doi.org/10.1016/j.compchemeng.2023.108161. https://www.sciencedirect.com/science/article/pii/S0098135423000303
18. Preskill, J.: Quantum computing in the NISQ era and beyond. Quantum **2**, 79 (2018). https://doi.org/10.22331/q-2018-08-06-79
19. Wilkins, A.: Record-breaking quantum computer has more than 1000 qubits. New Scientist (2023). https://www.newscientist.com/article/2399246-record-breaking-quantum-computer-has-more-than-1000-qubits/
20. Yang, M., Gao, F., Dai, W., Huang, D., Gao, Q., Shuang, F.: A scalable fully distributed quantum alternating direction method of multipliers for unit commitment problems. Adv. Quant. Technol. 2400286 (2024). https://doi.org/10.1002/qute.202400286. https://onlinelibrary.wiley.com/doi/abs/10.1002/qute.202400286
21. Zheng, X., Wang, J., Yue, M.: A fast quantum algorithm for searching the quasi-optimal solutions of unit commitment. IEEE Trans. Power Syst. **39**(2), 4755–4758 (2024). https://doi.org/10.1109/TPWRS.2024.3350382

Hybridizing Machine Learning and Optimization for Planning Satellite Observations

Romain Barrault[1](✉), Cédric Pralet[1], Gauthier Picard[1], and Eric Sawyer[2]

[1] DTIS, ONERA, Université de Toulouse, Toulouse, France
romain.barrault@onera.fr
romain.barrault@onera.fr
[2] CNES, Toulouse, France

Abstract. Planning the activities of an Earth observation satellite is a highly combinatorial task. It consists in regularly computing the sequence of observations to be performed by a satellite to collect images of candidate points of interest (POIs), while taking into account the time-dependent maneuvers required to point the satellite to the successive POIs. To solve such a recurrent optimization problem, we propose a novel approach that exploits offline learning techniques to approximate scheduling feasibility for sets of observation tasks. For this, we build a 0/1 neural network classifier whose inputs are related to the geographical positions of the POIs to be observed over the satellite orbit. We also learn hard capacity constraints limiting the number of observable POIs within orbit portions of various sizes. Finally, we introduce a hybrid algorithm, called **HySSEO**, optimizing the observation schedules based on a two-step process. The latter first searches for an optimal selection of POIs given the learned constraints, and then exploits this selection to bootstrap the search for an optimal schedule satisfying detailed time-dependent transition constraints. This hybrid optimization approach significantly improves the solution quality when compared to a standard scheduling approach.

Keywords: Hybrid AI · Neural Networks · Large Neighborhood Search · Scheduling · Earth Observation Satellite

1 Introduction

Earth observation satellites (EOSs) are vital for monitoring the planet, supporting diverse applications from climate studies to emergency response. *Agile* EOSs enhance responsiveness through advanced attitude control, enabling swift instrument reorientation towards targets of interest. However, planning the activities of agile EOSs presents significant challenges in terms of optimization. This explains why this problem has received attention over the last 25 years [26], leading to the use of various optimization techniques. The latter include, for example, greedy algorithms that iteratively insert observations one by one to build feasible plans, and several types of metaheuristics such as Large Neighborhood Search (LNS), where a current plan is iteratively optimized using destroy and repair operations. Among the state-of-the-art methods, we can cite an Adaptive LNS (ALNS) where six removal operators and three insertion operators are exploited [11]. Improvements of this algorithm using tabu search mechanisms have been proposed [9], as well as an extension that deals with the multi-satellite case by adding decisions related to the assignment of observations to satellites [8]. Another metaheuristic called GRILS for Greedy Randomized Iterated Local Search has also been recently developed and outperforms ALNS in the multi-satellite context [16]. In another direction, dynamic programming techniques have been studied to deal with cases where the profits collected depend on the time at which the observations are performed [15, 17]. All of these approaches include a critical feasibility constraint that assesses the temporal viability of observation schedules based on satellite agility models. For example, operational satellites may use detailed models that take into account the peculiarities of their on-board momentum control gyroscopes. Frequent evaluation of this constraint during the optimization process can significantly slow down solvers. Consequently, many approaches simplify the agility model by making assumptions such as constant maneuver times or constant angular velocity/acceleration to reduce computational complexity. However, these simplifications can lead to sub-optimal or even impractical schedules. There are also approaches that first search for solution plans using constant maneuver times, and then use a more complex time-dependent agility model [23].

But a key point is that none of the previous contributions exploit the fact that observation scheduling for EOSs is a highly recurrent optimization problem that is solved every day during several years of operation. To go beyond these existing approaches, we propose techniques that allow us to *learn* some aspects of the EOS scheduling problems to be solved regularly. The models learned during an offline phase are then used during an online phase, where EOS scheduling problems have to be solved for precise sets of candidate observations. In the literature, the use of learning methods for satellite scheduling problems has only been tested by a few authors, with attempts to build a decision policy that returns the next observation to insert in the current plan given the features of the candidate requests and the observations already selected [2, 10], and attempts to train a neural network that predicts the probability that an observation can be inserted in the plan of a given satellite [4]. The approach we propose is original compared to existing works because instead of learning an insertion heuristic or an insertion success estimator, we try to learn a so-called *scheduling feasibility function*. The latter is a Boolean function that takes as an input a set of POIs I to be observed and returns the value 1 if and only if the execution of all observations in I is considered as feasible from the point of view of the temporal constraints. More specifically, for the offline phase, the idea we promote is to learn a surrogate model that provides a good approximation of the temporal feasibility constraints, which is one of the main bottlenecks of the EOS schedulers. To do this, we first map the temporal feasibility problem to a binary classification problem for a neural network (NN). We also approximate capacity constraints providing upper bounds on the number of POIs that can be observed within orbit sections of different sizes. Obtaining these capacity constraints can be seen as a form of approximate knowledge compilation [22].

In the end, we adopt a hybrid AI and combinatorial optimization approach combining ML-based constraint modeling and classical decision algorithms [1, 12]. Globally, this paper brings several contributions with regard to the existing works:

(i) we show how a complex routing/scheduling problem (the visit of a set of POIs during some time windows) can be approximated as a simpler binary classification problem, by exploiting the geographical nature of the problem;
(ii) we propose an NN-based feasibility classifier that has a surprisingly low average error rate in terms of scheduling feasibility;
(iii) we propose an approximation of capacity constraints to help the optimization process for some POI distributions that are challenging for the NN classifier;
(iv) we develop a hybrid AI approach, HySSEO (*Hybrid Selection and Scheduling for Earth Observation*), where the feasibility constraints learned offline are exploited within an online optimization process. The online process is itself decomposed into a selection phase that searches for an optimal *set* of candidate POIs based on the learned feasibility constraints, and a scheduling phase where *sequences* of observations defining full solutions are computed. The global architecture obtained is shown in Fig. 1 (more details later on each component involved).

Section 2 provides background on the EOS scheduling problem. Section 3 details baseline resolution approaches for this problem. Section 4 presents the offline learning methods we propose to approximate the scheduling feasibility constraints. Section 5 details the online algorithms used to schedule EOS activities, including the hybrid AI approach proposed. Section 6 provides experimental results concerning the precision of the learned feasibility model and the quality of the schedules obtained. Section 7 concludes and provides perspectives on multi-satellite scenarios.

2 Problem Formalization

We consider a satellite in low Earth orbit that needs to collect acquisitions over points of interest on the ground during one pass over a specific area (during one orbit). Our goal is to select and schedule observations so as to maximize a total reward. The problem is overconstrained in the sense that capturing all candidate acquisitions is usually infeasible.

Space Discretization and Candidates POIs. For the sake of the learning approach mentioned in the introduction, we define the candidate observations from

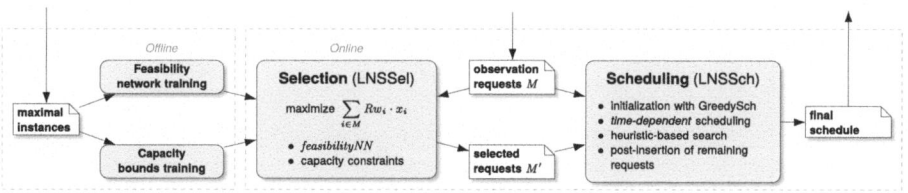

Fig. 1. HySSEO components: (i) training on maximal instances is performed offline; (ii) from the initial set of observation requests M, the selection component selects a subset M' by using the trained feasibility model and capacity constraints; (iii) finally M' is used in priority by an LNS-based scheduler, that may also add requests from $M \setminus M'$ if there is room for.

the point of view of the satellite (satellite-centered representation of the candidate observations), instead of using a fixed meshing of the Earth surface as usually done in the literature. This choice allows us to learn a unique scheduling feasibility model that can be reused for all the satellite revolutions over the Earth, whatever the precise longitude at which the satellite crosses the equatorial plane. Formally, as illustrated in Fig. 2, we build a grid of meshes between two latitudes $LatMin$, $LatMax$. To compute this grid for a North-to-South pass over all potential observation targets, we follow the process described below. The case of a South-to-North pass is symmetric. First, we consider different rolling angles (left-right rotations of the satellite allowing to capture areas on the left and right of its ground track). More precisely, we consider an angular step δ_r derived from the width of the field of view of the satellite sensor, and a maximum number of steps N_r related to the maximum observation angle usable to capture images. The set of candidate rolling angles is then $R = \{k \cdot \delta_r \mid k \in [-N_r..N_r]\}$. To define different positions of the satellite on its orbit, we consider an initial time t_0 at which the satellite reaches latitude $LatMax$ and a time-step δ_t derived from the length of the field of view of the sensor and the speed of the satellite on its orbit. We also define the number of time-steps N_t required to reach latitude $LatMin$. The set of candidate positions of the satellite on its orbit is then derived from the set of times $T = \{t_0 + k \cdot \delta_t \mid k \in [0..N_t]\}$, the orbital parameters of the satellite, and the space mechanic equations. From the previous sets of candidate rolling angles and candidate satellite positions, we build a grid of meshes by projecting, for each configuration, the field of view of the sensor on the Earth ellipsoid. This grid contains H rows and W columns, where $H = N_t + 1$ and $W = 2 \cdot N_r + 1$.

In the following, we denote by $\mathcal{M}$ the set of meshes obtained from this discretization process, and each potential POI of the problem is mapped to the center of a mesh in $\mathcal{M}$.

Optimization Problem. As an input, we consider a set of candidate observations $M \subseteq \mathcal{M}$. This set is strictly included in $\mathcal{M}$ when the end-users require images only on specific points on Earth. For each candidate observation $i \in M$, we have: (i) a position $P_i \in [1..H] \times [1..W]$ in the grid; (ii) a time window $[S_i, E_i]$, during which observation i can be performed, derived from a fixed maximum

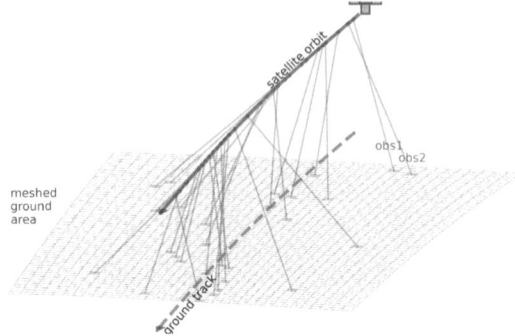

Fig. 2. Satellite-centered discretization of the field of view (in blue) and illustration of a set of selected meshes (in red) that can be successively observed by the satellite over its orbit. (Color figure online)

observation angle and from the ability of the satellite to point towards left-right and/or forward-backward directions; (iii) a reward Rw_i depending on the user requirements and the weather forecast (the higher the cloud coverage prevision, the lower the reward).

Additionally, the satellite must maneuver between successive observations in order to point its sensor towards the right direction at each step. In the following, we denote by tt the transition function such that $tt(i, j, t)$ returns the duration required by a maneuver from a configuration where the satellite is pointed to observation $i \in M$ to a configuration where the satellite is pointed to observation $j \in M$, when the maneuver starts at time t. This transition function is *time-dependent* (it depends on t) mainly due to the motion of the satellite on its orbit along time.

Definition 1 (EOSP). *The* Earth Observation Scheduling Problem *(EOSP) consists in finding a sequence of observations* $\sigma = [\sigma_1, \ldots, \sigma_K]$ *such that (i) each candidate observation in M appears at most once in σ; (ii) the successive observations can be performed during the allowed time windows; formally, the earliest start time of the first observation is $s_{\sigma_1} = S_{\sigma_1}$, the earliest start time of the kth observation is $s_{\sigma_k} = \max(S_{\sigma_k}, s_{\sigma_{k-1}} + tt(\sigma_{k-1}, \sigma_k, s_{\sigma_{k-1}}))$, and condition $s_{\sigma_k} \leq E_{\sigma_k}$ must be satisfied for every observation σ_k in σ; (iii) the total reward $(\sum_{i \in \sigma} Rw_i)$ is maximized.*

3 Baseline Search Strategies

In terms of Operations Research, the EOSP is a Time-Dependent Orienteering Problem with Time Windows (TD-OPTW) [21]. The latter problem is known to be NP-hard [7] and is usually addressed using non-exact methods, such as heuristic search, ant colony optimization [24], iterated local search [6], or large neighborhood search (LNS) [21]. The mapping between EOSP and TD-OPTW has been studied in previous works and attacked using large neighborhood search

Algorithm 1: GreedySch

Input: An EOSP p
Output: An admissible solution σ
$\sigma \leftarrow []$
$continue \leftarrow true$
while $|\sigma| \neq |M|$ and $continue$ **do**
  $i \leftarrow$ selectObs$()\sigma, M$
  if $i \neq \emptyset$ **then** $\sigma \leftarrow$ scheduleObs$()$i, σ,
  M **else** $continue \leftarrow false$
return σ

Algorithm 2: LNS to schedule

Input: An EOSP p, an admissible schedule s
Output: An admissible schedule s*
$s^* \leftarrow s$
while $duration < timeLimit$ **do**
  $s' \leftarrow$ repair$()$destroy$()$s
  if accept$()$s',s **then** $s \leftarrow s'$ **if**
  $\sum_{i \in s'} Rw_i > \sum_{i \in s^*} Rw_i$ **then**
  $s^* \leftarrow s'$
return s^*

specifically [11]. In this section, we detail two baseline search strategies that we consider in the following.

Greedy Scheduler. In order to solve a planning and scheduling problem such as EOSP or TD-OPTW, greedy algorithms are candidates of choice. Basically, greedy algorithms tackle problems by making the choice that seems the best at each decision step. While this usually fails to lead to the globally optimal solutions, it often creates solutions that have a quite good quality [3]. In the Space domain, for operational satellites, greedy algorithms are the baseline when it comes to scheduling tasks, because (i) they are fast (polynomial time in general), (ii) they can be guided by many efficient heuristics exploiting information about the observation tasks and the available time windows.

In the case of EOSP, Algorithm 1 sketches a greedy algorithm, looping over candidate observations. At each step, the idea is to select the best observation that can be added to the current schedule and to insert this observation at its best possible position in the sequence of observations. The process continues until all the observations are scheduled or there is no more place left in the current solution.

Selecting the next observation to schedule and its insertion position is a decision that strongly impacts the quality of the final solution. For this, the algorithm relies on two subfunctions that instantiate specific greedy heuristics. Function selectObs() first selects an observation from set M given the current solution σ. Here, we consider that the best observation is the observation $i \in M$ that is not already contained in σ and maximizes the ratio between the reward Rw_i and the additional time Δ_i required to maneuver between its predecessors and successors σ_k and σ_{k+1} at the best feasible insertion position in σ. By denoting as s_{σ_k} the earliest start time of observation σ_k in the current schedule and by $s_i = \max(S_i, s_{\sigma_k} + tt(\sigma_k, i, s_{\sigma_k}))$ the earliest time at which observation i can be performed after σ_k, this additional maneuver time is $\Delta_i = tt(\sigma_k, i, s_k) + tt(i, \sigma_{k+1}, s_i) - tt(\sigma_k, \sigma_{k+1}, s_k)$. The observation selection heuristic then chooses an observation in $\arg\max\{\frac{Rw_i}{\Delta_i} \mid i \in M, i \notin \sigma\}$, the idea being to favor observations that have the best "reward over maneuver time" ratio given the current schedule, or in other words the best yield. Function scheduleObs() positions the observation selected at the right place in the schedule. Note that selectObs() and scheduleObs() can be optimized by sharing some computations and data.

Large Neighborhood Search (LNS). LNS is a metaheuristic method employed in optimization problems, particularly those involving combinatorial optimization [18]. It is an advanced version of the simpler local search technique. Basically, local search begins with an initial solution to a problem and successively explores "neighboring" solutions generated by making minor adjustments to the current solution. In our case, the aim is to discover a solution that maximizes the sum of collected rewards. In LNS, a much larger set of neighboring solutions is explored at each step. In this extended neighborhood, the solution modifications can be more significant, potentially involving larger alterations of the current solution, but this wider neighborhood allows LNS to potentially escape from local optima and discover superior solutions overall.

An LNS algorithm for the EOSP is sketched in Algorithm 2. It mainly alternates between destroy() and repair() operations to find a better solution. The destroy() operation consists in deactivating some meshes from the current solution, and the repair() operation consists in refilling the solution plan through mesh activations. To decide whether the next destroy and repair operations have to work on the new solution obtained, we use function accept() that can be implemented in different ways. The simplest choice is to accept any solution after the repair process. Concerning the initialization, many metaheuristic methods start searching from good quality solutions found by a heuristic greedy algorithm. We also use this approach in our investigations.

4 Learning the Schedule Feasibility

For an EOS, schedules for successive orbits must be optimized several times a day over several years of operation. To solve this highly recurrent problem, we propose to learn, during an offline phase, a set of constraints called *feasibility constraints* that should be satisfied by any selection of observations.

4.1 Global Approach

To learn feasibility constraints during an offline phase, we combine two distinct learning approaches: (1) based on machine learning methods, we learn a global feasibility function *feasibilityNN* evaluating whether a set of POIs can be observed during a single satellite pass; (2) based on operations research methods, we learn hard capacity constraints limiting the number of POIs that can be observed within specific areas. For the second point, we consider a set of rectangular areas A in the grid of meshes, and for each area $a \in A$ we try to compute the maximum number of POIs that can be observed within a during a single satellite pass, referred to as $Capacity_a$.

During the online phase where a set of candidate observations M is available, the idea is to exploit the previous feasibility constraints to compute in a coarse-grain fashion an optimal selection of observations, before working on detailed scheduling decisions. Equations (1) to (4) give the corresponding coarse-grain optimization model, where Boolean variable x_i takes value 1 if the ith candidate mesh is selected, and value 0 otherwise. In Eq. (2), function *feasibilityNN* exploits the global feasibility function expressed as a neural network classifier to determine whether an assignment of the x_i variables is feasible. In Eq. (3), the hard capacity constraints enforce upper bounds about the number of selected meshes. As a result, the hybrid approach proposed combines on the one hand learning techniques through the offline computation of function *feasibilityNN* and bounds $Capacity_a$, and on the other hand a constraint-based optimization model that must be solved online, several times per day, to compute a plan for the satellite over the next decision horizon.

$$\text{maximize} \sum_{i \in M} Rw_i \cdot x_i \quad (1)$$

$$feasibilityNN(x_i, \ldots, x_{|M|}) = 1 \quad (2)$$

$$\forall a \in A, \sum_{i \in M \mid P_i \in a} x_i \leq Capacity_a \quad (3)$$

$$\forall i \in M, \, x_i \in \{0, 1\} \quad (4)$$

One of the main advantages here is that the optimization model proposed is focused on the selection aspect and does not directly use the time-dependent transition function *tt* that is associated with costly computations. As the calls to the transition function are the main bottleneck of the search methods in practice, building a surrogate model of the feasibility function is highly relevant. In the following, we successively detail how the global feasibility function *feasibilityNN* and the capacities $Capacity_a$ are learned.

4.2 Learning a Global Feasibility Function

From a Scheduling Feasibility Function to a Classification Problem. From a machine learning point of view, if we consider the grid of candidate meshes as an image where each pixel is black if the corresponding mesh is activated and white otherwise, our goal is to classify the images depending on whether they correspond to feasible sets of observations. On such an image, the satellite ground track would correspond to a straight vertical line crossing the rows in the middle. We will see later how we consider local and global features derived from this initial image.

To illustrate the approach, Fig. 3 provides images corresponding to sets of meshes for which it is possible to find feasible observation schedules given a realistic maneuver model. More specifically, Fig. 3(a) gives the image associated with the set of meshes activated in Fig. 2. In this case, the set of activated meshes is sparse and only a reduced number of meshes are observed due to the time consumed by the satellite maneuvers between the corresponding acquisition tasks. On the other side, Fig. 3(b) gives an example of an image where the set of activated meshes is dense. In this case, many contiguous meshes are activated and many small maneuvers are used, which is why the satellite is able to observe

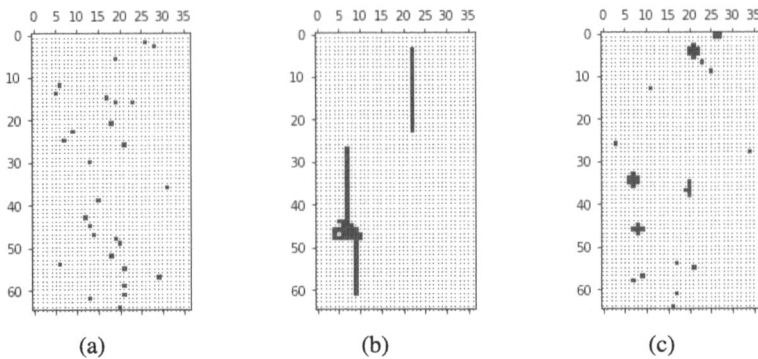

Fig. 3. Examples of maximum sets of meshes observable by the satellite during a single pass (mesh indices on the x-axis and y-axis): (a) sparse instance, (b) dense instance, (c) mixed instance

many meshes during a single pass. Figure 3(b) also shows standard patterns where the satellite observes strips of contiguous meshes that are parallel to its ground track. Last, Fig. 3(c) corresponds to a mixed instance involving both clusters of meshes and individual meshes spread over the observation area.

Input Features and NN Architecture. In our study, three kinds of features are considered to classify the images according to the feasibility of the mesh selection:

- *Image*: raw features corresponding to the activation of individual pixels in the mesh selection image;
- *Locals*: features of a set of predefined slices of the grid. Globally, we consider successive slices of height h separated by r rows, h and r being two parameters of the method. These slices are defined as $Slice_{k \cdot r+1, k \cdot r+h} = [k \cdot r + 1 .. k \cdot r + h] \times [1..W]$ for different values of k, so as to cover the whole grid. For each slice $Slice_{k \cdot r+1, k \cdot r+h}$, we compute two features: $nMeshesInSlice_k$ that represents the number of meshes activated in the slice, and $dispersionInSlice_k$ that approximates the total maneuver time required to successively visit the columns where meshes are activated in the slice. On the last point, if we denote by $Columns_k = [c_1, \ldots, c_Q]$ the indices of the successive columns where meshes are activated, we have $dispersionInSlice_k = \sum_{i \in [1..Q-1]} TI\_ttCols(c_{i+1} - c_i)$ where $TI\_ttCols(\delta)$ is a Time-Independent approximation on the minimum transition time required by any maneuver traversing δ columns. Such a transition time is non-linear in δ, especially for small maneuvers requiring an acceleration phase for the satellite;
- *Globals*: features representing global metrics over the set of selected meshes. Two global features are exploited: $nActive$, the total number of meshes activated in the grid, and $costMST$, the cost of a minimum spanning tree covering all the activated meshes, given that the cost of an edge between two activated meshes i and j is defined from a Time-Independent approximation $TI\_tt(i, j)$ of the (time-dependent) transition time between i and j. Note that the cost

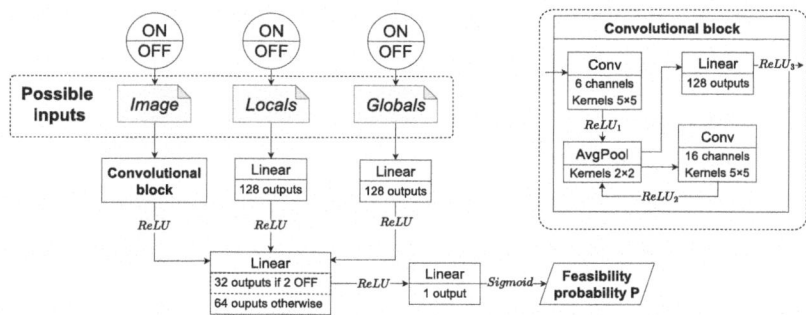

Fig. 4. Common architecture of designed neural networks

of a minimum spanning tree is also used as an efficient lower bound in works on traveling salesman problems with time windows [5].

Figure 4 displays the architecture shared by all neural networks designed in this work. The first feature to point out is that as a binary classifier, it returns a single number between 0 and 1 (thanks to the sigmoid activation function) embodying a probability; in this case, it is a feasibility probability for a given set of meshes. For the inputs, the three kinds of features listed before can be activated or deactivated, which gives us eight configurations ranging from a case where the neural network is empty to a case where all the features are used.

Feasibility Classifier and Generation of Training Instances. To train the NN classifier, we first build a set of feasible mesh selections $\mathcal{P}_{max}$ that are maximal in terms of inclusion. This means that for each maximal set $P^* \in \mathcal{P}_{max}$, each mesh selection $P \subseteq P^*$ is feasible and the scheduling algorithm estimates that it is not possible to activate one more mesh in P^* while keeping the feasibility of the selection. The examples provided in Figs. 3(a) to 3(c) are maximal positive instances obtained from a realistic time-dependent maneuver model. On this point, the maximal instances built are actually *approximately maximal* in the sense that the feasibility of a mesh selection is tested based on an approximate greedy algorithm that inserts the observations one by one in the current sequence of visits, each time at a position that is considered as the best one. Other efficient scheduling algorithms could be used [19], but it is worth mentioning that determining whether a single mesh can be added to a mesh selection while preserving the scheduling feasibility is NP-hard, due to the NP-hardness of the Traveling Salesman Problem with Time Windows [20]. This is why we only use approximate scheduling algorithms to estimate whether a set of mesh activations is maximal.

Then, for each maximum instance $P^* \in \mathcal{P}_{max}$, we generate N positive (resp. negative) instances by randomly deactivating (resp. activating) meshes. To better approximate the feasibility-infeasibility frontier, the number of positive (resp. negative) instances generated is higher for small numbers of meshes deactivations (resp. activations). Details about the instance generation protocol are given

4.3 Learning Capacity Constraints

We now describe how the areas A and the bounds $\{Capacity_a, a \in A\}$ used in Eq. (3) are defined. We recall that H and W respectively denote the height and width of the grid of meshes. The areas manipulated correspond to slices of meshes of various heights. More precisely, for each $h \in [1..H]$, we consider a slice $Slice_{1,h} = [1..h] \times [1..W]$ covering h successive rows and the entire grid width. Then, we solve an EOSP as defined in Definition 1 for a set M containing one candidate observation i per mesh in $[1..h] \times [1..W]$, with a unit reward $Rw_i = 1$. The best total reward $TotalRw_h^*$ found for area $a = Slice_{1,h}$ is then used to define capacity $Capacity_a$, that is $Capacity_a = TotalRw_h^*$. Due to the invariance of the problem along the grid rows, reward $TotalRw_h^*$ is exploited to define a maximum capacity not only over slice $Slice_{1,h}$, but also over all slices $Slice_{k+1,k+h} = [k+1..k+h] \times [1..W]$ of height h for $k \in [0..H-h]$. Note that for $h = H$, total reward $TotalRw_H^*$ gives the maximum number of meshes that can be selected over the whole grid.

From a theoretical point of view, the previous process guarantees that any solution of an EOSP problem defined over the grid must satisfy the capacity bounds, otherwise the optimal rewards found above would not be optimal. However, one difficulty is that solving the previous EOSP problem in NP-hard, because it contains the standard Traveling Salesman Problem with Time Windows as a special case [20]. As a result, finding the optimal total reward $TotalRw_h^*$ is not necessarily easy for large grid slices.

To overcome this difficulty, we exploit two main ideas. First, we use a non-exact LNS solver instead of a complete search engine to optimize the total reward. This LNS solver is run during an arbitrary CPU time set to h minutes. This implies that the capacities computed are not guaranteed to correspond to actual hard constraints, but it is worth mentioning that LNS is a state-of-the-art non-exact method for Time-Dependent Orienteering Problems with Time Windows [21]. Second, when optimizing the total reward for slice $Slice_{1,h+1}$, we reuse the solution found for slice $Slice_{1,h}$, in the spirit of a Russian doll search [25]. Basically, the best solution found for a slice $Slice_{1,h}$ selects customers within column range $[cmin_h..cmax_h]$. In practice, these customers are always placed around the ground track of the satellite because such customers have time windows that are large and contain the windows of customers placed near the borders of the grid. To simplify the resolution for height $h+1$, we consider only the customers located in area $[1..h+1] \times [cmin_h..cmax_h]$. For high values of h, this heuristic approach strongly simplifies the problem because the number of columns to take into account is quickly reduced to only 3 or 4.

5 Integrating the Learned Model Into a Scheduler

To integrate the trained models within an LNS solver for EOSP, we consider two successive LNS that both follow the generic scheme of Algorithm 2 provided before:

LNS for Selecting Observations using the surrogate model (LNSSel): in this case, the solutions manipulated are not full schedules but subsets of the set of candidate meshes M. Function destroy() of Algorithm 2 removes a portion of observations from the current solution with a priority for the deactivation of the observations that have a small ratio between the reward and the distance to the set of activated meshes, and function repair() adds as many observations as possible in a greedy manner while checking the feasibility with *feasibilityNN* and the capacity constraints. Since checking the feasibility using the surrogate is very fast, we can also afford multiple restarts until reaching a given time limit. A restart is performed after a fixed number of LNS iterations, and to perform a restart we generate a new initial solution containing 15 randomly chosen observations, enriched greedily with as many observations as possible.

LNS for solving an EOSP (LNSSch): in this case, solutions are sequences of observations from M. Function destroy() of Algorithm 2 randomly removes observations from the current sequence with a priority to observations close to the ones that just got removed, while function repair() adds and schedules as many observations as possible in a greedy manner, by calling the transition function. Here, each observation activation involves more computations than in LNSSel.

Using the Trained Models when Solving EOSP. Coming back to Fig. 1 given in the introduction, our hybrid approach is composed of an offline phase and an online phase. During the offline phase, the feasibility classifier is trained and capacity constraints are determined, as explained in Sect. 4. The online phase first selects a subset of good candidate observations M' using an LNS based on our *feasibilityNN* model and capacity constraints. Then, this subset is used as an input by an LNS-based scheduler that attempts to schedule as many observations from M' as possible, starting from a solution obtained by a greedy search procedure. After that, a second LNS-based scheduler attempts to add other meshes in $M \setminus M'$ to complete the plans. This hybrid offline/online algorithm is developed to answer two objectives. The first objective is to take advantage of the surrogate feasibility function *feasibilityNN* and capacity constraints to quickly compute good-quality solutions by exploring many observation selection strategies, instead of directly using complex time-dependent scheduling operations. The second objective is to challenge the robustness of the surrogate model *feasibilityNN*, by first searching for an optimal set of selected meshes $M' \subseteq M$ according to this model and then checking, based on the standard LNS algorithm for EOSP, whether all meshes in M' can actually be simultaneously observed. From a machine learning point of view, this objective is more challenging than just classifying some randomly generated instances.

Table 1. Error rates for the classifiers on each instance type: *dense* (D), *mixed* (M), *sparse* (S), *flat* (F), *borders* (B).

NN model	Image					G					{G,L5}					{G,L13}					CC				
Instance type	D	M	S	F	B	D	M	S	F	B	D	M	S	F	B	D	M	S	F	B	D	M	S	F	B
Error rate (%)	25	24	25	21	20	4.6	19	11	11	9.8	6.6	11	12	13	13	6.6	9.1	12	12	13	42	45	46	45	46

6 Experimental Evaluation

Now, we provide an experimental analysis on both the precision of the learned classifier and its capacity to select good sets of candidate observations within an EOSP solving process. The EOSP solvers are implemented in Java and executed on 20-core Intel(R) Xeon(R) CPU E5-2660 v3 @ 2.60 GHz, 62 GB RAM, Ubuntu 18.04.5 LTS, with an OpenJDK 11.0.9 JVM. The machine learning models have been developed and pre-trained using pytorch library in Python 3.11 [14]. These models have been serialized in ONNX and then loaded and called by our solvers using the ONNX Runtime for Java [13].

Generation of the Positive and Negative Instances. From each maximal set of meshes $P^* \in \mathcal{P}_{max}$, we generate 35 *positive instances* by deactivating m meshes belonging to P^* (random choice of the meshes deactivated). To better approximate the feasibility-infeasibility frontier, we generate more positive instances for small values of m: we generate $K_1 = 6$ random positive instances for $m = 1$, and for $m \in [2..10]$, we generate $K_m = round(5/6 \cdot K_{m-1})$ positive instances. Additionally, we generate $K_{>10} = 6$ positive instances where more than 10 meshes are deactivated. Similarly, from each maximal instance $P^* \in \mathcal{P}_{max}$, we generate 35 *negative instances* by randomly activating at least one mesh that does not belong to P^* (random choice of the meshes activated). Again, to better approximate the feasibility-infeasibility frontier, we generate more negative instances for small values of m, using the values of K_m introduced before. Last, each instance belongs to one out of five types: *dense* instances show a huge amount of activated meshes that are close to each other, implying small maneuvers; *mixed* instances have many isolated activated meshes with a few clusters of close meshes; *sparse* instances only have isolated meshes, and therefore a smaller number of observable meshes due to higher maneuver durations; *flat* instances show meshes displayed mostly horizontally, which is less easy to capture for the EOS; *borders* instances are characterized by a high amount of meshes near the borders of the meshing, which have smaller visibility time windows. We aim to cover realistic cases and determine the difficulty to observe a whole set of POIs based on the geometry of the instance.

Precision of the Learned Classifier. Several NN models have been designed for different input features. Figure 5 shows the results of several such models. Models Image and G (*Globals*) take as input only the corresponding features described in Sect. 4.2. {G,L5} and {G,L13} are based on global and local features computed

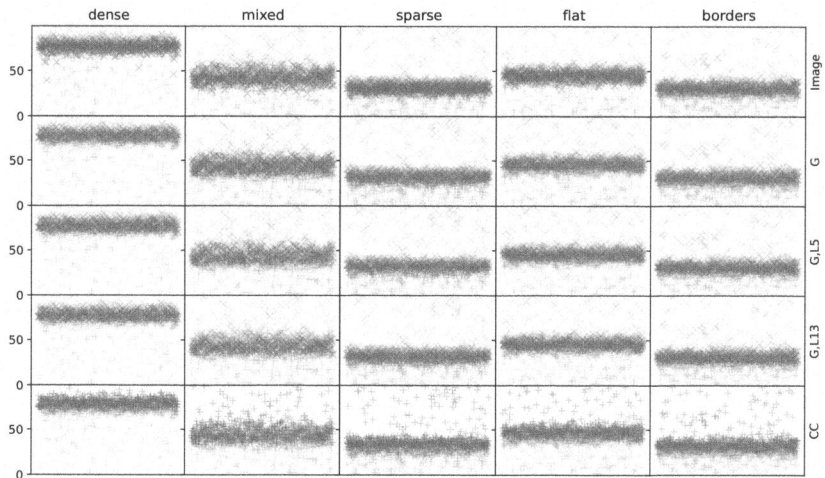

Fig. 5. Classification results for five classifiers (Image, G, {G,L5}, {G,L13}, and CC) on five types of instances (*dense, mixed, sparse, flat* and *borders*). On each graph, each dot corresponds to a test instance classified by the model, green if correctly classified and red otherwise. Negative instances are represented by crosses and positive instances by pluses. The x-axis is for the instance number and the y-axis is the amount of activated meshes in the instance. (Color figure online)

according to the description of Sect. 4.2 again. For the local features, {G,L5} considers successive slices of height $h = 5$ separated by $r = 1$ row, while {G,L13} considers successive slices of height $h = 13$ separated by $r = 6$ rows, implying much less NN inputs. Finally, a classifier based only on the capacity constraints (CC) claims that an instance is feasible if and only if it satisfies these constraints. The error rates of all these models are shown in Table 1, which indicates that models with global features perform best. Models with the image input do not show good performance in the current architecture, which is why our study is focused on the three other NN models (G, {G,L5}, {G,L13}). In addition, since capacity constraints are necessary feasibility conditions, they classify instances poorly on their own. On this point, only a minor part of the negative instances in the dataset are well classified by the CC approach, since activating less than 10 additional meshes rarely leads to a violation of the capacity constraints. Figure 6 provides the error rates of the models as a function of the number of meshes activated and deactivated in the positive and negative instances respectively. It shows that the G and {G,L5} models appear to be slightly more robust while {G,L13} gives fewer false positives. Finally, the CC classifier returns no false negative. This makes sense because as it classifies the instances based on necessary feasibility conditions, its only errors correspond to false positives.

Performances of HySSEO. Table 2 shows the results of the runs performed to evaluate the online algorithm. The learned model used in the LNSSel part of HySSEO varies between G, {G,L5}, {G,L13}, and all of the feasibility checks

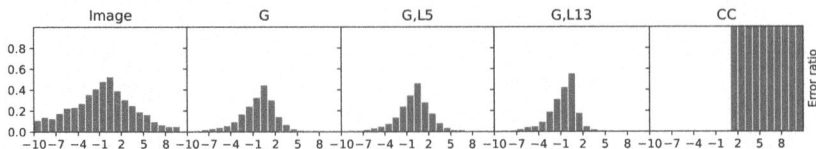

Fig. 6. Error rates of designed neural networks for different numbers of meshes added (positive value) or removed (negative values) from maximal instances.

used in this part also verify the capacity constraints learned offline. Indeed, preliminary results have shown that neural networks alone are not sufficiently constrained on dense instances and may return large sets of observations whose cardinality exceeds the capacity constraints, therefore models G, {G,L5}, and {G,L13} are combined with CC in all of our tests. For comparison, we also consider a last version of HySSEO that does not use any NN and only checks the capacity constraints, referred to as CC + {}. The algorithms are run for 1 and 5 minutes in order to compare their ability to converge faster or produce better results with enough time. For a fixed timespan, in the HySSEO algorithms, 16% of that timespan is used by LNSSel, 34% is used by LNSSched on the set of observations returned by LNSSel, and the remaining 50% of the computation time is used to optimize the insertion of other remaining requests. Table 2 shows the performance of the algorithms with $mean \pm std$ on 100 instances equally composed of dense, mixed and sparse instances, with 200 observation requests. The optimality gap is obtained by comparing to an LNSSched algorithm working on the full set of requested observations during a longer computation time of 1 hour. In the table, *nSites* is the number of observations involved in the final schedule, and LNS_1 (resp. LNS_2) refers to the first (resp. the second) use of LNSSched in HySSEO. The results show that the baseline LNS is beaten on average by all of our HySSEO algorithms, regardless of the allowed computation time. In addition, except for dense instances where CC + {} is highly efficient, using one of our three neural networks to compute feasibility checks improves the solution quality. In the experiments, the capacity constraint checks tend to return much too large request sets on mixed and sparse instances since it is blind with regards to the actual maneuver times required within a slice of meshes. However, we can note that configuration CC returns the best solutions after the first LNSSched, which is quite natural since it is easier to obtain a good solution when requests are less filtered. On the other hand, for the three NN-based HySSEO algorithms, the post-insertion process has a higher impact since the initial LNSSel phase is more constrained in this case. These three NN-based approaches show similar results in terms of solution quality. Figures 7 and 8 display the average inference time for feasibility checks in LNSSel and the average insertion time of a request into the current schedule in LNSSched, respectively. As expected, our learned model allows us to check the feasibility of a set of requests much faster than the feasibility of an actual observation insertion into a schedule. Note that the inference time for the first kind of checks slightly increases with the number

Table 2. Scheduling performance metrics.

Instance type		LNS	CC+{G}	CC+{G,L5}	CC+{G,L13}	CC+{ }	LNS	CC+{G}	CC+{G,L5}	CC+{G,L13}	CC+{ }
				1 minute					5 minutes		
Dense	nActive	n/a	61.6 ± 13.6	59.6 ± 11.3	60.6 ± 12.5	62.0 ± 14.2	n/a	61.5 ± 13.5	60.1 ± 11.8	60.6 ± 12.6	62.1 ± 14.2
	nSitesLNS₁	n/a	42.0 ± 9.0	41.8 ± 8.7	**42.1 ± 9.0**	41.5 ± 8.8	n/a	42.4 ± 8.9	42.4 ± 9.0	**42.4 ± 9.4**	41.9 ± 8.9
	gapLNS₁(%)	n/a	7.6 ± 3.9	8.3 ± 3.6	7.7 ± 4.2	7.2 ± 4.0	n/a	5.9 ± 2.7	6.6 ± 2.7	6.6 ± 3.5	6.1 ± 3.1
	nSitesLNS₂	**47.2 ± 9.1**	45.6 ± 9.1	45.9 ± 8.6	45.7 ± 9.2	45.6 ± 8.9	**47.8 ± 9.0**	46.8 ± 9.1	46.9 ± 9.0	47.0 ± 9.1	46.7 ± 8.7
	gapLNS₂(%)	5.5 ± 3.6	3.6 ± 2.7	3.9 ± 2.0	3.5 ± 2.2	**3.3 ± 2.0**	2.8 ± 2.2	2.0 ± 1.4	1.9 ± 1.5	2.0 ± 1.3	**1.8 ± 1.5**
Mixed	nActive	n/a	66.5 ± 4.9	59.3 ± 3.5	56.8 ± 4.0	97.5 ± 0.7	n/a	67.4 ± 4.0	60.2 ± 3.4	58.1 ± 4.2	97.5 ± 0.7
	nSitesLNS₁	n/a	46.4 ± 5.3	**47.5 ± 3.8**	47.4 ± 5.2	46.9 ± 2.7	n/a	46.5 ± 5.5	47.9 ± 4.0	**48.3 ± 4.5**	48.1 ± 2.9
	gapLNS₁(%)	n/a	18.8 ± 6.8	17.3 ± 6.7	17.8 ± 6.9	10.0 ± 3.9	n/a	17.6 ± 6.7	14.8 ± 5.7	15.3 ± 5.7	7.0 ± 3.2
	nSitesLNS₂	52.1 ± 3.3	54.0 ± 2.9	53.9 ± 3.0	**54.1 ± 3.0**	51.8 ± 2.8	53.4 ± 2.7	**54.8 ± 2.6**	54.3 ± 2.9	54.7 ± 2.7	53.5 ± 2.4
	gapLNS₂(%)	7.4 ± 3.9	4.8 ± 2.5	4.8 ± 2.4	**4.4 ± 2.3**	5.4 ± 3.2	4.0 ± 3.4	**1.7 ± 1.5**	2.2 ± 1.6	2.0 ± 1.6	2.4 ± 2.2
Sparse	nActive	n/a	49.2 ± 1.7	48.9 ± 2.1	46.3 ± 2.1	97.8 ± 0.4	n/a	49.1 ± 1.3	49.5 ± 2.2	46.7 ± 2.6	97.7 ± 0.5
	nSitesLNS₁	n/a	39.3 ± 4.2	43.1 ± 3.0	42.6 ± 2.8	**43.5 ± 1.6**	n/a	39.6 ± 3.8	43.9 ± 3.0	43.6 ± 2.9	**44.7 ± 1.9**
	gapLNS₁(%)	n/a	18.1 ± 7.6	13.1 ± 4.6	16.1 ± 4.6	7.7 ± 3.0	n/a	16.2 ± 7.1	11.1 ± 4.3	13.3 ± 3.8	5.2 ± 2.9
	nSitesLNS₂	46.8 ± 1.6	47.2 ± 1.7	47.5 ± 1.7	**47.9 ± 1.9**	46.5 ± 1.5	47.5 ± 1.4	47.7 ± 1.6	48.0 ± 1.3	**48.2 ± 1.5**	47.4 ± 1.8
	gapLNS₂(%)	6.6 ± 2.7	**3.1 ± 2.0**	4.3 ± 2.0	5.0 ± 2.5	5.0 ± 2.4	3.7 ± 2.0	2.1 ± 1.3	**2.0 ± 1.7**	2.5 ± 1.9	2.6 ± 2.1

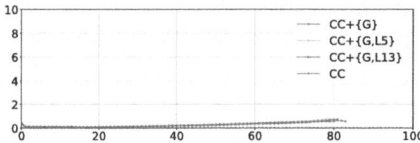

Fig. 7. Average inference time (in ms) for each number of activated meshes in the current solution.

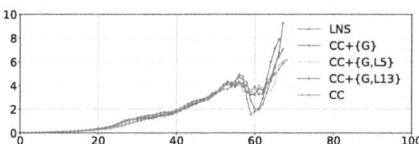

Fig. 8. Average insertion time (in ms) of an observation in the current schedule as a function of its current size.

of activated meshes due to the operations required to compute features such as *costMST*, while the second kind of checks is much slower since it requires testing each insertion position in the worst case. In Fig. 8, the behavior around 60 requests is due to the different kinds of instances involved in our dataset. More precisely, for many instances, the maximum number of selectable observations is around 60, and in this case testing all the possible insertion positions is long, while for other instances (typically the dense ones), the maximum number of selectable observations is much higher, and in this case finding a feasible insertion position in a schedule containing 60 observations is not that hard.

7 Conclusion

This paper proposes a novel approach to approximate, using a neural network, a feasibility constraint for a set of observation tasks by a satellite. The main idea is to map the feasibility check to a classification task, based on local and global features of the set of observations. We evaluated the classification performance of several combinations of features, and used them to select good subsets of candidate observation tasks upstream of a conventional scheduling process based on greedy search and large neighborhood search. We also proposed approximate capacity constraints that help discard unfeasible observation sets. The resulting feasibility classifiers have a small error rate and HySSEO is better than baseline methods in terms of solution quality. Computing local features for feasibility does not seem to improve the solution quality, and combining global features and capacity constraints is sufficient to provide results among the best.

One of the next steps is to apply this hybrid AI approach to plan the observations of several satellites. In this case, the problem to be solved is similar to a team orienteering problem, which is much harder to solve since it involves decisions on the allocation of observation tasks to the satellites. On this line, we believe that the mono-satellite techniques proposed can be relevant in multi-satellite scenarios, to explore large sets of observation dispatching decisions contrarily to traditional EOS planning methods that are slowed down by the computation of candidate maneuvers for many candidate observation insertion positions across all the satellites of a constellation. In another direction, we are also considering using our feasibility model as a constraint in constraint programming or linear programming frameworks, as in the Empirical Model Learning approach where neural networks can be encoded into standard optimization models [12].

References

1. Bengio, Y., Lodi, A., Prouvost, A.: Machine learning for combinatorial optimization: a methodological tour d'horizon. Eur. J. Oper. Res. **290**(2), 405–421 (2021). https://doi.org/10.1016/j.ejor.2020.07.063
2. Bensana, E., Verfaillie, G., Michelon-Edery, C., Bataille, N.: Dealing with uncertainty when managing an earth observation satellite. In: Fifth International Symposium on Artificial Intelligence, Robotic and Automation in Space (i-SAIRAS 1999), pp. 120–124 (1999)
3. Cormen, T.H., Leiserson, C.E., Rivest, R.L., Stein, C.: Introduction to Algorithms, 2nd edn. The MIT Press, Cambridge (2001)
4. Du, Y., Wang, T., Xin, B., Wang, L., Chen, Y., Xing, L.: A data-driven parallel scheduling approach for multiple agile earth observation satellites. IEEE Trans. Evol. Comput. **24**(4), 679–693 (2020)
5. Ducomman, S., Cambazard, H., Penz, B.: Alternative filtering for the weighted circuit constraint: comparing lower bounds for the TSP and solving TSPTW. In: Proceedings of the AAAI Conference on Artificial Intelligence, vol. 30, no. 1 (2016). https://doi.org/10.1609/aaai.v30i1.10434
6. Garcia, A., Arbelaitz, O., Vansteenwegen, P., Souffriau, W., Linaza, M.T.: Hybrid approach for the public transportation time dependent orienteering problem with time windows. In: Corchado, E., Graña Romay, M., Manhaes Savio, A. (eds.) Hybrid Artificial Intelligence Systems, pp. 151–158. Springer, Heidelberg (2010)
7. Golden, B.L., Levy, L., Vohra, R.: The orienteering problem. Naval Res. Logist. (NRL) **34**(3), 307–318 (1987)
8. He, L., Liu, X., Laporte, G., Chen, Y., Chen, Y.: An improved adaptive large neighborhood search algorithm for multiple agile satellites scheduling. Comput. Oper. Res. **100**, 12–25 (2018)
9. He, L., Weerdt, M., Yorke-Smith, N.: Time/sequence-dependent scheduling: the design and evaluation of a general purpose tabu-based adaptive large neighbourhood search algorithm. J. Intell. Manuf. **31**(4), 1051–1078 (2020)

10. Lam, J.T., Rivest, F., Berger, J.: Deep reinforcement learning for multi-satellite collection scheduling. In: Martín-Vide, C., Pond, G., Vega-Rodríguez, M.A. (eds.) Theory and Practice of Natural Computing, pp. 184–196. Springer (2019)
11. Liu, X., Laporte, G., Chen, Y., He, R.: An adaptive large neighborhood search metaheuristic for agile satellite scheduling with time-dependent transition time. Comput. Oper. Res. **86**, 41–53 (2017). https://doi.org/10.1016/j.cor.2017.04.006
12. Lombardi, M., Milano, M., Bartolini, A.: Empirical decision model learning. Artif. Intell. **244** (2016). https://doi.org/10.1016/j.artint.2016.01.005
13. ONNX Runtime developers: ONNX runtime (1.17.3). https://www.onnxruntime.ai (2024)
14. Paszke, A., et al.: An imperative style, high-performance deep learning library. In: Wallach, H., Larochelle, H., Beygelzimer, A., d' Alché-Buc, F., Fox, E., Garnett, R. (eds.) Advances in Neural Information Processing Systems, vol. 32, pp. 8024–8035. Curran Associates, Inc. (2019)
15. Peng, G., Dewil, R., Verbeeck, C., Gunawan, A., Xing, L., Vansteenwegen, P.: Agile Earth observation satellite scheduling: An orienteering problem with time-dependent profits and travel times. Comput. Oper. Res. **111**, 84–98 (2019). https://doi.org/10.1016/j.cor.2019.05.030
16. Peng, G.: Solving the agile Earth observation satellite scheduling problem with time-dependent transition times. IEEE Trans. Syst. Man Cybern. Syst. **52**(3), 1614–1625 (2022)
17. Peng, G., Song, G., Xing, L., Gunawan, A., Vansteenwegen, P.: An exact algorithm for agile earth observation satellite scheduling with time-dependent profits. Comput. Oper. Res. **120**, 104946 (2020). https://doi.org/10.1016/j.cor.2020.104946
18. Pisinger, D., Ropke, S.: Large neighborhood search. In: Gendreau, M., Potvin, J.-Y. (eds.) Handbook of Metaheuristics. ISORMS, vol. 272, pp. 99–127. Springer, Cham (2019). https://doi.org/10.1007/978-3-319-91086-4_4
19. Pralet, C.: Iterated maximum large neighborhood search for the traveling salesman problem with time windows and its time-dependent version. Comput. Oper. Res. **150**, 106078 (2023)
20. Savelsbergh, M.: Local search in routing problems with time windows. Ann. Oper. Res. **4**, 285–305 (1985)
21. Schmid, V., Ehmke, J.F.: An effective large neighborhood search for the team orienteering problem with time windows. In: Bektaş, T., Coniglio, S., Martinez-Sykora, A., Voß, S. (eds.) Computational Logistics, pp. 3–18. Springer (2017)
22. Selman, B., Kautz, H.: Knowledge compilation and theory approximation. J. ACM **43**(2), 193–224 (1996). https://doi.org/10.1145/226643.226644
23. Squillaci, S., Pralet, C., Roussel, S.: Comparison of time-dependent and time-independent scheduling approaches for a constellation of Earth observing satellites. In: 13th International Workshop on Planning and Scheduling for Space (IWPSS) (2023)
24. Verbeeck, C., Vansteenwegen, P., Aghezzaf, E.-H.: The time-dependent orienteering problem with time windows: a fast ant colony system. Ann. Oper. Res. (4), 481–505 (2017). https://doi.org/10.1007/s10479-017-2409-3
25. Verfaillie, G., Lemaître, M., Schiex, T.: Russian doll search for solving constraint optimization problems. In: Proceedings of the Thirteenth National Conference on Artificial Intelligence, AAAI 1996, vol. 1, pp. 181–187. AAAI Press (1996)
26. Wang, X., Wu, G., Xing, L., Pedrycz, W.: Agile earth observation satellite scheduling over 20 years: formulations, methods, and future directions. IEEE Syst. J. **15**(3), 3881–3892 (2021). https://doi.org/10.1109/JSYST.2020.2997050

Algorithm Configuration in Sequential Decision-Making

Luca Begnardi[1,2](✉), Bart von Meijenfeldt[1,2], Yingqian Zhang[1,2], Willem van Jaarsveld[1], and Hendrik Baier[1,2]

[1] Eindhoven University of Technology, Eindhoven, The Netherlands
{l.begnardi,b.m.v.meijenfeldt,YQZhang,W.L.v.Jaarsveld,
h.j.s.baier}@tue.nl
[2] Eindhoven Artificial Intelligence Systems Institute, Eindhoven, The Netherlands

Abstract. Proper parameter configuration of algorithms is essential, but often time-consuming and complex, as many parameters need to be tuned simultaneously and evaluation can be expensive. In this paper, we focus on sequential decision-making (SDM) algorithms, which are applied to problems that require a series of decisions to be taken sequentially, aiming for an optimal cumulative outcome for the agent. To do this, every time the agent needs to make a decision, SDM algorithms take the current state of the environment as input and provide a decision as output. We propose a taxonomy of algorithm configuration approaches for SDM and introduce the concept of Per-State Algorithm Configuration (PSAC). To perform PSAC automatically, we present a framework based on Reinforcement Learning (RL). We demonstrate how PSAC by RL works in practice by applying it to two SDM algorithms on two SDM problems: Monte Carlo Tree Search, to solve a collaborative order picking problem in warehouses, and AlphaZero, to play a classic board game called Connect Four. Our experiments show that, in both use cases, PSAC achieves significant performance improvements compared to fixed parameter configurations. In general, our work expands the field of automated algorithm configuration and opens new possibilities for further research on SDM algorithms and their applications. Code is available at: https://github.com/ai-for-decision-making-tue/Per-State_Algorithm_Configuration.

Keywords: algorithm configuration · sequential decision-making · reinforcement learning

1 Introduction

Proper parameter configuration is essential to ensure algorithms operate efficiently and effectively to solve specific problems. This is usually a time-consuming and difficult task, especially when many parameters need to be tuned or when the evaluation of parameter settings is expensive to compute. This algorithm configuration (AC) problem has been widely studied by the AI community [15]. Its approaches can be categorized as static or dynamic, depending on whether the parameters are set at the beginning of algorithm execution or changed at every iteration of the running algorithm [2,7]. Static approaches can be further divided into Per-Distribution AC, where the goal is to find a single set of parameters that works well across an entire distribution of problem instances, and Per-Instance AC (PIAC), which tries to find a function that maps each individual problem instance to a specific parameter configuration. The algorithm configuration literature mainly focuses on approaches for solving optimization problems, such as local search or genetic algorithms, which aim to find an optimal solution by cleverly exploring the space of solutions, typically by iteratively improving candidate solutions. In this paper, we focus on a different class of problems, namely sequential decision-making (SDM) problems. These model real-world problems where the environment changes over time and requires making a series of decisions that lead to a cumulative outcome. In contrast to the class of algorithms usually considered in algorithm configuration, SDM algorithms can work online by outputting one decision at a time instead of a complete solution (sequence of decisions). They must, therefore, be applied multiple times with different initial environment conditions to obtain a complete solution. This approach introduces a new dimension which should be taken into account by algorithm configuration methods. While SDM algorithms are often applied in practice with a fixed set of parameters for decision-making in a given problem instance, it could be beneficial to change them according to the current state of the instance at hand. The initial board position in a game of chess for example, when all the pieces are still available, is usually very different in terms of complexity from the board state in a position close to the end of the game, when most of the pieces have likely been captured. This can have a substantial impact on the behavior and performance of SDM algorithms, leading to the natural idea of setting algorithm parameters accordingly.

We formalize this as Per-State Algorithm Configuration (PSAC) and propose a framework based on Reinforcement Learning (RL) to learn how to do it in an automated way. We then present results that demonstrate the effectiveness of our method of applying PSAC on two popular SDM algorithms: Monte Carlo Tree Search (MCTS) [11] and AlphaZero (AZ) [18], a state-of-the-art combination of MCTS and deep learning. We test the first on a warehouse assignment problem and the second on the game of Connect Four. These results highlight the potential of our algorithm configuration approach to improve the performance of sequential decision-making algorithms.

Our contributions are as follows. (1) We extend the state-of-the-art taxonomy of AC approaches provided by [2] to be applicable to sequential decision-making problems, by introducing Per-State Algorithm Configuration (PSAC). (2) We propose a framework based on reinforcement learning to perform PSAC in an automated way. (3) We test automated as well as manual PSAC on use cases belonging to two different domains: games and real-world logistics.

2 Related Works

Algorithm Configuration. Most approaches to AC focus on determining the best set of parameters to apply to a certain algorithm, before it starts running. This can be done manually, by a human, or automatically, by a machine [2]. A common approach is to configure the algorithm based on the problem instance that it is trying to solve. This is known as Per-Instance Algorithm Configuration (PIAC). Recently, the idea of automatically adapting these parameters during the algorithm's execution has gained more interest. This approach is referred to as Dynamic Algorithm Configuration (DAC) and has two main lines of research: DAC by optimization and DAC by reinforcement learning. In particular, the second one was formalized in [7], and has then been applied to several classes of algorithms, including Genetic Algorithms (GA), Adaptive Large Neighborhood Search (ALNS), anytime planning algorithms, classical AI planning and deep learning algorithms such as Stochastic Gradient Descent (SGD) [2,6,14,20].

AC for Sequential Decision-Making is a less frequently explored concept in the algorithm configuration literature. Since SDM algorithms are widely used in the field of game AI, the tuning of their parameters has been tackled with a variety of approaches over the decades. The most common approach for algorithms expected to play only a single game (instance) has been constant per-instance configuration [3]. When algorithms are expected to face various unseen games, such as in General Game Playing or General Video Game Playing, a constant configuration across all games is common, tuned offline on a selection of games [8].

Related to dynamic AC, [19] proposed a method to tune the parameters of MCTS from scratch whenever facing a new game. The term DAC could also be used to describe a variety of search algorithm enhancements that work by acquiring certain information during each individual search, such as RAVE for MCTS agents [9] or the "killer heuristic" for Alpha-Beta agents [10], as well as some techniques specifically developed for time management of game-playing algorithms [12,22].

Per-state AC methods are much more rare; exceptions include time management techniques [4], which have not been generalized outside their game domains. In the field of classical motion planners, [21] also studied how to use RL in autonomous navigation systems to adapt planner parameters to the current state of the world. Since motion planners interact with an environment that can be defined by a Markov Decision Process (MDP), the authors design a meta-MDP on which RL operates. We are currently not aware of other instances of per-state algorithm configuration in the literature. We aim to generalize this idea to other SDM algorithms and application domains.

3 Algorithm Configuration in Sequential Decision-Making

We extend the existing taxonomy of algorithm configuration approaches by introducing the new concept of per-state configuration, and propose a framework of

Table 1. Taxonomy of approaches for SDM Algorithm Configuration

Approach	Objective
Per-Distribution (AC)	Constant parameter configuration
Per-Instance (PIAC)	Function that maps an instance to a configuration
Per-State (PSAC)	**Function that maps a problem state and instance to a configuration**
Dynamic (DAC)	Function that maps an algorithm state and a problem state and instance to a configuration

algorithm configuration for SDM (Table 1). One aspect to note is that while these approaches are reported as incremental, they are effectively independent of each other. For example, while in a warehouse optimization problem it might be interesting to configure an SDM algorithm depending on both the current state of the warehouse and the problem instance, so that it can be applied to multiple warehouses, when playing strategic games, it is often sufficient to specialize on a single problem instance, solely adapting to the encountered game states. Since the focus of the current study is on the novel concept of per-state configuration, in the remainder of this paper, we will consider the problem instance as fixed.

3.1 Formulation

To formalize PSAC, we first provide a formal definition of sequential decision-making algorithm.

Definition 1. *Sequential decision-making algorithm*

Given an instance $i \in I$ of a sequential decision-making (SDM) problem, which can be described as a Markov Decision Process (MDP) $\langle S, A, P, R \rangle$, a SDM algorithm $\mathcal{A}$ is applied, with parameters $\theta \in \Theta$, every time a decision needs to be made, taking the current state $s_t \in S$ as input and outputting the probability of taking the decision $a_t \in A$, $\mathcal{A}: S \times A \times \Theta \to \mathbb{R}_+$. The goal of a SDM algorithm is to maximize the expected cumulative return $R(T)$ over a time horizon T.

Popular SDM algorithms include simple myopic heuristics such as greedy algorithms, advanced search algorithms such as MCTS, and more sophisticated learning-based algorithms such as reinforcement learning.

As previously discussed, SDM algorithms are usually configured manually for given problem instances, and the selected parameter configuration θ is then kept fixed for every application of the algorithm to the same problem instance, disregarding its current state. To adapt the parameters to the current state of the environment, we need a function that maps s_t to the best set of parameters that should be applied to $\mathcal{A}$ to compute the next action a_t, with the objective of optimizing the cumulative outcome of applying these actions sequentially.

Definition 2. *Per-State Algorithm Configuration (PSAC)*
Given $\langle \mathcal{P}, \mathcal{A}, \Theta, \mathcal{F} \rangle$:

- A target instance i of a SDM problem $\mathcal{P}$, represented by the MDP $\langle S, A, P, R \rangle$.
- A **SDM algorithm** $\mathcal{A}$ that, if applied to state $s \in S$ with configuration parameters $\theta \in \Theta$, outputs the probability of taking decision $a \in A$, $\mathcal{A}: S \times A \times \Theta \to \mathbb{R}_+$.
- A space of **per-state configuration policies** $f \in \mathcal{F}$ with $f: S \times \Theta \to \mathbb{R}_+$ that map problem states to parameter configuration probabilities for $\mathcal{A}$.

The SDM algorithm $\mathcal{A}$ interacts with the MDP to collect episodes $\tau = \{s_t, a_t, r_t\}_{t=0}^T$, defining a distribution over episodes

$$P(\tau) = P_0(s_0) \prod_{t=0}^T \mathcal{A}(a_t | s_t, \theta) P(s_{t+1} | s_t, a_t) \tag{1}$$

where $P_0(s_0): S \to \mathbb{R}_+$ is a distribution over initial states.
If we let the configuration policy f select, at each timestep t, which θ should be applied to $\mathcal{A}$, we have

$$P(\tau) = P_0(s_0) \prod_{t=0}^T f(\theta_t | s_t) \mathcal{A}(a_t | s_t, \theta_t) P(s_{t+1} | s_t, a_t) \tag{2}$$

The goal is to find the configuration policy f that maximizes the expected cumulative return over $P(\tau)$:

$$G(f) = \mathbb{E}_{\tau \sim P(\tau)} \left[\sum_{t=0}^T \gamma^t r_t \right]. \tag{3}$$

3.2 Reinforcement Learning

Given the previous definition, maximizing the objective of Eq. 3 can easily be framed as a reinforcement learning problem, where the agent does not interact with the original MDP, but with a meta-MDP, which is an MDP defined by the tuple $\langle S, \Theta, P_\Theta, R \rangle$, where the states and rewards are the same as those of the original problem $\mathcal{P}$, the actions are the parameter configurations of algorithm $\mathcal{A}$, and the transition probability is now a function of the new action set $P_\Theta: S \times S \times \Theta \to \mathbb{R}_+$, which can be defined based on the original transition probability as $P_\Theta(s_{t+1} | s_t, \theta_t) = \sum_{a_t \in A} \mathcal{A}(a_t | s_t, \theta_t) P(s_{t+1} | s_t, a_t)$. Interacting with this new meta-MDP, we collect episodes $\tau_\Theta = \{s_t, \theta_t, r_t\}_{t=0}^T$, from which Eq. 2 can be updated as:

$$P(\tau_\Theta) = P_0(s_0) \prod_{t=0}^T f(\theta_t | s_t) P_\Theta(s_{t+1} | s_t, \theta_t) \tag{4}$$

The resulting framework is visualized in Fig. 1.

4 Benchmark Problems

To illustrate the generality of Per-State Algorithm Configuration (PSAC), we decided to focus on two different sequential decision-making problems, which are well-known and relevant for different audiences and pose different challenges: game playing and stochastic combinatorial optimization.

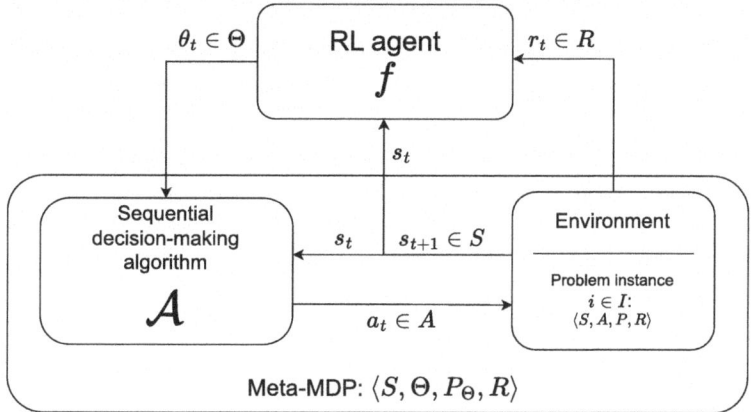

Fig. 1. Illustration of the PSAC framework. At each timestep t, the RL agent f interacts with the meta-MDP, observing the state of the current state and deciding which parameter configuration θ_t should be applied to the algorithm $\mathcal{A}$ (meta-level actions). Within the meta-MDP, the algorithm takes s_t and θ_t to compute the decision a_t that should be applied to the MDP (low-level actions).

4.1 Connect Four

We selected Connect Four (C4) as the game use case to strike the right balance between the computational resources required (e.g., each of the AZ policies [18] were trained on a minimum of 5 TPU years) and not being trivial to play well. C4 is a two-player game played on a vertically suspended grid with 6 rows and 7 columns. The players alternate between picking one of the 7 columns to place their disc in. The disc will then drop to the bottom unfilled row of the column. A player cannot play a disc in a column where all 6 rows are filled. Both players have different colored discs to differentiate theirs from their opponent's discs. The first player to connect four of their discs by placing them in a single line uninterrupted by any opponent discs wins the game. If all 7 columns are filled and neither player has managed to connect four discs, the game ends in a draw.

To create a single-player environment, we focus only on the learning of our agent and fix the opponent. In each episode, our agent will be randomly assigned to be the first or second player to move. We define the MDP as follows.

State. At each timestep t, the state s_t is the discs which are filling the slots in the 6×7 grid. Each slot can either be filled by a disc of the first player or second player, or be empty.

Actions. At each decision step, the player selects which column to drop one of its discs in. A column can only be selected if not all its rows are filled yet. After our action, if the episode is not finished, the opponent selects an action. If again the episode is not finished, the agent receives a new state.

Reward. The reward function is simply based on whether the game is won, lost or else, ongoing or a draw:

$$r(t) = \begin{cases} 1 & \text{if the agent played a winning move} \\ -1 & \text{if after the agent's move the opponent played a winning move} \\ 0 & \text{otherwise} \end{cases}$$

Objective. Since there are no intermediate rewards in this case, the objective can be both written as a summation or simply the reward of the last state: $R(T) = \sum_{t=0}^{T} r(t) = r(T)$.

4.2 Collaborative Order Picking

The collaborative human-robot order picking problem is a stochastic combinatorial optimization problem set in a modern warehouse, where human pickers and autonomous mobile robots (AMR) work side by side. AMRs handle the movement of goods, so that the pickers can focus on picking. The AMRs are usually associated with fixed sequences of pick-locations to visit, called pickruns, therefore, the task is to sequentially assign to each picker the next item to retrieve in order to further optimize the pick-rate. Since warehouses are highly stochastic, classical one-shot algorithms, which derive the solutions for the complete list of items, are not effective as they are not robust to random events occurring during plan execution. Therefore, the problem is a suitable benchmark for SDM algorithms such as MCTS.

We developed a discrete event simulation of the collaborative picking problem. The warehouse is represented as a rectangular graph, where nodes are discretized pick locations and edges are paths available between them. The simulated warehouse has a simple layout, with parallel shelves and picking aisles, without any cross-aisles. While humans can move freely, AMRs can only follow S-shaped routes and are not allowed to overtake other AMRs in the same lane. At the beginning of the simulation, every picker is in idling state. The system will then start assigning pickruns to AMRs, which are then dispatched and will immediately start traveling towards the first pick location in their list. The pickers are then also assigned to one of these pick locations depending on some configurable strategy and will start traveling to reach it. Both pickers and AMRs follow precomputed shortest paths to reach their destinations, taking into account the above-mentioned constraints, such as the S-routes. As soon as a picker and an AMR meet, the actual picking starts with a duration sampled from a parametrized distribution. Once the pick operation is over, the picker will receive a new picking task while the AMR will move to the next location. If the current pick location is the last in the pickrun, the AMR starts the drop-off operation, for which it has to reach one of the unloading stations located next to the bottom lane with need time sampled from a parametrized distribution. The AMR will then be assigned with a new pickrun and dispatched again. The speed of both AMRs and pickers is controlled by a random distribution every time a movement is processed. Finally, the system allows one to specify a set of distributions for pick durations. This could represent, for example, that items of different sizes are being picked. These different distributions can be applied at different times during the simulation, making the overall environment non-stationary.

We now define the MDP formulation.

State. The state s_t is the set of all the nodes in the graph. Each node contains features related to the controlled picker, its distance from the controlled picker, based on both AMRs and pickers moving capabilities, the presence of AMRs, free and busy pickers, and whether a picking operation is currently happening in the location. In addition, we have two global features, related to the current pick-time distribution, including a one-hot encoding of the current pick-time distribution $\mathcal{T} \in \mathbf{T}$ and the simulation time left before the next pick-time distribution change, since the shift points are fixed and known in advance.

Actions. At each decision step only one picker can be idle, therefore the set of actions contains all the locations which are the next pick-locations in at least one AMR's pickrun, plus the current location of the picker, which will trigger a waiting event to essentially postpone the decision if it is not a pick-locations included in a pickrun.

Reward. The reward function is simply based on whether an item was collected or not since the previous timestep:

$$r(t) = \begin{cases} 1 & \text{if an item was collected between timestep } t\text{-1 and timestep } t \\ 0 & \text{otherwise} \end{cases}$$

Objective. Since the reward function effectively counts the number of items collected during an episode, the objective is to maximize that number given a fixed duration T of the episodes: $R(T) = \sum_{t=0}^{T} r(t)$.

5 Experimental Setup

5.1 Algorithms

We applied PSAC to two popular SDM algorithms: MCTS and AZ. Below, we provide a brief description of how they work.

Monte Carlo Tree Search is one of the most commonly used SDM algorithms. For each encountered state, it builds a search tree incrementally and uses random sampling and simulations to estimate the values of actions and find the best one to apply. The UCT (i.e., Upper Confidence bounds applied to Trees) algorithm is a key component of MCTS, balancing exploration and exploitation by selecting actions that maximize the UCT value. The UCT is computed by $UCT = \frac{w_i}{n_i} + C\sqrt{\frac{\ln N}{n_i}}$, where w_i is the total reward of node i, n_i is the number of times node i has been visited, N is the search budget and C is a constant that balances exploration and exploitation. Every application of MCTS involves four main phases, which are repeated N times:

Selection: Starting from the root node, the algorithm traverses the tree by selecting promising nodes based on the UCT value.

Expansion: If the selected node is not a terminal state, new child nodes are added to the tree.

Simulation: A random simulation is run from the new node until either a terminal state is reached or a set amount of time, called rollout length l, has passed in the simulation, to estimate the outcome.

Backpropagation: The results of the simulation are used to update the values of the nodes along the path from the new node to the root.

At the end of the process, the algorithm returns the most visited action (from the root node) to be applied in the real environment.

AlphaZero is an algorithm that adapts the MCTS algorithm to be guided by neural networks (NN). In the selection phase, an adapted version of the UCT algorithm is used, which, in addition to the average reward and the number of node visits, also considers the prior probability of selecting actions. Furthermore, a prediction is used to estimate the outcome, rather than a simulation. Both the prior probability and the outcome estimate are predicted by NNs. The NNs are trained by playing against a copy of itself in a reinforcement learning setting, where actions are selected by MCTS (which is, in turn, dependent on the NNs) to align the networks with use in MCTS. AZ has shown impressive results across a range of board games [17].

In order to test our framework in a controlled way, we decided to apply AZ to C4 and MCTS to the collaborative picking use case.

5.2 Configuration Parameters

In each of our experiments, we chose to configure only a single parameter with limited range. Note that our theoretical framework does not limit the number of parameters that can be configured, but we made this choice in order to keep the analysis straightforward, while clearly showing the effectiveness of the approach.

In the case of C4, we believed that optimizing the exploration constant C would have the greatest impact. To ensure we can pick a wide variety of values for C without exploding the search space, we decided to pick our values for C on an exponential scale with base 2. The smallest value allowed is 2^{-7} and the largest 2^2 for a total of 10 values.

Since the collaborative order picking problem is highly stochastic and non-stationary, it is reasonable to assume that there should be a relation between the length of the rollouts, the amount of reward observed on average per timestep and the accuracy of the state-value estimation. Since the picking time can change, at different points in a single trajectory, the same value of rollout length l could potentially lead to collect very different amounts of reward. For example, if the average picking time is longer than the rollout length, no rewards will be observed, and the state cannot be evaluated properly. On the other hand, if the picking time is short and the rollout length is long, the stochasticity of the environment would make the observed rewards more and more uncertain, leading to an unreliable evaluation of the leaf state. For this reason, we chose the rollout length parameter l as the target of our experiment. In particular, in every rollout, the simulation will run for a number of simulation seconds between 5 and 100, with discrete steps of 5, for a total of 20 possible values.

All other parameters we kept fixed to values that seemed reasonable for the use case, as summarized in Table 2.

Table 2. Parameter Settings for the proposed use cases

Parameter	Collaborative Order Picking	Connect four
Search budget N	250	100
Discount factor γ	0.99	0.99
Exploration constant C	0.25	variable
Rollout strategy	Random	Value Function
Maximum number of chance nodes n_c	10	-

5.3 Simulation Setup

Connect Four. In C4 we used the standard dimensions of the grid, 6 rows and 7 columns. To make comparisons easier, we used the same AZ algorithm for our opponent and the PSAC implementation. We used the OpenSpiel library [13] to train the AZ algorithm. The AZ algorithm utilizes a residual NN with 64 nodes and a depth of 3 layers. We used the default values for all the training parameters except for the number of steps, which is set by default to run indefinitely. The parameters are as follows.

learning rate: 0.001, steps: 1000, N:100, replay buffer size:2^{16}, replay buffer reuse: 3, batch size: 2^{10}, C: 2, alpha: 1, and epsilon: 0.25.

One risk of having a fixed opponent is that the agent might learn to play well against this specific opponent but is not able to play well against different opponents, i.e. does not generalize well. To mediate this, at the start of each episode, we select an opponent from a set of potential opponents for the episode. Again, to ease the interpretation of results, we decided to align this with the action space of our Meta-MDP, meaning that our set of potential opponents will be AZ agents with C-values ranging from 2^{-7} to 2^2.

Another way to improve the trained agent's generalisation capabilities is to increase the diversity of the starting position. Learning how to play from a single starting position might result in the agent only learning how to play a fixed sequence of actions to win the game rather than more general patterns. We set the number of random moves at the start of each episode to 2, resulting in $7^2 = 49$ unique starting positions. We chose this number of random moves as a compromise between the variety of the starting positions and the balance of the starting position. For example, if we start with 4 random moves, positions exist in which the first player can force a win in 3 moves.

Collaborative Order Picking. For the collaborative order picking case we chose to keep the parameters of the simulation environment low in terms of the number of agents involved and the size of the warehouse to allow for more controllable experiments. In all experiments, the following parameters are the same: 8 aisles, 7 pick-locations per aisle, 4 pickers, 40 AMRs, and, on average, 10 picks per AMR pickrun. The travel times of pickers and AMRs are generated from a non-negative distribution [1] with, respectively, means 1.8 s and 0.8 s and standard deviations 0.4 s and 0.2 s.[1] The duration of the drop-off operation, from the moment the AMR reaches the drop-off location, is generated from a geometric distribution with mean of 50.0 s. Finally, the duration of the pick operation, from the moment both the picker and the AMR are in the right pick-location, is generated from non-negative distributions [1] $\{T_1, T_2, T_3, T_4\}$, with means $\{8.75, 17.5, 35.0, 70.0\}$s and standard deviations $\{1.25, 2.5, 5.0, 10.0\}$s. This is the only simulation parameter that will change in the experiments. Each simulation episode lasts 30 min of simulation time.

5.4 Deep Reinforcement Learning Algorithm

To implement PSAC we used the Proximal Policy Optimization (PPO) [16] algorithm. PPO has already been applied to many real-world SDM optimization use cases, including the collaborative human-robot order picking problem [5]. For the two domains, we used different NN backbones, depending on the input features, which are used both for the policy and the critic network. The policy network appends a softmax layer, representing the probability distribution over actions, i.e. the values of θ which can be applied to MCTS in the current timestep. The critic network appends a final layer to obtain a single value which represents the expected future reward in a state. In C4, we used a Tanh activation function since the values are bounded between -1 and 1, while in the Collaborative Order Picking use case the values can be real values, so no activation function is applied.

For C4, we used a residual NN backbone, due to the rectangular shape of the board. The input features are 4 binary 6×7 planes. The first three indicate whether a disc is on a slot on the board. The first plane considers the agent's own discs, the second the opponent's and the third whether the slots are unfilled. The fourth plane indicates whether the agent was the player to start. The residual backbone used two 3×3 residual blocks with 64 filters, batch normalization and a ReLu activation. After the two residual blocks, the output is flattened, and a single ReLu layer with 64 units is used.

---

[1] These times are applied every time the entities move from one location to another.

For Order Picking, the architecture consists of a multilayer perceptron (MLP) with one hidden layer of size 64, that independently processes the features of each node with the same set of weights to create node embeddings, which are then aggregated with a mean pooling layer. The two global features are also processed by two dedicated MLPs with hidden layers of size 64. The resulting embeddings are then concatenated and passed through the output layer.

The Tianshou library was used in this work for the implementation of PPO, and all the parameters of the algorithm are default, while PyTorch was used for implementing the NN architectures. All experiments were run with 16GB of RAM and without the use of GPU. The training and testing experiments for C4 were performed using 16 cores of an AMD EPYC(TM) Rome 7H12 CPU @ 2.6 GHZ and for the collaborative order picking problem on an Intel(R) Core(TM) i7-7700HQ CPU @ 2.80 GHz. The final configuration policy for C4 was trained for 800,000 timesteps (~18 h) and 500,000 timesteps (~8 h) for the collaborative order-picking problem.

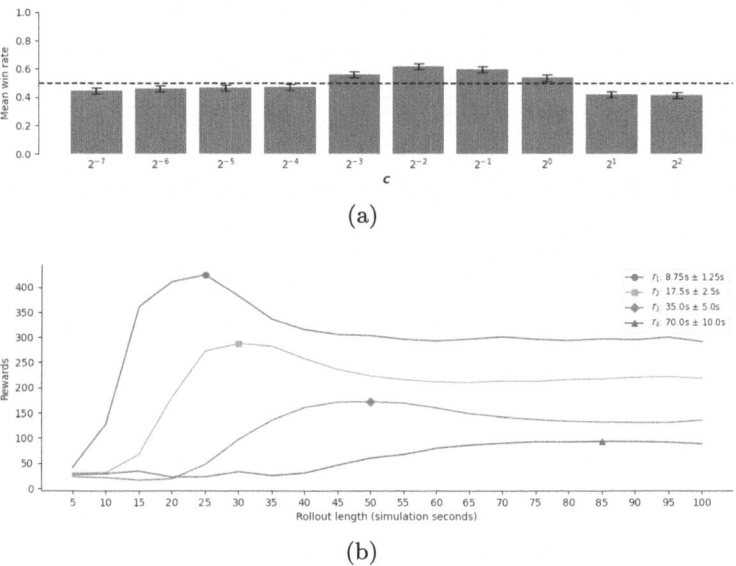

Fig. 2. Results of per-instance agents on Connect Four (a) and on the stationary environments of collaborative picking (b). (a) Mean win rate (draws counting as half wins) of a per-instance AlphaZero configuration vs. a range of AlphaZero opponents configured with $C \in \{2^{-7}, \ldots, 2^2\}$. Each opponent was played 200 times, half as first to act, half as second to act, resulting in each bar representing 2000 episodes. The whiskers represent 95% confidence intervals. (b) Average rewards over 10 episodes of MCTS applied with fixed values of l to four stationary environments.

5.5 Baselines

Figure 2 shows the results of the per-instance baseline experiments which were run to obtain the baselines for the two use cases.

In C4, per-instance agents were evaluated by running match-ups for each value of C of the action space. Figure 2(a) shows that the performance has an unimodal shape with the performance peaking at a value of 2^{-2} for C. Therefore, we used this value for the per-instance baseline agent.

For the collaborative picking use case, we manually created a configuration function, as follows: we first ran a grid search on stationary versions of the environment, where the picking time distribution $\mathcal{T} \in \mathbf{T}$ is fixed, to find the best value of l for each of those distributions. These values are highlighted in Fig. 2(b). The configuration function, which we will call *manual* PSAC from now on, as opposed to PSAC, which refers to the learned policy, picks, at each step, the value corresponding to the current distribution type $\mathcal{T}$. In addition, we also compared the results of the learned PSAC agent with a myopic greedy heuristic that assigns the pickers to the closest pick-location which is the next step in the pickrun of an AMR and is not currently assigned to another picker.

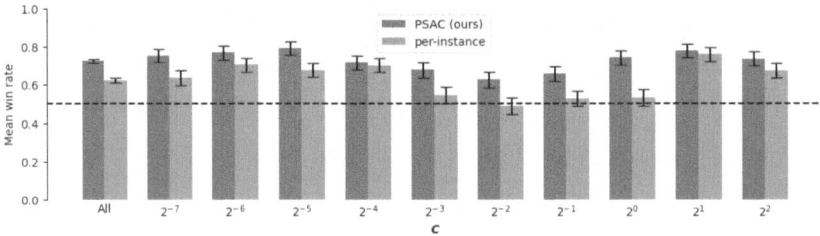

Fig. 3. Performance PSAC agent vs per-instance baseline. Mean win rate (draws counting as half wins) vs. AlphaZero opponents configured with different C. Each opponent was played 500 times, half as first to act, half as second to act. The whiskers represent 95% confidence intervals.

6 Results

Connect Four. To evaluate our approach, we have the PSAC agent and the optimized per-instance agent play against the same pool of opponents. The results are reported in Fig. 3. Our PSAC agent significantly outperforms the per-instance agent over all the matchups. Furthermore, the performance against the majority of specific AZ agents is significantly better as well, and, in all cases, better than that against the per-instance agent.

We also show in Fig. 4 the values for C that our PSAC agent selects. Surprisingly, Fig. 4(a) shows that the shape of the probability mass function is not uni-modal (as for the per-instance performances in Fig. 2(a)) but multi-modal.

Although it generally selects the same value for C as the agent per instance, at times it also prefers to select very small values (2^{-7} or 2^{-6}) or very large values (2^2). In Fig. 4(b) we can see the relation of the mean C versus the move number. At the very start of the episode, the mean values are low; around move 13, they peak and towards the end, they decrease again. We hypothesize that this is due to the first positions have been seen very frequently so AZ is quite confident that the best move is one of its most preferred moves, later on, the positions are seen less frequently and many alternatives seem promising so it will search more broadly, then near the end, there are fewer relevant alternatives again (some moves will be clearly losing so need not be considered) and the precise nature of the end game requires looking farther into fewer variations.

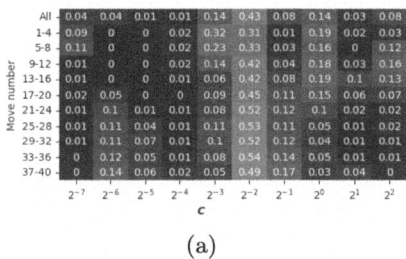
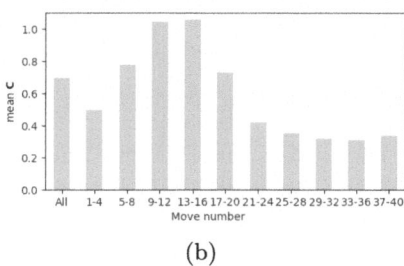

(a)   (b)

Fig. 4. Analysis of selected C values. Both plots are based on all the episodes of Fig. 3. (a) Probability mass function of the configured C vs. move number. (b) Mean C vs. move number.

Collaborative Order Picking. The rewards obtained running a grid search for the configuration of l in MCTS on the non-stationary environment, as well as running PSAC are reported in Table 3, together with the baselines discussed in the previous section. In addition, the results shown in Fig. 2(b) are also added, both to compare them with the greedy baseline and to highlight the best l parameters found for each pick-time distribution.

First, we can see that in every stationary environment, the best configuration of MCTS consistently outperforms the myopic heuristic. This is not true for the non-stationary environment, where every fixed configuration of MCTS performs worse than myopic. Secondly, manual PSAC clearly outperforms the myopic one, proving that per-state configuration of MCTS is necessary to achieve good results in this environment. Lastly, we consider the performance of PSAC. Learning how to configure the l parameter of MCTS during episodes, our method is able to considerably outperform every instance of MCTS with fixed parameters, as well as the myopic heuristic. While manual PSAC still provides the best result, Fig. 5 shows that PSAC learns a very similar behavior. Since manual PSAC requires manual labor and domain knowledge, it becomes impractical and less feasible to configure multiple parameters at the same time.

Table 3. Performances of per-instance MCTS, baseline heuristics and PSAC, in terms of means and standard deviations of rewards over 10 episodes.

Method	Environment				
	T_1: 8.75 ± 1,25	T_2: 17.5 ± 2.5	T_3: 35.0 ± 5.0	T_4: 70.0 ± 10.0	Non-stationary
MCTS, l : 5s	41.4 ± 34.65	30.2 ± 20.8	22.8 ± 17.17	26.9 ± 15.51	33.0 ± 24.52
MCTS, l : 10s	127.8 ± 65.5	31.0 ± 18.36	21.0 ± 23.94	28.9 ± 18.11	45.9 ± 22.51
MCTS, l : 15s	360.2 ± 13.14	67.3 ± 36.8	16.2 ± 11.91	34.1 ± 12.36	66.0 ± 18.41
MCTS, l : 20s	410.6 ± 15.87	180.3 ± 47.52	18.6 ± 14.93	22.3 ± 18.04	81.6 ± 18.56
MCTS, l :25s	**424.5 ± 13.57**	272.4 ± 8.49	46.9 ± 13.64	22.6 ± 15.25	89.9 ± 10.21
MCTS, l : 30s	381.5 ± 7.41	**287.4 ± 5.87**	96.8 ± 8.74	33.0 ± 18.57	109.1 ± 12.75
MCTS, l : 35s	335.6 ± 9.22	282.3 ± 5.68	135.7 ± 7.09	25.1 ± 13.41	118.4 ± 11.17
MCTS, l : 40s	315.1 ± 8.14	257.5 ± 4.25	160.6 ± 3.83	30.4 ± 17.54	122.0 ± 10.7
MCTS, l : 45s	305.4 ± 6.93	235.8 ± 4.42	171.1 ± 3.24	45.9 ± 11.27	127.5 ± 5.24
MCTS, l : 50s	303.2 ± 8.27	223.0 ± 6.16	**172.1 ± 3.7**	60.1 ± 2.43	129.6 ± 3.5
MCTS, l : 55s	296.0 ± 8.15	215.9 ± 4.5	169.2 ± 1.99	67.2 ± 2.71	130.5 ± 4.52
MCTS, l : 60s	292.6 ± 10.52	211.4 ± 5.28	159.7 ± 2.24	79.6 ± 3.69	136.1 ± 4.21
MCTS, l : 65s	296.6 ± 6.39	210.2 ± 3.68	148.6 ± 3.1	85.8 ± 2.64	133.7 + 3.63
MCTS, l : 70s	300.6 ± 15.27	213.2 ± 5.42	141.7 ± 1.95	89.7 ± 2.83	134.8 ± 3.57
MCTS, l : 75s	295.9 ± 7.62	212.5 ± 8.87	137.0 ± 2.45	92.4 ± 1.62	134.7 ± 3.41
MCTS, l : 80s	293.8 ± 8.28	216.3 ± 7.96	133.6 ± 2.94	92.2 ± 1.72	135.0 ± 2.83
MCTS, l : 85s	296.7 ± 12.53	217.2 ± 7.39	131.6 ± 1.56	**93.2 ± 1.54**	134.4 ± 4.45
MCTS, l : 90s	295.0 ± 5.31	220.2 ± 8.39	130.4 ± 3.38	92.6 ± 2.06	136.8 ± 4.69
MCTS, l : 95s	299.9 ± 12.49	221.6 ± 6.9	130.5 ± 3.11	91.4 ± 1.43	133.8 ± 4.49
MCTS, l : 100s	291.5 ± 8.24	218.8 ± 6.6	135.7 ± 2.93	88.6 ± 1.36	134.0 ± 3.1
Myopic heuristic	331.3 ± 10.87	245.0 ± 7.29	154.3 ± 3.74	86.3 ± 1.49	145.2 ± 5.1
Manual PSAC	-	-	-	-	**161.0 ± 6.72**
PSAC	-	-	-	-	154.45 ± 3.92

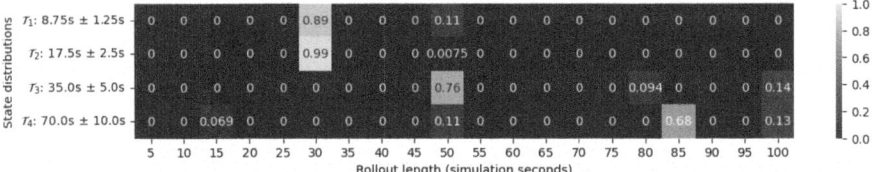

Fig. 5. Analysis of selected l values. Analysis of the l value selected by the PSAC agent for different T values in the non-stationary environment.

7 Conclusion

In this paper, we propose to adapt the state-of-the-art taxonomy of AC to SDM algorithms, introducing the concept of Per-State Algorithm Configuration. We then present a framework for doing that in an automated way with RL and show significant performance improvements for MCTS and AlphaZero on two different domains: a deterministic two-player game and a stochastic combinatorial optimization problem, inspired by a real-world use case. Future research directions include testing our method on a broader set of problems, including more complex games and large-scale real-world use cases as well as simpler white-box benchmarks, which would allow us to perform a deeper ablation analysis of the framework. Additionally, we plan to apply PSAC to configure multiple parameters simultaneously, to further improve the algorithm performances. Lastly, we intend to analyze more deeply in what type of states certain parameters are selected by PSAC. In general, our work extends the current field of automated algorithm configuration, opening up new possibilities for further research on sequential decision-making algorithms and their applications.

Acknowledgments. This research was made possible by TKI Dinalog and the Topsector Logistics and has received funding from the Ministry of Economic Affairs and Climate Policy (EZK) of the Netherlands and from the European Union's Horizon Europe Research and Innovation Programme, under Grant Agreement number 101120406. The paper reflects only the authors' view and the EC is not responsible for any use that may be made of the information it contains. This work used the Dutch national e-infrastructure with the support of the SURF Cooperative using grant no. EINF-9770.

References

1. Adan, I., van Eenige, M., Resing, J.: Fitting discrete distributions on the first two moments. Probab. Eng. Inf. Sci. **9**(4), 623–632 (1995). https://doi.org/10.1017/S0269964800004101
2. Adriaensen, S., et al.: Automated dynamic algorithm configuration. J. Artif. Intell. Res. **75**, 1633–1699 (2022). https://doi.org/10.1613/jair.1.13922
3. Baier, H., Kaisers, M.: Guiding multiplayer MCTS by focusing on yourself. In: IEEE Conference on Games, CoG 2020, Osaka, Japan, 24–27 August 2020, pp. 550–557. IEEE (2020). https://doi.org/10.1109/COG47356.2020.9231603
4. Baier, H., Winands, M.: Time management for Monte Carlo tree search. IEEE Trans. Comput. Intell. AI Games **8**(3), 301–314 (2016). https://doi.org/10.1109/TCIAIG.2015.2443123
5. Begnardi, L., Baier, H., van Jaarsveld, W., Zhang, Y.: Deep reinforcement learning for two-sided online bipartite matching in collaborative order picking. In: Asian Conference on Machine Learning, pp. 121–136. PMLR (2024)
6. Bhatia, A., Svegliato, J., Nashed, S.B., Zilberstein, S.: Tuning the hyperparameters of anytime planning: a metareasoning approach with deep reinforcement learning. In: Proceedings of the International Conference on Automated Planning and Scheduling, vol. 32, no. 1, pp. 556–564 (2022). https://doi.org/10.1609/icaps.v32i1.19842
7. Biedenkapp, A., Bozkurt, H.F., Eimer, T., Hutter, F., Lindauer, M.: Dynamic Algorithm Configuration: Foundation of a New Meta-Algorithmic Framework. Santiago de Compostela (2020)
8. Finnsson, H., Björnsson, Y.: Cadiaplayer: search-control techniques. Künstliche Intell. **25**(1), 9–16 (2011). https://doi.org/10.1007/S13218-010-0080-9

9. Gelly, S., Silver, D.: Combining online and offline knowledge in UCT. In: Ghahramani, Z. (ed.) Machine Learning, Proceedings of the Twenty-Fourth International Conference (ICML 2007), Corvallis, Oregon, USA, 20–24 June 2007. ACM International Conference Proceeding Series, vol. 227, pp. 273–280. ACM (2007). https://doi.org/10.1145/1273496.1273531
10. Huberman, B.J.: A program to play chess end games. Ph.D. thesis, Stanford University, USA (1968). https://searchworks.stanford.edu/view/2190328
11. Kocsis, L., Szepesvári, C.: Bandit based Monte-Carlo planning. In: Fürnkranz, J., Scheffer, T., Spiliopoulou, M. (eds.) ECML 2006. LNCS (LNAI), vol. 4212, pp. 282–293. Springer, Heidelberg (2006). https://doi.org/10.1007/11871842_29
12. Lan, L.C., Wu, T.R., Wu, I.C., Hsieh, C.J.: Learning to stop: dynamic simulation Monte-Carlo tree search. In: Proceedings of the AAAI Conference on Artificial Intelligence, vol. 35, no. 1, pp. 259–267 (2021). https://doi.org/10.1609/aaai.v35i1.16100
13. Lanctot, M., et al.: Openspiel: a framework for reinforcement learning in games. arXiv preprint arXiv:1908.09453 (2019)
14. Reijnen, R., Zhang, Y., Lau, H.C., Bukhsh, Z.: Online control of adaptive large neighborhood search using deep reinforcement learning. In: Proceedings of the International Conference on Automated Planning and Scheduling, vol. 34, pp. 475–483 (2024). https://doi.org/10.1609/icaps.v34i1.31507
15. Schede, E.: A survey of methods for automated algorithm configuration. J. Arti. Intell. Res. **75**, 425–487 (2022)
16. Schulman, J., Wolski, F., Dhariwal, P., Radford, A., Klimov, O.: Proximal Policy Optimization Algorithms (2017). arXiv:1707.06347
17. Silver, D.: A general reinforcement learning algorithm that masters chess, shogi, and go through self-play. Science **362**(6419), 1140–1144 (2018)
18. Silver, D., et al.: Mastering the game of go without human knowledge. Nature **550**(7676), 354–359 (2017)
19. Sironi, C.F., Liu, J., Winands, M.: Self-adaptive Monte Carlo tree search in general game playing. IEEE Trans. Games **12**(2), 132–144 (2020). https://doi.org/10.1109/TG.2018.2884768
20. Speck, D., Biedenkapp, A., Hutter, F., Mattmüller, R., Lindauer, M.: Learning heuristic selection with dynamic algorithm configuration. In: Proceedings of the International Conference on Automated Planning and Scheduling, vol. 31, pp. 597–605 (2021). https://doi.org/10.1609/icaps.v31i1.16008
21. Xu, Z., et al.: APPLR: adaptive planner parameter learning from reinforcement. In: 2021 IEEE International Conference on Robotics and Automation (ICRA), pp. 6086–6092 (2021). https://doi.org/10.1109/ICRA48506.2021.9561647
22. Ye, W., Abbeel, P., Gao, Y.: Spending thinking time wisely: accelerating MCTS with virtual expansions. In: Koyejo, S., Mohamed, S., Agarwal, A., Belgrave, D., Cho, K., Oh, A. (eds.) Advances in Neural Information Processing Systems, vol. 35, pp. 12211–12224. Curran Associates, Inc. (2022)

Self-supervised Penalty-Based Learning for Robust Constrained Optimization

Wyame Benslimane[✉] and Paul Grigas

Department of Industrial Engineering and Operations Research,
University of California-Berkeley, Berkeley, CA, USA
{wyame.benslimane,pgrigas}@berkeley.edu

Abstract. We propose a new methodology for parameterized constrained robust optimization, an important class of optimization problems under uncertainty, based on learning with a self-supervised penalty-based loss function. Whereas supervised learning requires pre-solved instances for training, our approach leverages a custom loss function derived from the exact penalty method in optimization to learn an approximation, typically defined by a neural network model, of the parameterized optimal solution mapping. Additionally, we adapt our approach to robust constrained combinatorial optimization problems by incorporating a surrogate linear cost over mixed integer domains, and a smooth approximations thereof, into the final layer of the network architecture. We perform computational experiments to test our approach on three different applications: multidimensional knapsack with continuous variables, combinatorial multidimensional knapsack with discrete variables, and an inventory management problem. Our results demonstrate that our self-supervised approach is able to effectively learn neural network approximations whose inference time is significantly smaller than the computation time of traditional solvers for this class of robust optimization problems. Furthermore, our results demonstrate that by varying the penalty parameter we are able to effectively balance the trade-off between sub-optimality and robust feasibility of the obtained solutions.

Keywords: Robust Optimization · Self-Supervised Learning · Learning to Optimize

1 Introduction

Decision-making under uncertainty is an essential aspect of various real-world applications, ranging from facility location planning, portfolio optimization, and inventory management to healthcare and logistics. Robust optimization (RO) [7] offers a powerful and computationally viable framework for addressing decision-making under uncertainty. Most approaches used for robust optimization rely on constructing a tractable reformulation based on a carefully designed uncertainty set [6,8], which is later fed into a standard optimization solver. Thus, solving robust optimization problems often involves specialized optimization algorithms like interior-point methods or other iterative algorithms that rely on first or second order information [7,10]. Although these methods have been well studied in theory, with strong convergence properties and theoretical guarantees, and applied in a variety of contexts, they face several limitations in terms of ease of computational efficiency and scalability to very large problems. These computational challenges are amplified even further when addressing demanding problems involving nonlinear functions, combinatorial constraints, and large-scale instances. The need for scalable robust optimization solutions, computed very efficiently with low latency, also stems from the need to solve real-time decision-making problems under time constraints in various applications. For instance, power systems require real-time adjustments to accommodate the fluctuating nature of renewable energy sources [28], while ride-sharing platforms need to efficiently dispatch drivers to meet dynamic customer demand and traffic conditions [17].

On the other hand, the use of machine learning for solving optimization problem instances has shown great potential to address some of the above challenges. Several models have been developed to combine the strength of data-driven approaches to solve decision-making problems efficiently [2,18]. While initial work in this area focused on improving the performance of unconstrained optimization [4,11,20], recent models have included methods for handling more general optimization problems [18]. For example, one line of work is to integrate constraints as a layer of a neural network using implicit differentiation [1,3,15,25]. Another line of work consists of using a surrogate model relying on custom loss functions to train models that approximate continuous constrained optimization solutions [13,19,22]. Still, the constrained problem formulations considered so far in this line of work have placed little focus on uncertainty, particularly as modeled with robust optimization.

While data-driven approaches have been used for robust optimization [9,26], most approaches focus on learning better uncertainty set representations. Our work adopts a different angle by directly learning the solution to the RO problem using the learning to optimize framework. By approximating the solution to a class of parametric robust optimization problems, we do not have to rely on classical iterative solvers that can be computationally expensive, especially for time-sensitive and large-scale problems. Furthermore, we rely on a self-supervised [22] penalty approach where we design a custom loss function based on the exact penalty method [12] to construct a neural network-based solver for the RO problem. This approach allows us to potentially achieve faster solution times and improved performance compared to traditional methods. Furthermore, to handle a wide range of constraints, we discuss how to adapt our architecture network to be able to handle both discrete and continuous feasible domains, broadening the applicability of our method to various problem settings.

To show the effectiveness and versatility of our method, we focus on three applications: multidimensional knapsack with continuous variables, combinatorial multidimensional knapsack with discrete variables, and an inventory management problem. The self-supervised learning method allows us to recover over

50% feasible solutions. By tuning the hyper-parameters in the loss function, our method guarantees a high feasibility level, while maintaining a reasonable suboptimality threshold. Furthermore, our method achieves significant computational speed-ups compared to traditional solvers, where we observe a minimum speed-up of 10 orders of magnitude when solving a single instance using the learned solver compared to classical solvers, emphasizing the practicality of our method for real-world optimization tasks.

2 Robust Constrained Optimization: Problem Setting and Self-supervised Learning Method

We consider the following family of robust optimization (RO) problems indexed by the instance parameter $z \in \mathcal{Z} \subseteq \mathbb{R}^{d_z}$:

$$x^*(z) = \mathrm{argmax}_{x \in \mathcal{X}} \; f_z(x) \qquad (1)$$
$$\text{subject to} \quad g_z(x, u) \leq 0 \quad \forall u \in \mathcal{U}(z)$$

Here, $\mathcal{X} \subseteq \mathbb{R}^{d_x}$ refers to the domain of the decision variables x, and $\mathcal{U} \subseteq \mathbb{R}^{d_u}$ refers to the domain of the uncertain parameters u. The objective function $f_z : \mathcal{X} \to \mathbb{R}$ depends on the instance parameter z, and is known with certainty given z. The constraint function $g_z : \mathcal{X} \times \mathcal{U} \to \mathbb{R}^m$, where m refers to the number of nominal constraints, depends on both the parameter z and on the uncertain parameters $u \in \mathcal{U}$. Let g_z^j refer to the jth constraint function for $j \in \{1, \ldots, m\}$. Without loss of generality, we assume that the uncertain parameters u occur in the constraints only.

Throughout the paper, we consider a more structured special case of the family of problem instances (1), which we refer to as "nominal-parameterized" instances. Specifically, in this special case, each z encodes a vector of uncertain parameters, $\hat{u}(z)$, which we refer to as the nominal values of the uncertain parameters. The version of (1) without robustness, i.e., with only constraints $g_z(x, \hat{u}(z)) \leq 0$ and $x \in \mathcal{X}$ is thought of as the nominal version of the problem. We then construct a norm-based uncertainty set based on considering all perturbations of the nominal values $\hat{u}(z)$ within a ball. Possible choices of the norm $\|\cdot\|$ include the ℓ_∞ norm corresponding to a box uncertainty set, and a quadratic norm corresponding to an ellipsoidal uncertainty set. This special class of problem instances is formalized in the below assumption, along with an additional assumption ensuring feasibility of (1).

Assumption 1. *The family of RO problems (1), parameterized by $z \in \mathcal{Z}$, satisfies:*

1. *For some fixed $\rho > 0$, there exists a mapping $\hat{u} : \mathcal{Z} \to \mathcal{U}$ to "nominal uncertain parameters" and $\mathcal{U}(z) = \{u : \|u - \hat{u}(z)\| \leq \rho\}$.*
2. *For all $z \in \mathcal{Z}$, the robust feasible set $\{x \in \mathcal{X} : g_z(x, u) \leq 0 \; \forall u \in \mathcal{U}(z)\}$ is non-empty.*

It is important to emphasize that we do not make any strong assumptions on the domain of decision variables $\mathcal{X}$, and our methodology allows for both convex and integer domains. These conditions in Assumption 1 are not overly restrictive and are standard in robust optimization.

Learning Methodology and Self-supervised Loss. Our methodology falls under the framework of "learning to optimize," whereby our goal is to approximate the mapping $z \mapsto x^*(z)$, specified by the family of RO problems (1), using machine learning. We use $h : \mathcal{Z} \to \mathcal{X}$ to denote such an approximation function, referred to as an optimization proxy/surrogate. The function h is selected from the hypothesis class $\mathcal{H}$ of candidate functions, which in all of our examples will be a class of neural network models. It is important that h returns outputs that are guaranteed to lie in the domain $\mathcal{X}$, and we elaborate on how to ensure this later in this paper. We consider a self-supervised setting whereby we have available a dataset $D = \{z_1, z_2, \ldots, z_n\}$ of realizations of instances of (1). Notably, we do not assume availability of $x^*(z_i)$, hence our methodology is *self-supervised*. Finally, we also assume availability of a loss function $L : \mathcal{X} \times \mathcal{Z} \to \mathbb{R}$, whereby $L(x, z)$ measures the quality of the solution x for the instance of (1) parameterized by z. Putting all of these ingredients together and applying the empirical risk minimization (ERM) principle of machine learning (a.k.a. sample average approximation) leads to the following training problem:

$$\min_{h \in \mathcal{H}} \frac{1}{n} \sum_{i=1}^{n} L(h(z_i), z_i) \tag{2}$$

The default supervised learning approach would treat this setting as a regression problem and choose a regression based loss function such as $L^{\text{SL}}(h(z), z) = \|h(z) - x^*(z)\|_2^2$. Instead of adopting a supervised learning approach, which requires access to pre-solved instances $x^*(z_i)$, we employ a penalty-based self-supervised loss function that directly encodes the problem structure by balancing optimality and feasibility with respect to the robust constraints. This approach is particularly useful in our setting where obtaining pre-solved instances can be computationally expensive due to the increased complexity of solving robust optimization counterparts over their nominal versions [6,7]. Inspired by the exact penalty method used to solve constrained optimization [12], we define our self-supervised loss function $L_\nu^{\text{SSL}} : \mathcal{X} \times \mathcal{Z} \to \mathbb{R}$, given a penalty parameter $\nu > 0$, as follows:

$$L_\nu^{\text{SSL}}(x, z) := -f_z(x) + \nu \sum_{j=1}^{m} \left[\max_{u \in \mathcal{U}(z)} g_z^j(x, u) \right]^+, \tag{3}$$

where $[\cdot]^+ := \max\{0, \cdot\}$ is the exact penalty function. The first term of the self-supervised loss function aims to maximize the objective function $f(\cdot)$, while the second term is a penalty that attributes a positive weight to any unsatisfied robust constraints.

Assumption 2. *The learning problem (2) is practically tractable in the sense:*

1. For all $z \in \mathcal{Z}$, the objective function $f_z(\cdot)$ is continuous and differentiable almost everywhere with respect to x on an open set containing $\mathcal{X}$.
2. For each $j \in \{1, \ldots, m\}$, the function $\bar{g}_z^j(x) = \left[\max_{u \in \mathcal{U}(z)} g_z^j(x, u)\right]^+$ is continuous and differentiable almost everywhere with respect to x on an open set containing $\mathcal{X}$.

Assumption 2 ensures practical tractability of our learning problem when using a neural network hypothesis class, since the required differentiability properties ensure that we can apply modern automatic differentiation frameworks such as PyTorch [23]. While part (1.) of the Assumption 2 is readily satisfied, part (2.), however, requires that the robust counterpart of the RO problem (1) has a tractable reformulation with differentiable constraints. Under the assumption that, for all $j \in \{1, \ldots, m\}$ and all fixed $x \in \mathcal{X}$, the constraint function $g_z^j(x, \cdot)$ is concave in the uncertain parameters u, the results of Ben-Tal [6] provide a framework for constructing such a tractable robust counterpart for a wide range of uncertainty sets and constraint functions.

Ensuring Domain Feasibility via Neural Network Structure. While the previously introduced self-supervised loss function is used to promote feasibility w.r.t. the robust inequality constraints, we adopt a different approach to ensure the variable domain constraint $x \in \mathcal{X}$. Indeed, the domain constraint $x \in \mathcal{X}$ is usually relatively much simpler, including simple continuous or discrete sets, than the robust inequality constraints. As such, specifically in the case of neural network hypothesis classes $\mathcal{H}$, we discuss how to directly engineer the network to automatically ensure that $h(z) \in \mathcal{X}$, for several important domains $\mathcal{X}$. For many typical convex domains, using a standard activation function at the last layer is sufficient to ensure feasibility of the output. For example, the ReLU activation can be used to ensure that $h(z) \in \mathcal{X} = \mathbb{R}_+^n$ and the sigmoid activation can be used to ensure that $h(z) \in \mathcal{X} = [0, 1]^n$.

The case of an mixed integer or discrete domain set $\mathcal{X}$ is more challenging. Related to this case is the problem of differentiating through a mixed integer program (MIP), which occurs in decision-focused learning and related areas [15,25,27]. Most of the literature addressing this or related challenges develop gradient-based approaches that are suited for differentiating through a generic MIP with complex constraints. In our case, we implicitly assume that the domain constraints represented by $\mathcal{X}$ are relatively simple and that the "complex" constraints are the robust ones that have been incorporated into the self-supervised loss function. As such, we adopt an approach based on the assumption that we can efficiently solve linear optimization problems over $\mathcal{X}$ and also a related family of "smoothed" quadratic optimization problems that enhance differentiability properties [27]. Our approach is inspired by [14], who propose using a linear integer optimization surrogate to address learning the solutions of a family of mixed integer nonlinear optimization problems, as well as the older literature on structured prediction methods [5,21].

To be precise, let us assume that $\mathcal{X} \subseteq \mathbb{R}^{d_x}$ is a non-empty compact mixed integer set. To ensure feasibility $h(z) \in \mathcal{X}$, we carefully design the last layer of our neural network models $h \in \mathcal{H}$. The previous layers can consist of any standard neural network architecture. The last layer should take in as input a vector $w \in \mathbb{R}^{d_x}$ and output one of two deterministic mappings: (i) $g_{\mathcal{X}}^{\text{train}}(w)$ at training time, and (ii) $g_{\mathcal{X}}^{\text{test}}(w)$ at testing/inference time. Let us first describe $g_{\mathcal{X}}^{\text{test}}$. We let $g_{\mathcal{X}}^{\text{test}}(w) \in \text{argmax}_{x \in \mathcal{X}} w^\top x$ be an arbitrary solution to a linear optimization problem over $\mathcal{X}$, given the cost vector $w \in \mathbb{R}^{d_x}$ determined by the computations at the earlier layers. Thus, one of our implicit assumptions about the simplicity of $\mathcal{X}$ is that it is tractable to solve linear optimization problems. We design $g_{\mathcal{X}}^{\text{train}}$ as an approximation of $g_{\mathcal{X}}^{\text{test}}$, with better differentiability properties. Specifically, let $\bar{\mathcal{X}} = \text{conv}(\mathcal{X})$ denote the convex hull of $\mathcal{X}$. Given some smoothing parameter $\gamma > 0$, we propose the following smooth $g_{\mathcal{X}}^{\text{train}}(w) = \text{argmax}_{x \in \bar{\mathcal{X}}} \left\{ w^\top x - \frac{\gamma}{2} \|x\|_2^2 \right\}$.

The second implicit assumption regarding $\mathcal{X}$ is that the above optimization problem is also readily solvable. Note that the above formulation is a continuous optimization problem over the convex hull $\bar{\mathcal{X}}$, and the mapping can be differentiated w.r.t. w by implicit differentiation or other techniques [27]. Inspired by [24], we further propose a learning-based approximation to $g_{\mathcal{X}}^{\text{train}}(w)$ to enhance the smoothness properties and ease of computation of evaluating the gradient. This approximation leads to a 2-phase approach, where the first phase trains a simple network approximating $g_{\mathcal{X}}^{\text{train}}(w)$ using a regression-based model. Once this model is trained, we freeze the weight of this network and use it as the last layer in the optimization predictive model, which we then train using our self-supervised approach.

Example 1. As a concrete example, consider the case where $x \in \mathcal{X} = \{0,1\}^{d_x}$ corresponds to binary integer constraints. To ensure feasibility with respect to $\mathcal{X}$ at testing time, we note that $g_{\mathcal{X}}^{\text{test}}(w) \in \text{argmax}_{x \in \mathcal{X}} w^\top x$ is given by the component-wise step function $[g_{\mathcal{X}}^{\text{test}}(w)]_i = \mathbf{1}(w_i > 0)$ where $\mathbf{1}(\cdot)$ is an indicator function equal to 1 if the argument is true and 0 otherwise.

Similarly the approximation function $g_{\mathcal{X}}^{\text{train}}(w)$ can also be expressed component-wise as $[g_{\mathcal{X}}^{\text{train}}(w)]_i = \min(1, \max(0, w_i/\gamma))$. The analytic expression can either be directly differentiated or approximated using a neural network, ensuring smoothness and facilitating efficient optimization through gradient-based methods.

3 Experimental Results

In this section, we present the numerical evaluation of our method, focusing on the following applications:

Application 1 (Multidimensional Knapsack Problem). We consider a multidimensional knapsack problem where the goal is to allocate a set of items, with a vector of values v_z and a matrix of weights W, across multiple knapsacks with capacities C_z. The items' weight matrix W belongs to a known uncertainty set $\mathcal{U}(z)$. The problem can be formulated as follows:

$$x^*(z) = \max_{x \in \mathcal{X}} v_z^\top x$$
$$\text{s.t.} \quad Wx \leq C_z, \quad \forall W \in \mathcal{U}(z).$$

We consider both the discrete case where $\mathcal{X} = \{0,1\}^{d_x}$ and the continuous case where $\mathcal{X} = [0,1]^{d_x}$. We generate a synthetic dataset $\mathcal{D} = \{(v_i, \hat{W}_i, C_i)\}_{i=1}^N$ for the knapsack problem, where $\hat{W}_i$ represents the nominal value of the weight matrix.

Application 2 (Inventory Management Problem). We consider an inventory management problem inspired by the classical newsvendor model, extended to a multi-retailer system with a centralized warehouse, with a formulation adopted from [16]. N retailers face uncertain demand $d_z = d_z^0 + Q_z u$, where d_z^0 is the expected demand, u is a k-dimensional vector of uncertainty factors, and Q_z is a matrix capturing the sensitivity of retailers' demand to these factors. The stocking decisions x aim to maximize profit across all retailers. Considering an auxiliary second stage variable $y(u)$ representing sales (as in [16]), the problem is formulated as the following two-stage adjustable robust optimization:

$$\begin{aligned}
\max_{x \in \mathcal{X}, P \in \mathbb{R}} \ & P \\
\text{s.t.} \quad & P \leq r_z^\top y(u) - c_z^{o\top} x \quad \forall u \in \mathcal{U}(z) \\
& y(u) = \min(x, d_z^0 + Q_z u) \ \forall u \in \mathcal{U}(z) \\
& \mathbf{1}^\top x \leq C_z
\end{aligned}$$

where the vector x is contained in $\mathcal{X} = \{x : 0 \leq x \leq c\}$, and r_z and c_z^o are the revenue and cost per unit sold respectively. Using a linear decision rule [10] for the second-stage decisions, i.e., $y(u) = Yu + y_0$, we derive the following one-stage robust problem:

$$\begin{aligned}
\max_{P \in \mathbb{R}, Y \in \mathbb{R}^{N \times k}, y_0 \in \mathbb{R}^N, x \in \mathcal{X}} \ & P \\
\text{s.t.} \quad & P \leq r_z^\top (Yu + y_0) - c_z^{o\top} x \ \forall u \in \mathcal{U}_z \\
& Yu + y_0 \leq x, \quad \forall u \in \mathcal{U}_z \\
& Yu + y_0 \leq d_z^0 + Q_z u, \quad \forall u \in \mathcal{U}_z \\
& \mathbf{1}^\top x \leq C_z.
\end{aligned}$$

We generate a synthetic dataset $\mathcal{D} = \{(r_i, c_i^0, d_i^0, Q_i, \hat{u}_i)\}_{i=1}^N$ for the inventory management problem, where $\hat{u}_i$ represents the nominal value of the uncertainty factor.

For each application, we use a synthetically generated dataset to train a two-layer fully connected neural network with problem-specific activation functions, as described in Sect. 2, to approximate the optimal solution of the parametric optimization problem. The dataset is split into training (70%), validation (15%), and testing (15%). At test time, we compute the optimal robust solution using the Gurobi solver and evaluate the performance of our method based on feasibility, optimality, and computational efficiency. To assess feasibility, we compute the average maximum constraint violation over the test set, defined as $\max_j [\bar{g}_z^j(h(z))]$, where $\bar{g}_z^j$ is given in Assumption 2, and report the percentage of feasible solutions produced by the learned model. For optimality, we measure regret defined as $(f_z^* - \hat{f}_z)/f_z^*$, where f_z^* is the optimal objective value obtained by Gurobi, and $\hat{f}_z$ is the objective value of the solution provided by the learned model. Finally, to evaluate computational efficiency, we compare the average

Table 1. Experimental Results for Constraint Optimization Problems

Application	Problem Size	Model & Penalty Coefficient	Optimality Regret	Max Constraint Violation	% Feasible Instances	Inference time	Solver time
Knapsack	$d_x = 50$	SL/	−1.41	3.93	0.88	**0.0023**	0.0279
	$m = 5$	SSL/1	**0.4**	0.36	57.6		
		SSL/10	0.64	0.082	95.6		
		SSL/20	0.63	0.088	98		
		SSL/50	0.70	**0.033**	**99**		
IP Knapsack	$d_x = 50$	SL/	−0.74	3.82	2	**0.0027**	0.1311
	$m = 5$	SSL/1	0.50	0.44	60		
		SSL/10	0.79	**0.06**	**94**		
		SSL/20	**0.22**	0.33	84		
		SSL/50	0.30	0.38	59		
Inventory	$d_x = N = 50$	SL/	−0.008	90.9	0	**0.0011**	0.3975
	$d_u = k = 5$	SSL/50	−1.8	0.065	22.6		
		SSL/100	−2.13	0.112	57.5		
		SSL/200	−1.94	0.0150	72.4		
		SSL/500	**−2.21**	**0.012**	**91.7**		

computational time required by both Gurobi and the learned solver reported in seconds.

Compared to the supervised learning approach, the solution provided by our method achieves a high level of feasibility, as shown in Table 1, whereby the percentage of feasible solutions returned by the self-supervised learning approach is consistently high across different applications and penalty coefficients. This highlights the effectiveness of our approach in generating feasible solutions, addressing a key limitation of supervised learning methods, which often produce infeasible solutions. Table 1 also highlights the trade-off between feasibility and optimality depending on the choice of the penalty coefficient used during training. Exploring this trade-off may be valuable in real-world applications, where the relative importance of feasibility and optimality depends on the specific requirements of the problem. Additionally, Table 1 demonstrates the computational efficiency of our method, achieving a speedup of over 10 orders of magnitude compared to a standard optimization solver like Gurobi. This efficiency can further be enhanced by the learned solver's ability to perform batch predictions, which may be extremely valuable when solving multiple problems simultaneously.

4 Conclusion

We present a self-supervised learning approach to solve optimization problems under uncertainty leveraging tractable reformulations of robust optimization problems. Our approach relies on penalty based self-supervised learning loss function to learn robust solutions, eliminating the need for pre-solved instances. Additionally, this work utilizes neural network models capable of outputting values for both integer and discrete variables, allowing our framework to handle a broad range of optimization problems. Numerical experiments validate the effectiveness of the proposed approach, revealing a trade-off between the feasibility and optimality of the solutions. Additionally, our method achieves significant computational speedup compared to traditional solvers which can be crucial in real-time decision-making applications.

Acknowledgments. This research was supported by NSF AI Institute for Advances in Optimization Award 2112533.

References

1. Agrawal, A., Amos, B., Barratt, S., Boyd, S., Diamond, S., Kolter, Z.: Differentiable convex optimization layers (2019). https://arxiv.org/abs/1910.12430
2. Amos, B.: Tutorial on amortized optimization (2023). https://arxiv.org/abs/2202.00665
3. Amos, B., Kolter, J.Z.: Optnet: differentiable optimization as a layer in neural networks (2021). https://arxiv.org/abs/1703.00443
4. Andrychowicz, M., et al.: Learning to learn by gradient descent by gradient descent (2016)
5. BakIr, G.: Predicting Structured Data. MIT Press (2007)
6. Ben-Tal, A., Den Hertog, D., Vial, J.P.: Deriving robust counterparts of nonlinear uncertain inequalities. Math. Program. **149**(1), 265–299 (2015)
7. Ben-Tal, A., El Ghaoui, L., Nemirovski, A.: Robust Optimization, vol. 28. Princeton University Press (2009)
8. Ben-Tal, A., Nemirovski, A.: Robust solutions of uncertain linear programs. Oper. Res. Lett. **25**(1), 1–13 (1999)
9. Bertsimas, D., Gupta, V., Kallus, N.: Data-driven robust optimization (2014). https://arxiv.org/abs/1401.0212
10. Bertsimas, D., Hertog, D.D.: Robust and adaptive optimization. Dynamic Ideas (2022). https://cir.nii.ac.jp/crid/1130295646841447468
11. Chen, T.: Learning to optimize: a primer and a benchmark. J. Mach. Learn. Res. **23**(189), 1–59 (2022)
12. Di Pillo, G., Grippo, L.: Exact penalty functions in constrained optimization. SIAM J. Control. Optim. **27**(6), 1333–1360 (1989)
13. Donti, P.L., Rolnick, D., Kolter, J.Z.: Dc3: a learning method for optimization with hard constraints (2021). https://arxiv.org/abs/2104.12225
14. Ferber, A., et al.: Surco: learning linear surrogates for combinatorial nonlinear optimization problems (2023). https://arxiv.org/abs/2210.12547
15. Ferber, A., Wilder, B., Dilkina, B., Tambe, M.: Mipaal: mixed integer program as a layer (2019). https://arxiv.org/abs/1907.05912
16. Iancu, D.A., Trichakis, N.: Pareto efficiency in robust optimization. Manage. Sci. **60**(1), 130–147 (2014)
17. Kim, S., Lewis, M.E., White, C.C.: Optimal vehicle routing with real-time traffic information. IEEE Trans. Intell. Transp. Syst. **6**(2), 178–188 (2005)
18. Kotary, J., Fioretto, F., Hentenryck, P.V., Wilder, B.: End-to-end constrained optimization learning: a survey (2021). https://arxiv.org/abs/2103.16378
19. Kotary, J., Vito, V.D., Christopher, J., Hentenryck, P.V., Fioretto, F.: Predict-then-optimize by proxy: learning joint models of prediction and optimization (2023). https://arxiv.org/abs/2311.13087

20. Li, K., Malik, J.: Learning to optimize (2016). https://arxiv.org/abs/1606.01885
21. Osokin, A., Bach, F., Lacoste-Julien, S.: On structured prediction theory with calibrated convex surrogate losses. Adv. Neural Inf. Proc. Syst. **30** (2017)
22. Park, S., Hentenryck, P.V.: Self-supervised primal-dual learning for constrained optimization (2022). https://arxiv.org/abs/2208.09046
23. Paszke, A., et al.: Pytorch: an imperative style, high-performance deep learning library (2019). https://arxiv.org/abs/1912.01703
24. Qi, M., Grigas, P., Shen, Z.J.M.: Integrated conditional estimation-optimization (2023). https://arxiv.org/abs/2110.12351
25. Vlastelica, M., Paulus, A., Musil, V., Martius, G., Rolínek, M.: Differentiation of blackbox combinatorial solvers (2020). https://arxiv.org/abs/1912.02175
26. Wang, I., Becker, C., Parys, B.V., Stellato, B.: Learning decision-focused uncertainty sets in robust optimization (2024). https://arxiv.org/abs/2305.19225
27. Wilder, B., Dilkina, B., Tambe, M.: Melding the data-decisions pipeline: decision-focused learning for combinatorial optimization (2018). https://arxiv.org/abs/1809.05504
28. Zamzam, A., Baker, K.: Learning optimal solutions for extremely fast ac optimal power flow (2019). https://arxiv.org/abs/1910.01213

Revisiting Pseudo-Boolean Encodings from an Integer Perspective

Hendrik Bierlee[1,2,3](✉), Jip J. Dekker[2,3], and Peter J. Stuckey[2,3]

[1] DTAI, KU Leuven, Leuven, Belgium
henk.bierlee@kuleuven.be
[2] Monash University, Melbourne, Australia
{jip.dekker,peter.stuckey}@monash.edu
[3] OPTIMA ITTC, Melbourne, Australia

Abstract. Traditionally, SAT encodings of complex constraints, such as linear constraints or more specifically PB constraints, are specified in terms of Boolean variables and clauses. However, often sets of related Boolean variables are either encodings of integer variables, or act as if they were. Furthermore, any encoding of linear constraints has to encode partial sums, and these are integers (even if the encoding does not explicitly notice this). By formally specifying the SAT encoding using integer variables and constraints, coupled with a procedure to encode this specification into SAT, we can gain some more insight into the encoding methods, and compose new ones. Experiments using these integer-driven encodings show that they can improve on standard approaches to encoding PB and integer linear constraints to SAT.

Keywords: Boolean Satisfaction · Pseudo-Boolean Equations · Encoding

1 Introduction

Boolean Satisfiability (SAT) is a powerful approach to solving combinatorial problems, but to use a SAT solver, we need to encode the problem into clauses. A common and important class of constraint to encode are linear constraints over Boolean literals, or *Pseudo-Boolean* (PB) constraints. In its normalized form, a PB constraint is $\sum_{i=1}^{n} q_i b_i \leq k$, where q_i and k are positive integer constants and b_i are Boolean literals. There are many competing methods used to encode a PB constraint, including the *Generalized Totalizer* (GT) [22], *Sequential Weight Counter* (SWC) [20], *Binary Decision Diagram* (BDD) [17], and the *Ripple Carry Adder* (RCA) encoding [26].

Each of these methods implicitly generates auxiliary encodings of some partial sums $\sum_{i \in J} q_i b_i$ where $J \subseteq \{1, \ldots, n\}$. For example, SWCs and BDDs encode a linear *decomposition* of the sum, such that: $y_j \geq \sum_{i=1}^{j} q_i b_i, 2 \leq j \leq n$, using encodings of $q_1 b_1 + q_2 b_2 \leq y_2$, and $y_j + q_{j+1} b_{j+1} \leq y_{j+1}, 2 \leq j < n$, finally adding $y_n \leq k$. Note that it is sufficient to enforce that y_j is greater than or equal to the partial sum, rather than equal to, since the overall inequality constraint is enforced by $y_n \leq k$. A binary tree decomposition of the sum, used by GTs, encodes the partial sums (assuming $n = 2^k$ for simplicity) as $y_i^1 \geq q_{2i-1}b_{2i-1} + q_{2i}b_{2i}, 1 \leq i \leq 2^{k-1}$ and $y_i^{j+1} \geq y_{2i-1}^j + y_{2i}^j, 1 \leq i \leq 2^{k-j}$, finally adding $y^k \leq k$. RCA encodings are free to decompose the sum in either way, and use binary encodings of partial sums.

In the original specifications of the encoders, apart from RCA, each set of Boolean variables which implicitly represents a partial sum is not explicitly treated as an integer variable y. In this paper, we show how to improve these encodings of PB constraints by recognising that all of these methods have one thing in common: they construct a **decomposition of addition constraints over *integers* representing the (lower bound of the) partial sums**. With this viewpoint, several advantages become evident. Importantly, any improvement on the integer decomposition automatically carries over to each of the PB encoders. The contributions of this work are:

- We show how all existing encodings can be generated from a simple integer viewpoint (Sect. 3). This allows their correctness and propagation consistency to be proved uniformly (Sect. 4.1).
- Existing PB encoding methods are automatically generalized to support integer linear constraints and side-constraints (Sect. 4.2).
- We show how to encode PB equations, $\sum_{i=1}^n q_i b_i = k$, more efficiently than by decomposing into two inequalities (Sect. 4.3).
- We show how different encodings can be applied to each individual decision variable *and* the partial sums (Sect. 4.4).

The remainder of the paper is organized as follows. In Sect. 2, we define the concepts used within the remainder of this paper. In Sect. 3, we introduce a simple encoding for the ternary inequality constraint over integers, which can recreate the GT, SWC, BDD and RCA encodings. In Sect. 4, we take advantage of the integer viewpoint by proposing multiple extensions. In Sect. 5, we evaluate the effect of these extensions on solver performance. We conclude our findings in Sect. 6. Note that related work is discussed throughout the paper.

2 Preliminaries

2.1 Constraint Programming

A *Constraint Satisfaction Problem* (CSP) instance, $P = (\mathcal{X}, \mathcal{D}, \mathcal{C})$, consists of a set of variables $\mathcal{X}$, with each $x \in \mathcal{X}$ restricted to taking values from some initial domain $D(x)$. For this paper, we assume domains are ordered sets of integers, and denote by $\text{lb}_D(x)$ and $\text{ub}_D(x)$ the least and greatest values in $D(x)$. We will use interval notation $[l, u]$ to represent the set of integers $\{l, l+1, \ldots, u\}$. A set of constraints $\mathcal{C}$ expresses relationships between the variables. An *assignment* of a CSP instance is a mapping of variables to values, which (if consistent with the domains and constraints) is a *solution* to the instance.

We say a domain D is *domain consistent* for constraint c over variables V if for all $v \in V$ and all $d \in D(v)$ there exist $d_{v'} \in D(v')$ for all $v' \in V \setminus \{v\}$, such that $(v = d) \wedge \bigwedge_{v' \in V \setminus \{v\}} (v' = d_{v'})$ is a solution of c. In other words, every value in every domain of $v \in V$ participates in a solution for c in D.

We say a domain D is $bounds(\mathbb{R})$ *consistent* for constraint c over variables V if for all $v \in V$ and for $d \in \{\text{lb}_D(v), \text{ub}_D(v)\}$ there exist **real** numbers $d_{v'}$, where $\text{lb}_D(v') \leq d_{v'} \wedge d_{v'} \leq \text{ub}_D(v')$ for all $v' \in V \setminus \{v\}$ such that $v = d \wedge \bigwedge_{v' \in V \setminus \{v\}} v' = d_{v'}$ is a (real) solution of c. That is to say, every lower and upper bound for $v \in V$ participates in a real solution for c within the bounds of D.

2.2 Satisfiability

A SAT problem can be considered a special case of a CSP, where the domain for all variables x is $D(x) \in \{0, 1\}$, representing the values *false* and *true* . A *literal* is either a Boolean variable x or its negation $\neg x$. We extend the negation operation to operate on literals, *i.e.* $\neg b = \neg x$ if $b = x$ and $\neg b = x$ if $b = \neg x$. We use the notation $b = v$ where b is a literal and $v \in \{0, 1\}$ to encode the appropriate form of the literal, *i.e.* if $v = 1$ it is equivalent to b and if $v = 0$ it is equivalent to $\neg b$. The notation $b \neq v$ is defined similarly to encode $\neg(b = v)$. A *clause* is a disjunction of literals. In a SAT problem P, the constraints $\mathcal{C}$ are clauses. A partial assignment θ maps each Boolean literal b to either true $\theta(b) = \{1\}$ or false $\theta(b) = \{0\}$, or unknown $\theta(b) = \{0, 1\}$.

2.3 Encoding CP to SAT

Given an integer x with (possibly non-contiguous) domain $D(x) = \{d_1, d_2, \ldots, d_m\}$, a variable encoding method maps x to a set of Boolean *encoding* variables. We use semantic brackets to name the Booleans, where the Boolean $[\![f]\!]$ is true iff the formula f holds. Two Booleans with different names $[\![f]\!]$ and $[\![g]\!]$ refer to the same underlying literal iff $[\![f]\!] \equiv [\![g]\!]$.

The *interpretation* of the assignment of a variable encoding returns the assignment of the original encoded integer. This can require a variable encoding to be associated with a *consistency constraint* to ensure that the interpretation correctly maps to a single integer.

The *order* encoding of x, denoted $x{:}\mathbb{O}$, introduces $m - 1$ encoding variables $[\![x \geq v]\!], v \in \{d_2, \ldots, d_m\}$. $[\![x \geq v]\!]$ is true iff x is assigned a value greater or equal to v. We extend our semantic brackets notation of the order encoding to express other relations on x in terms of these encoding variables:

$$[\![x \geq v]\!] \equiv 1, v \leq d_1 \tag{1a}$$

$$[\![x \geq v]\!] \equiv [\![x \geq d_{i+1}]\!], d_i < v \leq d_{i+1} \tag{1b}$$

$$[\![x \geq v]\!] \equiv 0, v > d_m \tag{1c}$$

$$[\![x > v]\!] \equiv [\![x \geq v+1]\!] \tag{1d}$$

$$[\![x < v]\!] \equiv \neg[\![x \geq v]\!], [\![x \leq v]\!] \equiv \neg[\![x \geq v+1]\!] \tag{1e}$$

We can map a partial assignment θ on the Booleans encoding integer variable x to any value a where $arg\,max_{d \in D(x)}(\theta(\llbracket x \geq d \rrbracket)) \leq a < arg\,min_{d \in D(x) \cup \{d_m+1\}}$ $(\theta(\llbracket x < d \rrbracket))$. No value for a may exist, unless the following *Implication Chain* (IC) constraint is enforced:

$$\bigwedge_{i=3}^{m} \llbracket x \geq d_i \rrbracket \to \llbracket x \geq d_{i-1} \rrbracket \qquad (2)$$

Other important integer variable encodings are the direct, the binary and mixed radix encodings [9]. The *binary* encoding of x, denoted $x{:}\mathbb{B}$, introduces n encoding variables $\llbracket \text{bit}(x,k) \rrbracket$, $k \in 0..n-1$. Then, $\llbracket \text{bit}(x,k) \rrbracket$ is true iff the k-th most significant bit in the two's complement binary representation of the value assigned to x is true, i.e., in C notation: $\llbracket \text{bit}(x,k) \rrbracket$ =(x >> k) & 1. For simplicity, we assume x is known to be non-negative, extensions to two's complement representations are well understood. The number of bits required $n = \lceil log(d_m + 1) \rceil$) is given by the upper bound of x. Let $\mathcal{B}(x) = 0..n-1$ and let $\mathcal{R}(x)$ be the set of representable integers for the binary encoding of x, $\mathcal{R}(x) = 0..2^n - 1$.

The *consistency constraint* for the binary encoding enforces that the Boolean representation can only represent values in the initial domain $D_i(x)$. It is the SAT encoding of the constraints (a) $x \geq d_1$ if $d_1 > 0$, (b) $x \leq d_m$ if $d_m < 2^{n-1}$ (c) $x \neq d$ for each $d_1 < d < d_m, d \notin D_i(x)$. Constraints (a) and (b) are encoded using lexicographic decompositions, while (c) can be simply encoded as a clause $\vee_{k \in \mathcal{B}(x)} \llbracket \text{bit}(x,k) \rrbracket \neq \text{bit}(d,k)$. For highly sparse domains, more efficient approaches are possible for (c).

2.4 Pseudo-Boolean Constraints

A PB constraint has the form $\sum_{i=1}^{n} q_i b_i \,\#\, k$ where q_i, k are integer constants, b_i are Boolean literals and $\# \in \{<, \leq, =, \geq, >\}$. A *normalized* PB constraint requires that k is positive, $1 \leq q_i \leq k, \forall 1 \leq i \leq n$ and the comparator $\#$ is $\leq$. Since a PB constraint can always be normalized to one or two normalized PB inequalities, we only consider normalized PB constraints, until Sect. 4.3.

3 PB Encodings as Integer-Based Decompositions

All encodings for the PB constraint $\sum_{i=1}^{n} q_i b_i \leq k$ decompose the constraint into ternary constraints that implicitly or explicitly encode integer partial sums as auxiliary integer variables. The difference between the encodings is the shape of the decomposition, the domains of the auxiliary variables, and their choice of variable encoding (order for GT, SWC and BDD, binary for RCA).

The first step of the integer viewpoint is to introduce *principal* integer variables x_i with domains $D(x_i) = \{0, q_i\}$ to represent the PB terms $x_i = q_i b_i$. This allows us to represent the PB constraint as the integer linear constraint, $\sum_{i=1}^{n} x_i \leq k$. Next, we create an arbitrary binary tree with leaf nodes labelled x_i

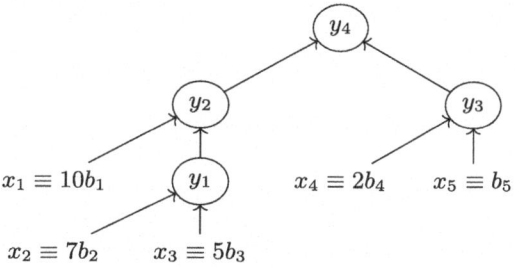

Fig. 1. A possible tree decomposition of $10b_1 + 7b_2 + 5b_3 + 2b_4 + b_5 \leq 15$

that sums up the n original terms using $n-1$ internal nodes. Each node would be labelled by a new auxiliary variable $y_i, 1 \leq i \leq n-1$ that encodes the partial sum of the tree rooted at the internal node i. The final variable y_{n-1} is also denoted as y_{root}.

Each node i creates a constraint $l(i) + r(i) = y_i$, $l(i)$ is y_l if the left child of node i is an internal node l and x_l if the left child of node i is the leaf node x_l. $r(i)$ is defined similarly. Note that we can weaken any of these ternary equalities to an inequality while still encoding the PB constraint. Hence, we can encode $l(i) + r(i) \leq y_i$ for any auxiliary y_i, since the PB constraint will hold if and only if all inequalities are satisfied. The overall PB constraint is then enforced by adding a constraint $y_{root} \leq k$ or equivalently $y_{root} = k$ if each path from root to leaf has a weakening inequality.

The introduced variables can be restricted to domain $[0, k]$ since there is no solution where any of them takes a value above k. To recreate different encodings, we may restrict the domains further. For example, in the GT encoding for a non-root internal node i, we can define $D(y_i) = \{u + v \mid u \in D(l(i)), v \in D(r(i)), u + v \leq k\}$. For the root variable $D(y_{root}) = \{k\}$.

Example 1. The tree decomposition shown in Fig. 1 decomposes the PB constraint as $x_2 + x_3 = y_1$, $x_1 + y_1 = y_2$, $x_4 + x_5 = y_3$, $y_2 + y_3 = y_4$, and $y_4 \leq 15$. The GT decomposition makes further changes: it weakens the equalities to inequalities, omits the final constraint, and has domains $D(y_1) = \{0, 5, 7, 12\}$, $D(y_2) = \{0, 5, 7, 10, 12, 15\}$, $D(y_3) = \{0, 1, 2, 3\}$, and $D(y_4) = \{15\}$. □

Once a tree decomposition has been constructed, the encoding of the PB constraint can be reduced to the repeated encoding of *ternary (in)equality* constraints. The RCA encoding of $x + y = z$ assumes the variables x, y and z are binary encoded, then encodes the following simple constraints:

$$\begin{aligned}
c_1 &\equiv [\![\text{bit}(x, k)]\!] \wedge [\![\text{bit}(y, k)]\!] \\
c_{k+1} &\equiv ([\![\text{bit}(x, k)]\!] + [\![\text{bit}(y, k)]\!] + c_k \geq 2), \quad k \in 1..n-1 \\
[\![\text{bit}(z, 0)]\!] &\equiv [\![\text{bit}(x, k)]\!] \oplus [\![\text{bit}(y, k)]\!] \\
[\![\text{bit}(z, k)]\!] &\equiv [\![\text{bit}(x, k)]\!] \oplus [\![\text{bit}(y, k)]\!] \oplus c_{k-1}, \quad k \in 1..n-1
\end{aligned} \quad (3)$$

Here, c_k is the carry bit from the kth addition, and $\oplus$ is exclusive-or.

The following equation defines the ternary inequality constraint encoding, for $x + y \leq z$ assuming each of x, y and z is an order encoded integer variables with appropriate domains:

$$\bigwedge_{v \in D(x)} \bigwedge_{w \in D(y)} (\llbracket x \geq v \rrbracket \wedge \llbracket y \geq w \rrbracket) \rightarrow \llbracket z \geq v + w \rrbracket \tag{4}$$

Although other encodings of the ternary inequality constraint can be devised, we will show how Eq. (4) is the shared structure between the GT, SWC, and BDD encoding. In the following subsections, we first describe the existing encoding method from the literature. Then, we show a particular decomposition of a PB constraint into ternary inequalities and the domains of the auxiliary integer variables. Finally, we prove how an order encoding of the integer variables and encoding of the ternary inequality constraints (by Eq. (4)) produces an equivalent set of clauses as the original encoding methods given the same PB constraint. The only difference is that some trivially assignable literals (*e.g.* by unit propagation) of the original encoding may already be removed in our method.

3.1 Generalized Totalizer

The GT encoding [22] constructs a complete binary tree bottom-up. Each node Z is associated with a set of auxiliary variables z_v, where v corresponds to the lower bound of the partial sum of the subtree. Each leaf node represents a PB term $q_i b_i$ with singleton set $\{z_{q_i}\}$, where $z_{q_i} \equiv b_i$. Every internal node Z with child nodes L and R contains an auxiliary Boolean variable z_v for every possible sum of the values of the auxiliary variables of child nodes L and R, where those greater than k are all represented by $k+1$. That is, $Z = \{z_{\min(v+w,k+1)} \mid l_v \in L, r_w \in R\}$. The GT is encoded by adding a unary clause $\neg a_{k+1}$ for the root node, and for every internal node Z:

$$l_v \rightarrow z_v \quad l_v \in L \tag{5a}$$
$$r_w \rightarrow z_w \quad r_w \in R \tag{5b}$$
$$(l_v \wedge r_w) \rightarrow z_{\min(v+w,k+1)} \quad l_v \in L, r_w \in R \tag{5c}$$

Clearly, the nodes of the GT encoding imply a higher-level structure, and has been formalized before as an equivalent PB-based decomposition [25]. We will now define the integer-based GT encoding. The auxiliary Boolean variables of every node Z are represented by a single auxiliary integer variable y_i of a binary tree decomposition of the following ternary inequality constraints:

$$\bigwedge_{i=1}^{n-1} l(i) + r(i) \leq y_i \tag{6}$$

where $D(y_i) = \{v + w : v \in D(l(i)), w \in D(r(i)), v + w \leq k\}$, except $D(y_{root}) = \{k\}$. We can show this equivalent to the GT encoding.

Theorem 1. *Encoding Eq. (6) is, after unit propagation, equivalent to GT.*

3.2 Sequential Weight Counter

In the SWC encoding [20], a series of n weight counters is introduced. The i-th counter enforces $\sum_{i'=1}^{i} q_{i'} b_{i'} \leq k$ by introducing auxiliary variables $z_{i-1,j}$ for $1 \leq j \leq k$, where $z_{i,j}$ represents $\sum_{i'=1}^{i} q_{i'} b_{i'} \geq j$. We consider $z_{0,j} \equiv 0, 0 \leq j \leq k$. We encode the i-th counter with the following clauses:

$$z_{i-1,j} \to z_{i,j} \quad 1 \leq j \leq k \tag{7a}$$

$$\neg (z_{i-1,k+1-q_i} \wedge b_i) \tag{7b}$$

$$(z_{i-1,j} \wedge b_i) \to z_{i,j+q_i} \quad 1 \leq j \leq k - q_i \tag{7c}$$

$$b_i \to z_{i,j} \quad 1 \leq j \leq q_i \tag{7d}$$

From the integer perspective, the same encoding can be described using a linear tree that introduces $n+1$ auxiliary integer variables $w_i \in [-k, 0]$ except $w_0 = 0$ and $w_n = -k$, and the following ternary inequalities.

$$\bigwedge_{i=1}^{n} x_i + w_i \leq w_{i-1} \tag{8}$$

These inequalities are derived from $x_i + y_{i-1} \leq y_i$ where $y_i \in [0, k]$, which is equivalent from an integer perspective to $x_i - y_i \leq -y_{i-1}$ assuming $w_i = -y_i$. However, encoding $x_i + y_{i-1} \leq y_i$ would lead to a subtle difference in clauses compared to the original SWC encoding.

Theorem 2. *Encoding Eq. (8) is, after unit propagation, equivalent to SWC.*

3.3 Binary Decision Diagram

A BDD [14] is a binary, rooted, directed, acyclic graph which models a Boolean function. Every node is associated with a selector variable which determines whether a path follows either the node's zero or one edge. When the path reaches a terminal node $z_\top$ or $z_\bot$, the function returns zero or one, respectively. In ordered BDDs, selector variables appear in the same order on all paths from the root, organizing the BDD into one layer per selector variable. A quasi-reduced QOBDDs has no isomorphic sub-BDDs. Fully reduced ROBDDs are quasi-reduced and have no *identity* nodes, where all edges target the same child.

Construction algorithms exist which produce ROBDDs that model satisfaction of PB constraints [4]. The layer i models the decision variable b_i using multiple nodes (i, j), each associated with an auxiliary variable $z_{i,j}$. The set of edges E contains $((i, j), (i', j'), p)$ with p equal to zero (one) iff there is a zero (one) edge from node (i, j) to node (i', j'). The ROBDD is encoded as follows:

$$z_{1,1} \wedge \neg z_\bot \wedge z_\top \tag{9a}$$

$$z_{i,j} \to z_{i',j'} \quad (i,j),(i',j'),0) \in E \tag{9b}$$

$$(b_i \wedge z_{i,j}) \to z_{i'',j''} \quad (i,j),(i',j'),1) \in E \tag{9c}$$

Before moving to the integer perspective, we establish two general properties of BDDs. First, an ROBDD can be converted to a QOBDD by replacing any *long* edges $((i, j), (i', j'), p)$ where $i' - i > 1$ by chains of $i' - i - 1$ identity nodes, so that $i' = i + 1$ for all edges. This requires no additional variables or clauses, as the new identity node variables are all equivalent to $z_{i,j}$.

Second, we adapt the PB intervals of the above construction algorithm. A path from root node $(1, 1)$ to node (i, j) is associated with a partial assignment of $b_1, b_2, \ldots, b_{i-1}$, which yields a partial sum. Then, $W_{i,j}$ is the set of all partial sums given by all possible paths from the root node to node (i, j). In a QOBDD, the sub-QOBDD, rooted at the node (i, j), covers all PB constraints $w + \sum_{i'=i}^{n} q_i b_i \leq k$ in the interval $\min W_{i,j} \leq w \leq \max W_{i,j}$. Hence, intervals on the same layers do not overlap, as that would produce identical sub-ROBDD, contradicting the isomorphism property.

In the integer perspective, a single integer variable y_i represents each layer i of the QOBDD as a partial sum $\sum_{j=1}^{i} x_j \leq y_i$. This is enforced by the following linear tree of ternary inequalities:

$$\bigwedge_{i=1}^{n} x_i + y_{i-1} \leq y_i \qquad (10)$$

The domain values 0 and q_i of x_i are represented by the zero and one edges at layer i, respectively. The j-th domain value $d_{i,j}$ of y_i is represented by node (i, j) and is equal to $\max W_{i,j}$, except for $y_0 = 0$ and $y_n = k$. The domain values can also be recursively calculated given QOBDD edges E, bottom-up:

$$d_{i,j} = \max \{d_{i-1,j'} + pq_i : ((i-1, j'), (i, j), p) \in E\} \qquad (11)$$

Two interesting notes on this specification are that

1. some non-reduced BDDs cannot be represented by our encoding of the ternary inequality constraints, and
2. other encodings of BDDs exist, which would require different constraint encodings of the ternary inequalities.

Theorem 3. *Encoding Eq. (10) is, after unit propagation, equivalent to BDD.*

4 Taking Advantage of the Integer Viewpoint

There are various advantages of specifying the PB encodings in terms of integer ternary inequalities. First, we will show that by proving the propagation consistency of the ternary inequality encoding, we can also prove the propagation consistency of the overall encoding of the PB constraint. Second, since the input of the specification is a integer linear constraint representing a PB constraint, we can support integer linear variables and constraints directly. This also applies to integer variables, which arise from detecting sets of literals that act like integers due to existing side-constraints. Third, we can encode equality constraints more efficiently by replacing each ternary inequality in the decomposition for a ternary equality. Fourth, we can create encodings where the choice of each variable representation is mixed: using order encoding for small domain sizes to benefit from strong propagation, and binary encodings for those with large domain sizes to keep the encoding small.

4.1 Proving Consistency of All Encodings

Instead of proving consistency on every encoding individually, we can prove the propagation consistency on the ternary inequality, and use this to prove domain consistency of every decomposition. Surprisingly, the encoding of $x + y \leq z$, given by Eq. (4), does not maintain bounds($\mathbb{R}$) consistency:

Example 2. Consider the domains $D(x) = \{0, 4, 8\}$, $D(y) = \{0, 7, 10\}$, and $D(z) = \{0, 4, 7, 8, 10, 11\}$ then suppose we have $[\![y \geq 7]\!]$ and $\neg[\![z \geq 10]\!]$ ($z \leq 9$). We would hope to propagate $x \leq 2$, or equivalently $\neg[\![x \geq 4]\!]$, through the clause $\neg[\![x \geq 4]\!] \vee \neg[\![y \geq 7]\!] \vee [\![z \geq 11]\!]$. However, we do not have $\neg[\![z \geq 11]\!]$. The issue is that there is no IC to propagate $[\![z \geq 11]\!] \rightarrow [\![z \geq 10]\!]$. □

If we assume the upper bound of z satisfies an IC, we can prove the following:

Lemma 1. *Encoding $x + y \leq z$ with Eq. (4) ensures that lower bounds on x and y are propagated correctly to z (or cause failure), and when an upper bound u on z satisfies an IC (i.e. $[\![z \leq u]\!] \rightarrow [\![z \leq u+1]\!]$), then propagation of upper bounds on x and y are correctly propagated, and these upper bounds also satisfy ICs.*

A consequence of the above theorem is that the decompositions of a PB constraint into a tree of ternary integer inequalities enforces bounds($\mathbb{R}$) consistency (or, equivalently for this constraint, domain consistency) by satisfying an implicit IC on upper bounds without explicitly encoding the clauses in Eq. (2):

Theorem 4. *The decomposition of a PB constraint $\sum_{i=1}^{n} q_i b_i \leq k$ into ternary inequalities encoded using Eq. (4) enforces domain consistency of the constraint.*

Note that the counterexample of Example 2 does not conflict with Theorem 4, since it cannot occur, unless we branch on intermediate literals, in this case by setting $\neg[\![z \geq 10]\!]$. Note, the proof given for this result for GT by Joshi et al. [22] seems to never mention the need for enforcing ICs, which suggests it may actually be incomplete.

4.2 Using Integer Decision Variables

An integer linear constraint $\sum_{i}^{n} q_i x'_i \leq k$ over order encoded integer variables x'_i and constants q_i, k can be converted to a PB constraint by splitting out every integer linear term $q_i x'_i$ into multiple PB terms using $\sum_{v \in D(x'_i)} q_i [\![x'_i \geq v]\!]$. However, encoding the resulting PB constraint leads to a high degree of redundancy, since it introduces a GT leaf node, SWC counter, or BDD layer per domain value, for every integer linear term.

Since our ternary inequalities are defined over integers, the PB encoding methods are easily extended to integer linear encodings. As in Sect. 3, we use views $x_i \equiv q_i x'_i$, where $D(x_i) = \{q_i v : v \in D(x'_i)\}$ to establish a linear constraint with unit coefficients $\sum_{i=1}^{n} x_i \leq k$. Encoding the ternary inequalities results in just a single leaf node, counter, or layer per term. In particular, BDDs are thus generalized to *Multi-valued Decision Diagram* (MDDs), as every node on layer i will have one edge per domain value of x_i (see *e.g.* [5]).

Even if constraints are not explicitly modelled using integer variables, we can use the knowledge of other constraints to treat sets of literals as such. In particular, we look for IC and *At-Most-One* (AMO) constraints, where a set of literals follow similar behaviour to order and direct encoded integer. Bofill et al. first explored how the presence of AMO constraints can lead to improved PB encodings [13]. We find that using our encodings the IC constraints can be similarly exploited. This results in novel encodings for GT and SWC, and, for BDD, this results in the same encoding as [2].

For some PB constraints, it might be efficient to introduce additional integer variables. Given a PB constraint $\sum_{i=1}^{n} q_i b_i \leq k$, suppose that there are shared coefficients $q_1 = q_i, 2 \leq i \leq l$ (w.l.o.g. we assume the first l coefficients are the same). We can replace the term $q_1 b_1 + \cdots + q_1 b_l$ by an integer term $q_1 x$ where $D(x) = \{0, 1, \ldots, l\}$ and encode the constraint $b_1 + \cdots + b_l \leq x$ using a cardinality network (*e.g.* using [3]). This performs the addition in a more compact way, and generates new auxiliary variables for x that recognize symmetric situations, and has been shown to improve encodings [5]. Note that the integer term $q_1 x$ is simply a view on the x variable so that we do not need to introduce additional Booleans to encode it [2].

Another use case of integer decision variables is to re-use a (non-constant) integer variable at the root of one constraint's decomposition as a node for another, as has been done for GT [21].

4.3 Encoding Equality Constraints

Rather than split an equality constraint $\sum_{i=1}^{n} q_i b_i = k$ into two inequalities $\sum_{i=1}^{n} q_i b_i \leq k$ and $\sum_{i=1}^{n} q_i b_i \geq k$, it would be preferable to encode the equality directly. We can do so straightforwardly by decomposing the equality into integer equalities of the form $x + y = z$, and then generating an encoding of these directly. In fact, the unweighted totalizer encoding for cardinality constraints (where all $q_i = 1$) supports directly encoding both lower- and upper bound, and consequently supports equality as well [6]. Since domain consistency is NP hard (by reduction to subset sum [15]) we will only enforce bounds($\mathbb{R}$) consistency, which is equivalent to the consistency enforced by splitting into two inequalities.

The encoding simply reuses the PB encoding methods for $\sum_{i=1}^{n} q_i b_i \leq k$, with the following changes. Instead of decomposing to ternary inequalities $x + y \leq z$, we decompose to ternary equalities $x + y = z$. The domain computation for each introduced integer variable remains the same. The given ROBDD should model the equality constraint, which can be achieved by changing the base case of the construction algorithm [4].

The encoding of the ternary equality constraint $x+y = z$ is given by the usual encoding of $x + y \leq z$ from Eq. (4), with the following encoding on $x + y \geq z$:

$$\bigwedge_{v \in D(x)} \bigwedge_{w \in D(y)} (\llbracket x \leq v \rrbracket \wedge \llbracket y \leq w \rrbracket) \to \llbracket z \leq v + w \rrbracket \tag{12}$$

The advantage of the equality encoding over splitting into two inequalities is that it reuses the same variables for the intermediate sums, rather than building two copies of the intermediate sums, after we normalize the second inequality.

Example 3. Consider encoding the equation $10b_1 + 7b_2 + 5b_3 + 2b_4 + b_5 = 15$, using the same tree decomposition as in Example 1. The constraints are $x_2 + x_3 = y_1$, $x_1 + y_2 = y_2$, $x_4 + x_5 = y_3$, $y_2 + y_3 = y_4$, $y_4 = 15$, with the same domains as shown in Example 1. We encode each ternary equation as defined above, *e.g.* $y_2 + y_3 = y_4$ generates (trivial) clauses such as $\neg \llbracket y_2 \geq 5 \rrbracket \vee \neg \llbracket y_3 \geq 2 \rrbracket \vee \llbracket y_4 \geq 7 \rrbracket \equiv true$ from $(\llbracket y_2 \geq 5 \rrbracket \wedge \llbracket y_2 \geq 2 \rrbracket) \to \llbracket y_4 \geq 7 \rrbracket$ and more interesting clauses such as $\llbracket y_2 \geq 7 \rrbracket \vee \llbracket y_3 \geq 3 \rrbracket \vee \neg \llbracket y_4 \geq 8 \rrbracket$ from $(\llbracket y_2 \leq 5 \rrbracket \wedge \llbracket y_3 \leq 2 \rrbracket) \to \llbracket y_4 \leq 7 \rrbracket$, which is equivalent to just $\llbracket y_2 \geq 7 \rrbracket \vee \llbracket y_3 \geq 3 \rrbracket$. □

We can prove results analogous to Theorem 1 and Theorem 4 about the encoding of $x + y \geq z$ generated by Eq. (12).

Theorem 5. *The decomposition of a pseudo-Boolean constraint $\sum_{i=1}^{n} q_i b_i \geq k$ into ternary inequalities encoded using Eq. (12) enforces domain consistency of the constraint.*

4.4 Mixed Integer Variable Encodings of the Decompositions

By Theorem 4, order encoding the integer variables in a decomposition enforces domain consistency on the inequality constraint. However, the number of clauses in Eq. (4) grow cubically with the domains of the variables. This makes the encoding less effective as domain sizes grow. Alternatively, a binary encoding of integer variables creates fewer clauses, but trades in the consistency guarantees.

Depending on the chosen decomposition, the domains of the different auxiliary variables vary in size. In the SWC, the domains of (non-constant) auxiliary variables are of the same size. In the GT, the size of auxiliary domains grows at each subsequent layer. In the BDD decomposition, the auxiliary domain values directly correspond with the nodes at each layer of a given QOBDD. They start with one node at the root layer, grow in number until the middle layer, and then shrink again until the terminal layer. Since the decomposition is agnostic as to which variable encoding is used for which variable, we can choose the order encoding for domains up to size c, and binary for larger domains. We will denote this simple heuristic by $\mathbb{O} \leq c$.

Consider the equality constraint $x + y = z$ in a decomposition of an inequality constraint. Suppose we are given a pre-determined choice of order or binary encoding for each variable indicated by :$\mathbb{O}$ or :$\mathbb{B}$, respectively. So far, these choices have always been $x{:}\mathbb{O} + y{:}\mathbb{O} = z{:}\mathbb{O}$, which we have encoded using Eq. (4) after relaxing to an inequality constraint. If instead the choices are $x{:}\mathbb{B}+y{:}\mathbb{B}=z{:}\mathbb{B}$, we can directly apply the RCA encoding of Eq. (3) for ternary equalities. Unfortunately, for mixed encodings such as $x{:}\mathbb{O}+y{:}\mathbb{B}=z{:}\mathbb{B}$, often there are no (efficient) encodings. However, there are efficient encodings for $x{:}\mathbb{O} \# x{:}\mathbb{B}, \# \in \{\leq, \geq, =\}$ [12]. We can introduce $x{:}\mathbb{B}$ as an additional integer variable and decompose the constraint to $x{:}\mathbb{O} \leq x{:}\mathbb{B} \wedge x{:}\mathbb{B} + y{:}\mathbb{B} = y{:}\mathbb{B}$. If the decomposition contains a ternary inequality constraint $x{:}\mathbb{B} + y{:}\mathbb{B} \leq z{:}\mathbb{B}$, then to apply the RCA encoding, we rewrite it to an equality constraint. To avoid removing solutions, we change the auxiliary variable's domain $D(z) = \min(\mathrm{lb}_D(z), \mathrm{lb}_D(x) + \mathrm{lb}_D(y))..\mathrm{ub}_D(z)$.

Example 4. Consider the PB constraint $8b_1 + 6b_2 + 3b_3 + 3b_4 + 2b_5 \leq 10$. After constructing a BDD, we decompose using Eqs. (10) and (11) and order encode all variables with domain size up to 3 (*i.e.* $\mathbb{O} \leq 3$):

$x_1{:}\mathbb{O} \leq y_1{:}\mathbb{O}$ where $D(x_1{:}\mathbb{O}) = \{0,8\}, D(y_1{:}\mathbb{O}) = \{1,8\}$

$x_2{:}\mathbb{O} + y_1{:}\mathbb{O} \leq y_2{:}\mathbb{O}$ where $D(x_2{:}\mathbb{O}) = \{0,6\}, D(y_2{:}\mathbb{O}) = \{2,7,8\}$

$x_3{:}\mathbb{O} + y_2{:}\mathbb{O} \leq y_3{:}\mathbb{B}$ where $D(x_3{:}\mathbb{O}) = \{0,3\}, D(y_3{:}\mathbb{B}) = \{5,7,8,10\}$

$x_4{:}\mathbb{O} + y_3{:}\mathbb{B} \leq y_4{:}\mathbb{O}$ where $D(x_4{:}\mathbb{O}) = \{0,3\}, D(y_4{:}\mathbb{O}) = \{8,10\}$

$x_5{:}\mathbb{O} + y_4{:}\mathbb{O} \leq 10$ where $D(x_5{:}\mathbb{O}) = \{0,2\}$

This requires further decomposition of the two mixed encoding constraints:

$x_3{:}\mathbb{O} \leq x_3{:}\mathbb{B}$ where $D(x_3{:}\mathbb{O}) = D(x_3{:}\mathbb{B}) = \{0,3\}$

$y_2{:}\mathbb{O} \leq y_2{:}\mathbb{B}$ where $D(y_2{:}\mathbb{O}) = \{2,7,8\}, D(y_2{:}\mathbb{B}) = \{2,7,8\}$

$x_3{:}\mathbb{B} + y_2{:}\mathbb{B} = y_3{:}\mathbb{B}$ where $D(y_3{:}\mathbb{B}) = 2..10$

$x_4{:}\mathbb{O} \leq x_4{:}\mathbb{B}$ where $D(x_4{:}\mathbb{O}) = D(x_4{:}\mathbb{B}) = \{0,3\}$

$x_4{:}\mathbb{B} + y_3{:}\mathbb{B} = y_4{:}\mathbb{B}$ where $D(y_4{:}\mathbb{B}) = 2..10$

$y_4{:}\mathbb{B} \geq y_4{:}\mathbb{O}$ where $D(y_4{:}\mathbb{O}) = \{8,10\}$

□

To support binary encoded integer decision variables (see Sect. 4.2), representing a term $x_i = q_i x'_i{:}\mathbb{B}$ is not without overhead (unlike $x_i = q_i x'_i{:}\mathbb{O}$). We use efficient encodings which are pre-computed for a given coefficient and domain [11]. By using common sub-expression elimination of terms, we avoid introducing the same encoding twice [16]. To encode decompositions of equality constraints (see Sect. 4.3), we apply $x{:}\mathbb{O} = x{:}\mathbb{B}$ [12].

5 Experimental Evaluation

In this section, we evaluate the GT and BDD decomposition, and apply each practical extension. First, we test the PB encoding in its base form by converting each integer decision variable into PB terms using the order encoding. The next configuration ($+\mathbb{O}$) supports the integer linear constraint directly (Sect. 4.2).

Next, we add direct support for equality constraints (+eq), rather than splitting each in two inequalities (Sect. 4.3). Finally, we mix in the binary encoding for (auxiliary) integer variables (+$\mathbb{O} \leq c$) using two cut-offs (Sect. 4.4), as well as a uniform binary approach (+$\mathbb{B}$). We omit the SWC decomposition because its constant size auxiliary variable domains are not as interesting for the mixed encoding, and due to space limitations.

As baseline, we include various configurations of three established SAT encoders which can encode integer linear constraints. Savile Row (version 1.10.1) implements the fundamental PB encodings that we study in this paper, as well as binary encoding approaches such as *Generalized n-Level Modulo Totalizer* (GMTO) and *Global Polynomial Watchdog* (GPW) [7]. These PB encodings are generalized for AMO side-constraints, allowing for principal integer variables, the GT encoding is enhanced by a BDD-like reduction algorithm [13]. Equality constraints are separately encoded as two inequalities. Picat-SAT (version 3.5) primarily uses the binary encoding with RCA and ad-hoc optimization [27,28]. Fun-sCOP (version 20230601-13h09m) hybridizes order and binary encodings of the principal integer variables [24]. Yet, its binary encoding employs PB encodings such as BDD, which still order encode the auxiliary variables uniformly. Consequently, there was no difference between their binary and mixed approach. We have omitted PBLib [23], another common PB encoding library, as it does not support side constraints or integer variables, and we have found that it is not competitive in an integer setting.

All encodings are solved using the same executable of CaDiCaL (v2.1.0) [8] with a 3 GB memory limit and a time limit of 60 seconds. For two different integer linear problems, we generate three instance sets. We compare the number of solved instances, followed by their average solve time, in parentheses. The encoding size is shown as the number of thousands of variables and clauses, with the number of memory- and timeouts (if any) indicated by a superscript, or by—if all failed to encode. Our implementation of the encoding methods, the generated benchmark instances, and the benchmark scripts are available [10].

5.1 Multidimensional Bounded Knapsack Problem

In the *Multidimensional Bounded Knapsack Problem* (MBKP) we decide for N item types how many x_i to pack (up to B) such that a minimum profit $\sum_{i=1}^{N} x_i p_i \geq P$ is reached. Each item is restricted by M dimensions of weight with $\bigwedge_{i=1}^{M} \sum_{i=j}^{N} x_j w_{i,j} \leq W_i$. Adapted from [18], instances can be generated by first generating C coefficient sets for $w_{i,j}$ and p_i, sampling uniformly from $[1, Q]$. Then, for each coefficient set, we choose S capacity sets for $1 \leq s \leq S$ with a capacity factor $f = \frac{s}{S}$: generating $W_i = f \sum_{j=1}^{N} B w_{i,j}, 1 \leq i \leq M$, and $P = (1-f) \sum_{j=1}^{N} B p_j$. As f increases, the constraints become less strict, and instances turn from unsatisfiable to satisfiable. Since the low and high ends of f are trivial, we normalize f to be within $0.4 \leq f \leq 0.6$. We generate one PB instance set (where B=1), and two integer linear instance sets with different Q.

Table 1. Results for 3 sets of 100 MBKP instances ($C = 4, S = 25, 0.4 \leq f \leq 0.6$).

(N, B, M, Q)	$(50, 1, 25, 50)$		$(10, 10, 200, 50)$		$(10, 10, 200, 250)$	
	solved	vars./cl.	solved	vars./cl.	solved	vars./cl.
$GT + \mathbb{B}$	84 (6.0)	15/76	85 (11.5)	36/224	74 (16.5)	52/332
$GT + \mathbb{O} \leq 25$	88 (5.7)	14/54	**94 (10.9)**	39/238	87 (14.9)	58/337
$GT + \mathbb{O} \leq 75$	**89 (5.7)**	14/54	93 (11.8)	43/252	**91 (16.6)**	58/340
$GT + \mathbb{O}$	67 (3.5)	84/7852	0	701/159425	—	—
GT	68 (3.9)	84/7852	0	985/164102	—	—
$BDD + \mathbb{B}$	78 (3.3)	23/99	78 (10.1)	40/234	68 (12.0)	56/343
$BDD + \mathbb{O} \leq 25$	75 (8.9)	26/113	93 (13.4)	42/244	86 (15.1)	58/334
$BDD + \mathbb{O} \leq 75$	76 (10.2)	29/114	90 (13.8)	43/252	88 (16.1)	58/335
$BDD + \mathbb{O}$	75 (6.4)	351/694	61 (17.4)	775/7983	0	2723/28503
BDD	74 (5.2)	351/694	0	8313/16588	0	30312/60561
Savile Row (GT)	82 (2.3)	61/5946	0	610/108981[56]	—	—
Savile Row (SWC)	74 (13.9)	795/1561	9 (13.9)	2244/22459	0	10566/105809[30]
Savile Row (BDD)	86 (0.8)	230/446	44 (2.8)	582/5535	40 (2.2)	2019/19390
Savile Row (GPW)	87 (3.5)	22/117	74 (12.1)	83/288	72 (13.1)	111/394
Savile Row (GMTO)	87 (5.1)	20/43	89 (8.0)	50/179	84 (10.9)	63/234
Fun-sCOP	83 (0.2)	232/667	58 (8.5)	1252/3512	44 (9.0)	5405/15092
Picat-SAT (RCA)	81 (6.5)	10/62	70 (16.7)	27/211	47 (19.6)	42/359

In the results shown in Table 1, the effects of using integer decision variables ($+\mathbb{O}$) is difficult to compare due to encoding memory- and timeouts. Using the uniform binary encoding ($+\mathbb{B}$) improves results in every case, presumably because the auxiliary domains are considerable. A relatively small cut-off value, with only a minority of all domain values order encoded, pushes the number of solved instances for both GT and BDD on all instance set, with the former outperforming the control encoders. In experiments not shown here, a larger cut-off value of 150 or 300 was less effective.

5.2 Multidimensional Bounded Subset Sum Problem

The *Multidimensional Bounded Subset Sum Problem* (MBSSP) is similar to MBKP, but requires packed items to sum up to an exact capacity using equality constraints. We decide for item types $1 \leq i \leq N$ how many x_i items to select (up to bound B) such that an exact subset sum k_i is reached for each $1 \leq j \leq M$ dimension given the weight $q_{i,j} \in [1, Q]$ of the item j in the dimension i. That is, $\bigwedge_{i=1}^{M} \sum_{j=1}^{N} q_{i,j} x_j = k_i$. We generate instances by uniformly sampling all $q_{i,j}$. Then, to guarantee a solution exists, we uniformly sample an assignment a_j for every x_j, yielding feasible values for $k_i = \sum_{j=1}^{N} q_{i,j} a_j$ for $1 \leq i \leq M$. Again, we generate one PB and two integer linear instance sets.

Table 2. Results for 3 sets of 100 MBSSP instances.

(N, B, M, Q)	$(40, 1, 15, 50)$		$(12, 10, 8, 50)$		$(12, 10, 8, 250)$	
	solved	vars./cl.	solved	vars./cl.	solved	vars./cl.
GT + eq + $\mathbb{B}$	100 (2.5)	7/34	100 (1.2)	2/14	**100 (1.4)**	3/19
GT + eq + $\mathbb{O} \leq 25$	100 (2.4)	6/55	100 (7.0)	2/58	99 (10.3)	3/253
GT + eq + $\mathbb{O} \leq 75$	100 (2.3)	6/55	100 (9.4)	3/59	100 (11.7)	3/253
GT + eq + $\mathbb{O}$	6 (36.8)	32/5794	0	38/22574	0	$103/177961^{96}$
GT + $\mathbb{O}$	4 (46.4)	68/6446	0	81/23596	0	$251/207563^{99}$
GT	2 (37.9)	68/6446	0	101/25768	0	$291/196827^{97}$
BDD + eq + $\mathbb{B}$	99 (9.8)	10/44	**100 (1.1)**	2/14	**100 (1.4)**	3/20
BDD + eq + $\mathbb{O} \leq 25$	—	—	100 (6.1)	3/59	100 (10.2)	3/254
BDD + eq + $\mathbb{O} \leq 75$	—	—	100 (8.1)	3/60	100 (10.1)	3/254
BDD + eq + $\mathbb{O}$	35 (35.7)	116/458	7 (42.6)	49/1024	0	192/4045
BDD + $\mathbb{O}$	29 (35.3)	232/458	5 (34.3)	99/1024	1 (17.3)	385/4048
BDD	32 (36.3)	232/458	0	1037/2068	0	4117/8225
Savile Row (GT)	**100 (0.2)**	47/3190	0	67/15607	—	—
Savile Row (SWC)	33 (48.3)	597/1166	1 (24.7)	269/2728	0	1322/13412
Savile Row (BDD)	34 (28.0)	153/294	0	75/725	0	289/2816
Savile Row (GPW)	98 (12.4)	19/89	15 (25.6)	9/31	9 (25.2)	11/42
Savile Row (GMTO)	94 (13.7)	18/39	11 (31.7)	5/18	8 (32.1)	6/23
Fun-sCOP	26 (38.3)	156/442	5 (25.6)	155/435	1 (46.7)	688/1930
Picat-SAT (RCA)	81 (16.1)	5/29	100 (1.6)	2/13	100 (1.8)	2/19

The results in Table 2 show an improvement when using integer variables (+$\mathbb{O}$) on non-PB instances. The support for equality constraints predictably halves the number of auxiliary variables, and makes a positive effect on solver performance overall. Again, the binary encoding approach is better than the order-based ones, and equals the best control for two out of three instance sets. Mixing in the order encoding does not improve performance for MBSSP, as it perhaps suits the equality constraints less.

6 Conclusion and Future Work

In this paper, we have shown that by thinking of encodings of PB constraints as computing partial sums, we can reframe the fundamental GT, SWC, and BDD encodings as integer decompositions. But once we have the *integer viewpoint*, any improvement on the decomposition automatically improves each encoding method. We extend the decomposition by supporting linear constraints, equality constraints, and mixed encodings for each individual integer variable. We show how each extension practically improves each method across two types of benchmarks and compared to three strong baseline encoders.

In future work, additional PB encodings may be understood from the same viewpoint. Some methods (*e.g.* Sorting Networks [17]) may also use an implicit order encoding for their auxiliary variables, while others may use the binary or direct encoding (*e.g.* alternative BDD encodings [1]). Further extensions of the decomposition (*e.g.* by improving its encoding) could also be of interest.

Acknowledgments. This research was partially funded by the Australian Government through the Australian Research Council Industrial Transformation Training Centre in Optimisation Technologies, Integrated Methodologies, and Applications (OPTIMA), Project ID IC200100009, and H. Bierlee was also partially funded by the European Research Council (ERC) under the EU Horizon 2020 research and innovation programme (Grant No 101002802, CHAT-Opt).

Disclosure of Interests. The authors have no competing interests to declare that are relevant to the content of this article.

Appendix 1: proofs

Proof (Theorem 1).
We show that the clauses generated by the integer tree decomposition are (after closing under unit propagation) the same as those for the GT encodings. We consider a single node Z with child nodes L and R from the GT encoding and corresponding integer constraint $l(i)+r(i) \leq y_i$. The binary tree of the decomposition is the same as in the GT encoding, where $[\![y_i \geq v]\!] \equiv z_v$, $[\![l(i) \geq v]\!] \equiv l_v$ and $[\![r(i) \geq v]\!] \equiv r_v$. Encoding the ternary inequality of Eq. (6) yields the following clauses.

$$\bigwedge_{v \in D(l(i))} \bigwedge_{w \in D(r(i))} ([\![l(i) \geq v]\!] \wedge [\![r(i) \geq w]\!]) \rightarrow [\![y_i \geq v+w]\!]$$

$$\equiv \left(\bigwedge_{v \in D(l(i))} ([\![l(i) \geq v]\!] \wedge [\![r(i) \geq 0]\!]) \rightarrow [\![y_i \geq v]\!] \right)$$

$$\wedge \left(\bigwedge_{w \in D(r(i))} ([\![l(i) \geq 0]\!] \wedge [\![r(i) \geq w]\!]) \rightarrow [\![y_i \geq w]\!] \right)$$

$$\wedge \left(\bigwedge_{v \in D(l(i)) \setminus \{0\}} \bigwedge_{w \in D(r(i)) \setminus \{0\}} ([\![l(i) \geq v]\!] \wedge [\![r(i) \geq w]\!]) \rightarrow [\![y_i \geq v+w]\!] \right)$$

The case decomposition is correct since 0 is the lower bound of every internal variable y_i, except for y_{root}, which is fixed to k. We show these are the same clauses as generated for a GT node. The literals $[\![l(i) \geq 0]\!] = 1$ and $[\![r(i) \geq 0]\!] = 1$ by Eq. (1a). We can skip iterations where $[\![y_i \geq 0]\!] = 1$, since the resulting clause is satisfied. In the GT encoding, the unary clause $\neg a_{k+1}$ will fix $z_{k+1} = 0$ if present in its child node Z, by $z_{k+1} \to a_{k+1}$. This propagates through the entire tree, so that $z_{k+1} \equiv 0$ for all nodes. Consequently, the first two conjuncts are equivalent to the binary clauses Eqs. (5a) and (5b) of GT since $D(l(i)) = \{v \mid l_v \in L \setminus \{l_{k+1}\}\} \cup \{0\}$ and $D(r(i)) = \{v \mid r_v \in R \setminus \{r_{k+1}\}\} \cup \{0\}$:

$$\left(\bigwedge_{v \in D(r(i)) \setminus \{0\}} [\![l(i) \geq v]\!] \to [\![y_i \geq v]\!] \right) \wedge \left(\bigwedge_{w \in D(l(i)) \setminus \{0\}} [\![r(i) \geq w]\!] \to [\![y_i \geq w]\!] \right)$$

The remaining (third case) clauses are, after unit propagation, equivalent to the ternary clauses of the GT decomposition. Since, for any $v + w > k$, the literal $[\![y_i \geq v + w]\!] \equiv 0$, which is simply falsified (leading to a binary clause, rather than a ternary clause including z_{k+1}).

For the root node if $v + w \leq k$ then the literal $[\![y_i \geq v + w]\!] \equiv 1$ by Eq. (1a), and so the resulting clauses are trivial; otherwise, they agree with those in GT. Note that the GT decomposition adds Boolean variables $\{a_v \mid v \leq k\}$ for the root node, but since they only appear positively, they can be trivially assigned to 1 (by SAT preprocessing).

Proof (Theorem 2). Encoding each ternary inequality $x_i + w_i \leq w_{i-1}$ yields the following clauses:

$$\bigwedge_{v \in D(x_i)} \bigwedge_{u \in D(w_i)} ([\![x_i \geq v]\!] \wedge [\![w_i \geq w]\!]) \to [\![w_{i-1} \geq v + u]\!]$$

$$\equiv \bigwedge_{v \in D(x_i)} \bigwedge_{u \in D(w_i)} ([\![x_i \geq v]\!] \wedge [\![w_{i-1} < v + u]\!]) \to [\![w_i < u]\!]$$

We split the conjunctions:

$$\left(\bigwedge_{u=-k}^{0} ([\![x_i \geq 0]\!] \wedge [\![w_{i-1} < 0 + u]\!]) \to [\![w_i < u]\!] \right) \tag{13a}$$

$$\wedge \left(([\![x_i \geq q_i]\!] \wedge [\![w_{i-1} < q_i - k]\!]) \to [\![w_i < -k]\!] \right) \tag{13b}$$

$$\wedge \left(\bigwedge_{u=-k+1}^{-q_i} ([\![x_i \geq q_i]\!] \wedge [\![w_{i-1} < q_i + u]\!]) \to [\![w_i < u]\!] \right) \tag{13c}$$

$$\wedge \left(\bigwedge_{u=-q_i+1}^{0} ([\![x_i \geq q_i]\!] \wedge [\![w_{i-1} < q_i + u]\!]) \to [\![w_i < u]\!] \right) \tag{13d}$$

When we equate $[\![w_i < u]\!] \equiv z_{i,-u+1}$, for $1 \leq i \leq n, -k \leq u \leq 0$ we can show that the four cases above correspond to the four cases in the SWC encoding. The first case, Eq. (13a), $[\![x_i \geq 0]\!] \equiv 1$, and $[\![w_i - 1 < -k]\!] \equiv 0$, and the resulting clauses are equivalent after unit propagation to Eq. (7a) and removing the trivial clause (when $u = -k$). In the second case, Eq. (13b), the clauses are equivalent once we note that $[\![w_i < -k]\!] \equiv 0$. The third case Eq. (13c), produces exactly the clauses from Eq. (7c), generated in reverse order, i.e. $\bigwedge_{u=-k+1}^{-q_i} \left((b_i \wedge z_{i-1,-q_i-u})\right) \to z_{i,-u} \equiv \bigwedge_{l=1}^{k-q_i} \left((b_i \wedge z_{i-1,k-q_i-l}) \to z_{i,k-l}\right)$, where $l = u + k$. Finally, Eq. (13d) corresponds to Eq. (7d), as we have $[\![w_{i-1} < q_i + u]\!] \equiv 1$ since $q_i + u \geq q_i - q_i + 1 > 0$ (which is the upper bound of w_{i-1}).

Proof (Theorem 3).
The clauses for $x_i + y_{i-1} \leq y_i$ are:

$$\bigwedge_{v \in D(x_i)} \bigwedge_{d_{i-1,j} \in D(y_{i-1})} ([\![x_i \geq v]\!] \wedge [\![y_{i-1} \geq d_{i-1,j}]\!]) \to [\![y_i \geq d_{i-1,j} + v]\!]$$

Clearly, for every edge $((i-1,j),(i,j'),p) \in E$, there is a clause with antecedents $[\![x_i \geq v]\!]$ (where $v = pq_i$) and $[\![y_{i-1} \geq d_{i-1,j}]\!]$, which corresponds to the clause from that same edge in the original encoding. It remains to be shown that the clause's consequent $[\![y_i \geq d_{i-1,j} + v]\!]$ matches the literal $[\![y_i \geq d_{i,j'}]\!]$ of the target node (i,j'). If $d_{i-1,j} = \max W_{i-1,j}$, and there is a p edge $(i-1,j)$ and (i,j'), then $\max W_{i,j'-1} < d_{i,j} + v \leq \max W_{i,j'}$. By the domain construction of Eq. (11), we have $d_{i,j'-1} < w \leq d_{i,j'}$. Consequently, $[\![y_i \geq w]\!] \equiv [\![y_i \geq d_{i,j'}]\!]$ by Eq. (1b). Where $v = 0$, the clause simplifies to $[\![y_{i-1} \geq d_{i-1,j}]\!] \to [\![y_i \geq d_{i-1,j}]\!]$ by Eq. (1a).

Proof (Lemma 1). No assumptions are made on the domains, except that all are non-empty (otherwise the problem would be trivially unsatisfiable). However, we do know that the clause $([\![x \geq v]\!] \wedge [\![y \geq w]\!]) \to [\![z \geq v + w]\!]$ exists if and only if $v \in D(x)$ and $w \in D(y)$.

Lower Bound Propagation. Given a fixed lower bound v for x and lower bound w for y, then since these bounds are in $D(x)$ and $D(y)$ respectively, we have $[\![x \geq v]\!]$ and $[\![y \geq w]\!]$ hold. Hence, the clause $[\![x \geq v]\!] \wedge [\![y \geq w]\!] \to [\![z \geq v+w]\!]$ will propagate the correct bound (or cause failure if $v + w > ub(z)$).

Upper Bound Propagation. Suppose we have a fixed upper bound u for z, then we have $[\![z \leq u]\!] \equiv \neg[\![z \geq u']\!], d_i^z \leq u < u' = d_{i+1}^z$ for some adjacent domain values $d_i^z, d_{i+1}^z \subseteq D(z)$. So $\neg[\![z \geq u']\!]$ holds. By the IC assumption, $\neg[\![z \geq u'']\!], u'' > u'$ also holds.

Suppose we have a fixed lower bound $w \in D(y)$ for y (which might be the original lower bound), then we know that $[\![y \geq w]\!]$ holds. Propagation should enforce that $x \leq u - w$. Now $[\![x \leq u - w]\!] \equiv \neg[\![x \geq u - w + 1]\!] \equiv \neg[\![x \geq v]\!], d_j^x < u - w + 1 \leq v = d_{j+1}^x$ for some adjacent domain values $d_i^x, d_{j+1}^x \subseteq D(x)$. Consider the clause $[\![x \geq v]\!] \wedge [\![y \geq w]\!] \to [\![z \geq v + w]\!]$. Since $v + w \geq (u - w + 1) + w = u + 1 \geq u'$, we have that $\neg[\![z \geq v + w]\!]$ holds by the IC assumption, and the clause propagates $\neg[\![x \geq v]\!]$. Note also, for any value $v' > v, v' \in D(x)$ we have the clause $[\![x \geq v']\!] \wedge [\![y \geq w]\!] \to [\![z \geq v' + w]\!]$, which propagates $\neg[\![x \geq v']\!]$ again by the IC assumption on z. This proves that upper bounds are correctly propagated, and any propagated upper bound also satisfies the IC assumption.

Proof (Theorem 4). First, domain consistency on linear inequality constraints is equivalent to bounds($\mathbb{R}$) consistency [15]. Next, the (non-zero) lower bounds on the leaf integers $x_i = q_i b_i$ are set by the fact that $[\![x_i \geq q_i]\!] \equiv b_i$. Lemma 1 shows that the lower bounds are correctly propagated up the tree. For the top most inequality $l(root) + r(root) \leq y_{root}$, the IC is satisfied automatically since y_{root} has a singular domain, thus all literals mentioning y_{root} are either true or false by definition. This means upper bounds propagate to the children of the root correctly and satisfy the IC. By induction, using Theorem 1 all upper bounds propagated by the tree are correct and always satisfy the IC. Finally, if an upper bound propagated on the leaf integer $x_i = q_i b_i$ removes value q_i, then it sets $\neg[\![x_i \geq q_i]\!] \equiv \neg b_i$.

Proof (Theorem 5). The fact that the combined encodings using Eq. (4) and Eq. (12) enforce bounds ($\mathbb{R}$) consistency of $\sum_{i=1}^{n} q_i b_i = k$, follows from its equivalence to bounds($\mathbb{R}$) consistency (or equivalently domain consistency) of the two inequalities $\sum_{i=1}^{n} q_i b_i \leq k$ and $\sum_{i=1}^{n} q_i b_i \geq k$ [19].

References

1. Abío, I., Gange, G., Mayer-Eichberger, V., Stuckey, P.J.: On CNF encodings of decision diagrams. In: Quimper, C.-G. (ed.) CPAIOR 2016. LNCS, vol. 9676, pp. 1–17. Springer, Cham (2016). https://doi.org/10.1007/978-3-319-33954-2_1
2. Abío, I., Mayer-Eichberger, V., Stuckey, P.J.: Encoding linear constraints with implication chains to CNF. In: Pesant, G. (ed.) CP 2015. LNCS, vol. 9255, pp. 3–11. Springer, Cham (2015). https://doi.org/10.1007/978-3-319-23219-5_1
3. Abío, I., Nieuwenhuis, R., Oliveras, A., Rodríguez-Carbonell, E.: A parametric approach for smaller and better encodings of cardinality constraints. In: Schulte, C. (ed.) CP 2013. LNCS, vol. 8124, pp. 80–96. Springer, Heidelberg (2013). https://doi.org/10.1007/978-3-642-40627-0_9
4. Abío, I., Nieuwenhuis, R., Oliveras, A., Rodríguez-Carbonell, E., Mayer-Eichberger, V.: A new look at BDDs for pseudo-boolean constraints. J. Artif. Intell. Res. **45**, 443–480 (2012). https://doi.org/10.1613/jair.3653
5. Abío, I., Stuckey, P.J.: Encoding linear constraints into SAT. In: O'Sullivan, B. (ed.) CP 2014. LNCS, vol. 8656, pp. 75–91. Springer, Cham (2014). https://doi.org/10.1007/978-3-319-10428-7_9
6. Bailleux, O., Boufkhad, Y.: Efficient CNF encoding of boolean cardinality constraints. In: Rossi, F. (ed.) CP 2003. LNCS, vol. 2833, pp. 108–122. Springer, Heidelberg (2003). https://doi.org/10.1007/978-3-540-45193-8_8
7. Bailleux, O., Boufkhad, Y., Roussel, O.: New encodings of pseudo-boolean constraints into CNF. In: Kullmann, O. (ed.) SAT 2009. LNCS, vol. 5584, pp. 181–194. Springer, Heidelberg (2009). https://doi.org/10.1007/978-3-642-02777-2_19
8. Biere, A., Fazekas, K., Fleury, M., Heisinger, M.: CaDiCaL, Kissat, Paracooba, Plingeling and Treengeling entering the SAT Competition 2020. In: Balyo, T., Froleyks, N., Heule, M., Iser, M., Järvisalo, M., Suda, M. (eds.) Proceedings of SAT Competition 2020 – Solver and Benchmark Descriptions. Department of Computer Science Report Series B, vol. B-2020-1, pp. 51–53. University of Helsinki (2020)
9. Biere, A., Heule, M., van Maaren, H., Walsh, T. (eds.): Handbook of Satisfiability, 2 edn. IOS Press (2021)

10. Bierlee, H., Dekker, J.J.: Pindakaas: CPAIOR-25 (submission) (2024). https://doi.org/10.5281/zenodo.14500064
11. Bierlee, H., Dekker, J.J., Lagoon, V., Stuckey, P.J., Tack, G.: Single constant multiplication for SAT. In: Dilkina, B. (ed.) Integration of Constraint Programming, Artificial Intelligence, and Operations Research - 21st International Conference, CPAIOR 2024, Uppsala, Sweden, 28–31 May 2024, Proceedings, Part I. Lecture Notes in Computer Science, vol. 14742, pp. 84–98. Springer, Heidelberg (2024). https://doi.org/10.1007/978-3-031-60597-0_6
12. Bierlee, H., Gange, G., Tack, G., Dekker, J.J., Stuckey, P.J.: Coupling different integer encodings for SAT. In: Schaus, P. (ed.) Proceedings of the 19th International Conference on the Integration of Constraint Programming, Artificial Intelligence, and Operations Research (CPAIOR 2022). LNCS, vol. 13292, pp. 44–63. Springer, Heidelberg (2022). https://doi.org/10.1007/978-3-031-08011-1_5
13. Bofill, M., Coll, J., Nightingale, P., Suy, J., Ulrich-Oltean, F., Villaret, M.: SAT encodings for pseudo-boolean constraints together with at-most-one constraints. Artif. Intell. **302**, 103604 (2022). https://doi.org/10.1016/j.artint.2021.103604
14. Bryant, R.: Graph-based algorithms for boolean function manipulation. IEEE Trans. Comput. **C-35**(8), 677–691 (1986). https://doi.org/10.1109/TC.1986.1676819
15. Choi, C.W., Harvey, W., Lee, J., Stuckey, P.J.: Finite domain bounds consistency revisited. In: Sattar, A., Kang, B. (eds.) AI 2006. LNCS (LNAI), vol. 4304, pp. 49–58. Springer, Heidelberg (2006). https://doi.org/10.1007/11941439_9
16. Cocke, J.: Global common subexpression elimination. In: Northcote, R.S. (ed.) Proceedings of a Symposium on Compiler Optimization, Urbana-Champaign, Illinois, USA, 27-28 July 1970, pp. 20–24. ACM (1970). https://doi.org/10.1145/800028.808480
17. Eén, N., Sörensson, N.: Translating pseudo-boolean constraints into SAT. J. Satisf. Boolean Model. Comput. **2**(1–4), 1–26 (2006). https://doi.org/10.3233/sat190014
18. Han, B., Leblet, J., Simon, G.: Hard multidimensional multiple choice knapsack problems, an empirical study. Comput. Oper. Res. **37**(1), 172–181 (2010). https://doi.org/10.1016/j.cor.2009.04.006
19. Harvey, W., Stuckey, P.: Improving linear constraint propagation by changing constraint representation. Constraints **8**(2), 173–207 (2003)
20. Hölldobler, S., Manthey, N., Steinke, P.: A compact encoding of pseudo-boolean constraints into SAT. In: Glimm, B., Krüger, A. (eds.) KI 2012. LNCS (LNAI), vol. 7526, pp. 107–118. Springer, Heidelberg (2012). https://doi.org/10.1007/978-3-642-33347-7_10
21. Jabs, C., Berg, J., Järvisalo, M.: Core boosting in sat-based multi-objective optimization. In: Dilkina, B. (ed.) Integration of Constraint Programming, Artificial Intelligence, and Operations Research - 21st International Conference, CPAIOR 2024, Uppsala, Sweden, 28–31 May 2024, Proceedings, Part II. LNCS, vol. 14743, pp. 1–19. Springer, Heidelberg (2024). https://doi.org/10.1007/978-3-031-60599-4_1
22. Joshi, S., Martins, R., Manquinho, V.: Generalized totalizer encoding for pseudo-boolean constraints. In: Pesant, G. (ed.) CP 2015. LNCS, vol. 9255, pp. 200–209. Springer, Cham (2015). https://doi.org/10.1007/978-3-319-23219-5_15
23. Philipp, T., Steinke, P.: PBLib – a library for encoding pseudo-boolean constraints into CNF. In: Heule, M., Weaver, S. (eds.) SAT 2015. LNCS, vol. 9340, pp. 9–16. Springer, Cham (2015). https://doi.org/10.1007/978-3-319-24318-4_2

24. Soh, T., Le Berre, D., Banbara, M., Tamura, N.: SCOP: sat-based constraint programming system. In: Proceedings of XCSP3 Competition 2018 (XCSP18), pp. 93–94 (2018)
25. Vandesande, D., Wulf, W.D., Bogaerts, B.: Qmaxsatpb: a certified maxsat solver. In: Gottlob, G., Inclezan, D., Maratea, M. (eds.) Logic Programming and Nonmonotonic Reasoning - 16th International Conference, LPNMR 2022, Genova, Italy, 5–9 September 2022, Proceedings. LNCS, vol. 13416, pp. 429–442. Springer, Heidelberg (2022). https://doi.org/10.1007/978-3-031-15707-3_33
26. Warners, J.P.: A linear-time transformation of linear inequalities into conjunctive normal form. Inf. Process. Lett. **68**(2), 63–69 (1998). https://doi.org/10.1016/S0020-0190(98)00144-6
27. Zhou, N.-F., Kjellerstrand, H.: The Picat-SAT compiler. In: Gavanelli, M., Reppy, J. (eds.) PADL 2016. LNCS, vol. 9585, pp. 48–62. Springer, Cham (2016). https://doi.org/10.1007/978-3-319-28228-2_4
28. Zhou, N.-F., Kjellerstrand, H.: Optimizing SAT encodings for arithmetic constraints. In: Beck, J.C. (ed.) CP 2017. LNCS, vol. 10416, pp. 671–686. Springer, Cham (2017). https://doi.org/10.1007/978-3-319-66158-2_43

Multi-task Representation Learning for Mixed Integer Linear Programming

Junyang Cai[✉][iD], Taoan Huang[iD], and Bistra Dilkina[iD]

University of Southern California, Los Angeles, CA, USA
{caijunya,taoanhua,dilkina}@usc.edu

Abstract. Mixed Integer Linear Programs (MILPs) are highly flexible and powerful tools for modeling and solving complex real-world combinatorial optimization problems. Recently, machine learning (ML)-guided approaches have demonstrated significant potential in improving MILP-solving efficiency. However, these methods typically rely on separate offline data collection and training processes, which limits their scalability and adaptability. This paper introduces the first multi-task learning framework for ML-guided MILP solving. The proposed framework provides MILP embeddings helpful in guiding MILP solving across solvers (e.g., Gurobi and SCIP) and across tasks (e.g., Branching and Solver configuration). Through extensive experiments on three widely used MILP benchmarks, we demonstrate that our multi-task learning model performs similarly to specialized models within the same distribution. Moreover, it significantly outperforms them in generalization across problem sizes and tasks.

Keywords: Deep Learning · Mixed Integer Linear Programming · Multi-task Learning · Graph Neural Networks

1 Introduction

Many real-world **problem domains**, such as path planning [48], scheduling [9,19], and network design [12,27], fall into the category of combinatorial optimization (CO) and are generally NP-hard to solve. Designing efficient algorithms for CO problems is both important and challenging. Mixed Integer Linear Programs (MILPs) provide a versatile framework for modeling and solving various CO problems. MILPs involve optimizing a linear objective function subject to linear constraints, with some variables restricted to integer values. Significant advancements in MILP solvers, such as Gurobi [22] and SCIP [5], have been achieved by leveraging techniques like Branch-and-Bound (BnB) [38], complemented by a suite of heuristics to enhance performance.

Recent advancements in machine learning (ML) offer new avenues to improve MILP solvers. ML methods, by learning from complex historical distributions, can enhance both exact solvers like BnB and heuristic solvers. For exact solvers, ML techniques can predict **tasks** like which node to expand [37,51], which variable to branch on [7,21,32], which cut to apply [47,52], or how to schedule heuristics [11,25,33]. For heuristic solvers, ML approaches can predict solutions directly [14,29,34] or integrate Large Neighborhood Search into MILP solvers [28,50]. These ML-driven enhancements hold the promise of bridging gaps in traditional solvers.

While learning-based methods for MILPs have demonstrated success in single-task settings, they face limitations in generalizing across multiple domains and tasks. Current approaches often rely on specialized models tailored to individual tasks and domains, resulting in computationally expensive training pipelines and the need for carefully curated datasets. This lack of generalization hampers real-world applicability. To address this, researchers have begun to explore single models capable of addressing multiple problem domains, drawing inspiration from foundation models. However, these efforts have largely been restricted to single-task settings [16,30,40]. Despite these breakthroughs, an important observation remains unaddressed: MILPs from the same problem domain often share structural and characteristic similarities that can be leveraged for multi-task learning.

This paper presents a unified multi-task learning framework for MILP solving, designed to learn shared representations across multiple tasks within a problem domain. Our approach involves a two-step training process: first, we train a shared representation layer for MILPs alongside fixed, randomly initialized task-specific output layers; then, we fine-tune the task-specific layers while keeping the shared representation layer fixed. We evaluate our framework across three tasks—**Backdoors** (root-node branching) [7], **Predict-and-Search** (PaS) [29], and **Solver Configurations**—and three common problem domains: Combinatorial Auction (CA) [39], Maximal Independent Set (MIS) [53], and Minimum Vertex Cover (MVC) [15].

Our main findings and contributions are as follows:

- **Multi-task Learning Framework:** We introduce a unified multi-task learning framework for MILPs that leverages a shared representation layer and task-specific output heads, enabling efficient adaptation on unseen tasks.
- **Generalization on Size:** Our multi-task models achieve competitive results compared to task-specific models within the same distribution and show superior generalization on more significant problem instances.
- **Cross-Task Generalization:** Through cross-task evaluations, we multi-task train on two tasks and fine-tune on the third, which consistently outperforms specialized models trained on the third task.

The remainder of this paper provides background, related work, methodology, experimental results, and discussions.

2 Background

This section defines MILPs and provides background knowledge on our three tasks: BACKDOOR, PAS, and CONFIGURATION.

2.1 Mixed Integer Linear Programming

A Mixed Integer Linear Program (MILP) $P = (A, b, c, I)$ is defined as:

$$\min\{c^T x \mid Ax \leq b,\ x \in \mathbb{R}^n,\ x_j \in \{0,1\}\ \forall j \in I\},$$

where $A \in \mathbb{R}^{m \times n}$, $b \in \mathbb{R}^m$, $c \in \mathbb{R}^n$, and $I \subseteq \{1, ..., n\}$ is the set of indices for binary variables. The objective is to minimize $c^T x$ by finding a feasible assignment for x that satisfies the constraints. MILP solvers rely heavily on Branch-and-Bound (BnB) [38] that constructs a search tree to find feasible solutions with minimum costs. This process involves repeatedly solving LP relaxations of the MILP and branching on integer variables that are fractional in the LP solution, creating subproblems until all integrality constraints are met.

A key aspect of MILP solvers is their vast configuration space, with parameters spanning integer, continuous, and categorical values, influencing nearly every step of the BnB process. While the solvers' default settings aim for robust performance across heterogeneous MILP benchmarks, there is significant potential to improve configuration settings for specific distributions of instances by selecting the solver parameters effectively (CONFIGURATION).

2.2 Backdoors for MILP

Initially introduced for Constraint Satisfaction Problems [56], backdoors were later generalized to MILPs [13]. In the context of MILPs, strong backdoors are defined as subsets of integer variables such that branching exclusively on these variables yields an optimal integral solution. Research [18] has further demonstrated speedups in MILP solving times by prioritizing branching backdoor variables instead of branching exclusively on them.

Given a MILP instance $P = (A, b, c, I)$, a pseudo-backdoor (BACKDOOR) of size $K \ll |I|$ is a small subset $B \subset I$ of binary variables, with $|B| = K$, whose variables are prioritized for branching to improve solver performance [7,17]. We guide the solver's decision-making process by assigning higher branching priority to the variables in B at the start of the tree search. This adjusted branching order influences the solver's primal heuristics and enhances the overall pruning efficiency in the BnB procedure.

2.3 Predict-and-Search

Predict-and-Search (PAS) [23] is a primal heuristic that leverages the prediction of the optimal solutions to guide the search process. Given a MILP instance, $P = (A, b, c, I)$, let $p_\theta(x_i \mid P)$ denote the predicted probability for each binary variable $x_i \in I$. PAS identifies near-optimal solutions by exploring a neighborhood informed by these predictions. Specifically, it selects k_0 binary variables X_0 with the smallest $p_\theta(x_i \mid P)$ and k_1 binary variables X_1 with the largest $p_\theta(x_i \mid P)$, ensuring X_0 and X_1 are disjoint ($k_0 + k_1 \leq q$). Variables in X_0 are fixed to 0, and those in X_1 are fixed to 1 in a sub-MILP. However, PAS allows up to $\Delta \geq 0$ of these fixed variables to be flipped during solving. Formally, let $B(X_0, X_1, \Delta) = \{x : \sum_{x_i \in X_0} x_i + \sum_{x_i \in X_1}(1 - x_i) \leq \Delta\}$ be the neighborhood defined by X_0, X_1, and Δ, and let D represent the feasible region of the original MILP. PAS then solves the following optimization problem: $\min c^T x$ s.t. $x \in D \cap B(X_0, X_1, \Delta)$. Restricting the solution space to $D \cap B(X_0, X_1, \Delta)$ simplifies the problem, enabling the solver to find high-quality feasible solutions to the original MILP more efficiently.

3 Related Work

This section provides an overview of related work in learning techniques for MILP solving, focusing on learning to branch, solution prediction, and algorithm configuration. Additionally, it highlights research about generalization in the context of ML-guided solving.

3.1 Machine Learning for MILP Solving

There has been extensive research leveraging machine learning techniques to improve MILP solvers. Common approaches represent MILPs as bipartite graphs [21] and employ graph neural networks (GCN [14,21,26,29,34,47,55] and GAT [7,28]) to learn various MILP decisions. Learning methods are diverse, some commonly used are imitation learning [24,47,51], contrastive learning [7,28,29,41], and reinforcement learning [8,50,52]. For a comprehensive survey on machine learning for MILP solving, we refer readers to [49]. Here, we focus on our selected three tasks.

Learning to Branch: Several studies [3,21,32,41,44] have explored learning to branch by imitating strong branching heuristics and predicting scores or ranking variables. Still, these approaches require solver-specific implementations and multiple test-time inferences. In contrast, backdoor approaches [7,17] focus on predicting branching variables at the root node, treating the solver as a black box with a single inference. The first work to use ML-guided techniques for identifying effective backdoors is [17], which employs a scorer model and a classifier model trained on data collected via biased sampling methods from [13]. Building on this, [7] utilizes a contrastive learning model to generate backdoors, leveraging a novel Monte Carlo tree search-based data collection approach introduced in [35].

Solution Prediction: The goal is to predict partial assignments of high-quality feasible solutions in a MILP to guide the search. [14,54] identifies variables that remain unchanged across near-optimal solutions and searches within their neighborhood. [45] and [34] propose fixing predicted variables and letting the MILP solver optimize the rest as a warm start. However, if predictions are inaccurate, fixing variables can lead to low-quality or infeasible solutions. [23] introduces PaS, which searches for solutions within a predefined neighborhood of the prediction, improving feasibility and quality. [29] extends PaS using contrastive learning and novel optimization-based methods for handling low-quality or infeasible solutions.

Instance-Specific Configuration: Introduced by [31], this approach extracts features from problem instances and uses G-means clustering to group similar instances for configuration selection. Hydra-MIP [57] improved this by incorporating features from short solver runs before selecting configurations for full runs. In the NeurIPS 2021 ML4CO competition [20], participants successfully applied ML-guided regression methods to choose the best configurations [55]. [26] learns MILP similarities related to solution costs and uses K-nearest neighbors to select configurations at inference time. Different from choosing the best from a set of configurations, we propose a contrastive learning approach that learns to generate a new configuration by learning to discriminate between good and bad ones.

3.2 ML-Guided Solving Generalization

Despite advancements in ML-guided solving, generalizing learned models across tasks and problem domains remains a key challenge. Most ML methods are tailored to specific problem classes, limiting their applicability in real-world scenarios. Recently, some works have addressed this issue. [30] introduced Distributional MIPLIB, the first multi-domain library for advancing ML-guided MILP methods and exploring cross-domain generalization. [40] proposed MILP-Evolve, leveraging large language models to generate diverse MILP classes, showing strong generalization when trained on a large dataset. [16] introduced GOAL, a generalist model for various combinatorial optimization problems, but it has not been applied to MILP domains. Additionally, multi-task learning has been applied in other CS problems like computer vision [43] and natural language processing [10], but no one has applied it to the CO domain.

4 Multi-task Representation Learning

We aim to learn MILP embeddings that are effective across different learning tasks and instances within the same problem domain, and that can be easily fine-tuned for new tasks. We share the same network architecture for different tasks to enable multi-task training with task-specific layers attached. Training data from other tasks is processed alternately by batch in every epoch.

However, this alternating training strategy often leads to competition between different tasks, causing significant oscillations in the loss curves and a strong bias in testing performance towards one of the tasks. To address this issue, we introduced a two-step training strategy. First, we train the shared network architecture while keeping the task-specific layers fixed at randomly initialized weights. We use three randomly initialized task-specific layers to enhance the robustness of each task. This ensures the shared network architecture learns a general MILP embedding that is not overly specialized to a single random initialization. Second, we fine-tune the task-specific layers for each task while keeping the shared network architecture fixed. This step ensures that the standard MILP embedding benefits each specific task without disrupting the shared representation.

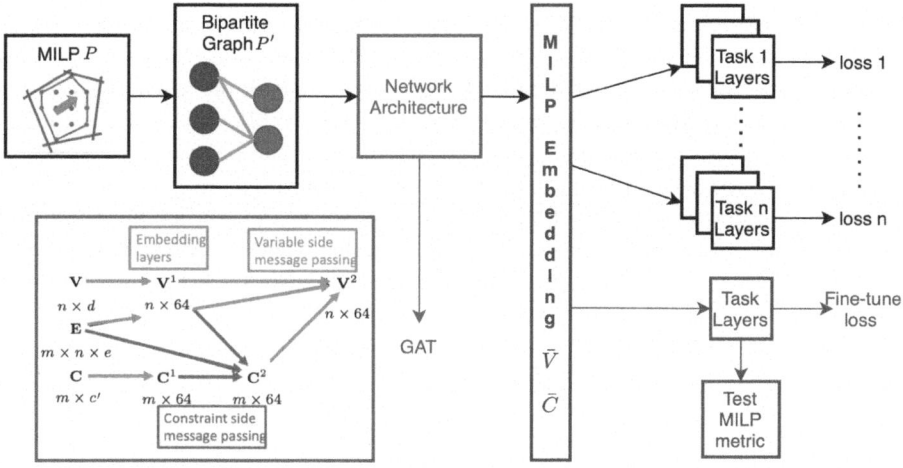

Fig. 1. This figure illustrates the multi-task learning framework for Mixed Integer Linear Programs (MILPs). Step 1 (red) involves training a network architecture with a graph representation of MILPs to produce general MILP embeddings for multiple random task-specific layers. In Step 2 (purple), task-specific layers are fine-tuned for different downstream tasks, optimizing task-specific losses while preserving a shared representation and evaluating using task-specific metrics. In our settings, MILP embedding $(\bar{V}, \bar{C}) = (V^2, C^2)$ and number of tasks $n = 2$. (Color figure online)

The pipeline is shown in Fig. 1. With this multi-task training pipeline, we achieve a standard representation embedding of MILPs that can be generalized across different tasks within the same problem domain. It is worth noting that this pipeline is adaptable to **any graph representation of MILPs, any size-invariant network architecture**, and **any loss function**. This paper focuses on the bipartite graph representation, Graph Attention Networks, and contrastive loss. In the following subsections, we explain the data collection process, bipartite graph representation, network architecture, loss function, and applying learned networks for the tasks: BACKDOOR, PAS, and CONFIGURATION. We choose these three MILP tasks because there are existing works [7,29] that have utilized the same architecture and loss function, which allows us to effectively explore a multitask model without starting entirely from scratch. Additionally, these three tasks require only one-time inference with static features.

4.1 Data Collection and Data Representation

Since we use contrastive loss, carefully collecting positive and negative samples for practical training is crucial. For BACKDOOR, we follow [7] and employ the

Monte Carlo Tree Search (MCTS) algorithm proposed in [35] to generate candidate backdoors. Positive samples are selected as the backdoors with the shortest runtimes, while negative samples are chosen as the backdoors with the longest runtimes. For CONFIGURATION, we use SMAC3 [42] to collect candidate configurations. Positive samples are the configurations with the best feasible solution found, while negative samples are the configurations with the worst feasible solution found. For PAS, we follow [29] and collect a set of optimal or near-optimal solutions as positive samples. We identify low-quality solutions for negative samples by finding the worst feasible solutions that differ from the positive samples in, at most, 10% of the binary variables.

We represent the MILP $P = (A, b, c, I)$ as a bipartite graph, following the approach in [21]. The resulting bipartite graph, denoted as $P' = (G, V, C, E)$, consists of a graph G with two types of nodes: variable nodes and constraint nodes. An edge (i, j) exists between a variable node i and a constraint node j if the variable i appears in the constraint j with a nonzero coefficient, i.e., $A_{ji} \neq 0$. Constraint, variable, and edge features are represented as matrices $V \in \mathbb{R}^{n \times d}$, $C \in \mathbb{R}^{m \times c'}$, and $E \in \mathbb{R}^{m \times n \times e}$, respectively. This bipartite graph representation ensures that the MILP encoding is invariant to permutations of variables and constraints. Additionally, it enables the use of predictive models designed for graphs of varying sizes, allowing deployment on problems with differing numbers of variables and constraints. The features utilized include 15 variable features (e.g., variable types, coefficients, upper and lower bounds, and root LP-related features), four constraint features (e.g., constant terms and senses), and one edge feature (the coefficients).

4.2 Network Architecture and Contrastive Loss

Our network architecture is a Graph Attention Network (GAT) [6], which takes the bipartite graph P' as input and outputs the MILP embeddings (V^2, C^2). To enhance the modeling capacity and manage interactions between nodes, embedding layers are employed to adjust the sizes of the feature embeddings to $V^1 \in \mathbb{R}^{n \times L}$ and $C^1 \in \mathbb{R}^{m \times L}$. Subsequently, the GAT performs two rounds of message passing. In the first round, each constraint node in C^1 attends to its neighboring variable nodes via an attention mechanism with H attention heads, producing updated constraint embeddings C^2. Similarly, each variable node in V^1 attends to its neighboring constraint nodes in the second round, yielding updated variable embeddings V^2 using a separate set of attention weights. The GAT is designed to learn a shared MILP embedding representation (V^2, C^2) that is generalizable across various tasks and instances. In the experiments, embedding vector size L and number of attention heads H are set to 64 and 8.

The task-specific layers vary depending on the task. For BACKDOOR and PAS tasks, only variable features are required during prediction. In these cases, the task-specific layers consist of a multi-layer perceptron with a sigmoid activation function, which outputs a score between 0 and 1 for each variable. For CONFIGURATION, we have additional layers combining variable and constraint features, mapping the embedding size to the number of configuration parameters, and finally, sigmoid and softmax functions are used to generate scores between 0 and 1 for numerical and categorical parameters.

The model is trained or fine-tuned using a contrastive loss function that scores parameters by learning to emulate superior samples while avoiding inferior ones. Given a set of MILP instances $\mathcal{P}$ for training, let $\mathcal{D} = \{(S_+^P, S_-^P) : P \in \mathcal{P}\}$ represent the set of positive and negative samples for all training instances. Let $p_\theta(P)$ denote the model's prediction for an instance P with network parameters θ. Using dot products for similarity, the InfoNCE [46] contrastive loss is defined as:

$$\mathcal{L}(\theta) = \sum_{(S_+^P, S_-^P) \in \mathcal{D}} \frac{-1}{|S_+^P|} \sum_{a \in S_+^P} \log \frac{\exp(a^\top p_\theta(P)/\tau)}{\sum_{a' \in S_-^P \cup \{a\}} \exp(a'^\top p_\theta(P)/\tau)},$$

where τ is a temperature hyperparameter set to 0.07 in the experiments, following [28].

4.3 Applying Learned Network

During testing, given a MILP instance, we convert it to a bipartite graph and inference one-time with the network to output a score vector with one score for each variable/parameter.

- For BACKDOOR, the binary variables with the highest scores are greedily selected as the predicted backdoors based on a user-defined backdoor size K.
- For PAS, X_0 and X_1 are selected greedily based on the predictions, and a constrained optimization problem is solved using the hyperparameters k_0, k_1, and Δ.
- For CONFIGURATION, categorical parameters are set to the option with the highest score, while numerical parameters use the output score.

Full details of bipartite graph features, GAT network architecture, and hyperparameter settings are provided in the Appendix.

5 Experiments

This section introduces the setup for empirical evaluation and presents the results. The code and Appendix are available at https://github.com/caidog1129/MILP_multitask.

5.1 Experiment Setup

Benchmark Problems and Instance Generation:
We evaluate our approach on three NP-hard benchmark problems widely used in existing studies [21,23]: combinatorial auction (CA), minimum vertex cover (MVC), and maximum independent set (MIS). Previous works [7,29] have demonstrated promising results in predicting backdoor variables and PaS assignments on these problem domain benchmarks. MVC and MIS instances are generated using the Barabási–Albert random graph model [2], while CA instances are generated based on arbitrary relations described in [39]. This paper focuses on the formulation that includes only binary variables as it aligns with the common problem domains in our three tasks. Still, our multi-task learning approach works on the general MILP problem domain.

Table 1. Instance sizes for the benchmarks: Combinatorial Auctions (CA), Maximum Independent Set (MIS), and Minimum Vertex Cover (MVC) across the tasks BACKDOOR, PaS, and CONFIGURATION. The table distinguishes between small (S) and large (L) instances. For CA, the instance sizes are characterized by the number of items and bids. For MIS and MVC, the instance sizes are described by the average degree and the number of nodes.

Benchmarks	Description	BACKDOOR	PaS	CONFIGURATION
CA-S	(# items, # bids)	(175, 850)	(2000, 4000)	(2000, 4000)
CA-L		(200, 1000)	(3000, 6000)	(3000, 6000)
MIS-S	(avg degree, # nodes)	(4, 1250)	(5, 6000)	(4, 3000)
MIS-L		(4, 1500)	(5, 9000)	(5, 6000)
MVC-S	(avg degree, # nodes)	(5, 1500)	(5, 6000)	(4, 3000)
MVC-L		(5, 2000)	(5, 9000)	(5, 6000)

We utilize Distributional-MIPLIB [30] to generate 200 training small (S) instances and 100 test S instances for each task. We also generated another 100 test large (L) instances for each task to test the model's generalizability. We train and fine-tune the model on S instances and test on both S and L instances. Each task involves different problem sizes: BACKDOOR focuses on improving optimal solving time for smaller instances (with optimal solutions found in approximately hundreds of seconds), while PaS and CONFIGURATION address more challenging instances (solving optimally takes over an hour), aiming to achieve better primal solutions within a runtime cutoff. Table 1 provides detailed information on the generated instances for each problem domain and task.

Baselines and Approaches:
We compare our approach with baseline solvers: *Gurobi* [22] (default settings) for BACKDOOR and PaS, and *SCIP* [5] (default configuration) for CONFIGURATION. *Gurobi*, the state-of-the-art commercial solver, and *SCIP*, the best open-source solver, provide a diverse evaluation framework, demonstrating that our multi-task learning framework is not tied to a single solver.

For CONFIGURATION, we include an additional baseline, *SMAC*, which uses SMAC3 [42] to perform configuration space search in 5 rounds per test instance. We do not choose *Gurobi* for CONFIGURATION because we cannot find a set of working configurations from previous literature or our experiments. We do not compare with other learning-based methods [26], as their approach requires an extensive database of pre-explored configurations to select similar instances, whereas our task generates new configurations specific to each instance.

The model trained specifically on a single task is denoted as *Single-task*. We have three single-task models, each focusing on a specific task. For simplicity, we use consistent notation to indicate which task each model is associated with. On the other hand, the model trained using our multi-task learning framework is referred to as *Multi-task* (*Multi-task-BAPAS*, *Multi-task-BACONFIG*, and *Multi-task-PASCONFIG*), where the suffix indicates the two tasks used in multi-task training). For *Single-task*, we train on 200 instances for each task. For *Multi-task*, we first train on 200 instances from each task with multiple random task-specific layers fixed; then, we fine-tune the task-specific layer from scratch on the same 200 instances used in the first step. For new tasks, we use the same 200 instances for both *Single-task* training and fine-tuning new task-specific layers.

Evaluation Metrics:
We evaluate performance on 100 test instances using the following metrics:

1. **Solve Time:** The time in seconds required for the solver to find the optimal solution to the MILP. This metric evaluates the efficiency of solving optimization problems.
2. **Primal Gap (PG):** [4] Defined as the normalized difference between the primal bound v and the best-known objective value v^*:

$$\text{PG} = \frac{|v - v^*|}{\max(|v^*|, \epsilon)},$$

where $\epsilon = 10^{-8}$ avoids division by zero. This metric applies when v exists and $vv^* \geq 0$.
3. **Primal Integral (PI):** [1] The integral of the primal gap over runtime $[0, t]$. PI reflects both the quality and speed of finding solutions.

We evaluate solve time for the BACKDOOR, while for PAS and CONFIGURATION, we measure the primal gap and primal integral within a runtime cutoff. We do not collect primal gap or primal integral for the BACKDOOR, as doing so would interfere with the solving process, particularly for instances that can be solved in hundreds of seconds.

Hyperparameters:
Experiments are conducted on 2.4 GHz Xeon-2640v3 CPUs with 64 GB memory. Training is performed on an NVIDIA V100 GPU with 112 GB memory. We use Gurobi 10.0.2 and SCIP 9.0.0 as solvers. For training and fine-tuning, the Adam optimizer [36] is used with a learning rate of 1×10^{-4}. The batch size is set to 32,

Table 2. Same-Task Performance: Solve time (seconds) for BACKDOOR and Primal Integral for PAS averaged over 100 test instances for each benchmark. We compare the performance of *Gurobi*, *Single-task* (trained on BACKDOOR or PAS), and *Multi-task-BAPAS*. Results include the mean, standard deviation, and the number of instances each approach wins. The best-performing entries are highlighted in bold for clarity.

Benchmarks	Approaches	BACKDOOR Solve Time			PAS Primal Integral		
		Mean	Std Dev	Wins	Mean	Std Dev	Wins
CA-S	Gurobi	252.66	122.06	5	14.27	5.28	8
	Single-task	228.46	**114.83**	44	9.43	5.11	45
	Multi-task-BAPAS	**219.23**	115.13	**51**	**9.36**	**4.99**	**47**
MIS-S	Gurobi	150.98	204.42	29	16.33	4.01	0
	Single-task	141.76	204.02	**39**	**3.00**	1.18	**59**
	Multi-task-BAPAS	**133.18**	**190.68**	32	3.35	**1.16**	41
MVC-S	Gurobi	81.06	**113.24**	7	1.27	0.60	0
	Single-task	75.83	122.63	**53**	**0.52**	**0.09**	43
	Multi-task-BAPAS	**74.65**	113.81	40	**0.52**	0.12	**57**
CA-L	Gurobi	831.43	464.74	20	19.53	6.70	7
	Single-task	784.77	511.40	33	14.14	5.86	24
	Multi-task-BAPAS	**740.08**	**409.45**	**47**	**9.88**	**4.88**	**68**
MIS-L	Gurobi	489.25	840.36	22	30.73	8.34	0
	Single-task	433.18	734.15	32	50.83	8.27	0
	Multi-task-BAPAS	**401.23**	**666.22**	**46**	**2.23**	**0.74**	**100**
MVC-L	Gurobi	461.59	1062.18	30	3.44	1.88	0
	Single-task	413.63	735.18	29	1.19	0.52	30
	Multi-task-BAPAS	**393.69**	**711.34**	**41**	**0.98**	**0.40**	**70**

and training and fine-tuning are run for 1000 epochs (training converges in less than 12 h, and fine-tuning converges in less than 1 h). The number of parameters in the shared layers and task-specific layers is roughly a ratio of 10:1. During testing, the model with the best validation loss is selected. Runtime cutoffs are set to 1000 s for PAS and 900 s for CONFIGURATION instances. For the backdoor task, we follow [7] to select backdoor size K. For the PAS task, we use k_0, k_1, and Δ as described in [29]. For the CONFIGURATION task, we select 15 parameters related to the solving process (e.g., branching, cut selection, LP relaxation) based on SCIP 9.0.0, following [26].

Additional details on instance generation, data collection, configuration parameters, and hyperparameter settings are provided in the Appendix.

Our experiments aim to answer the following research questions:

- Does our newly proposed approach for learning to generate configurations for MILP solvers outperform default solvers and other baseline methods?

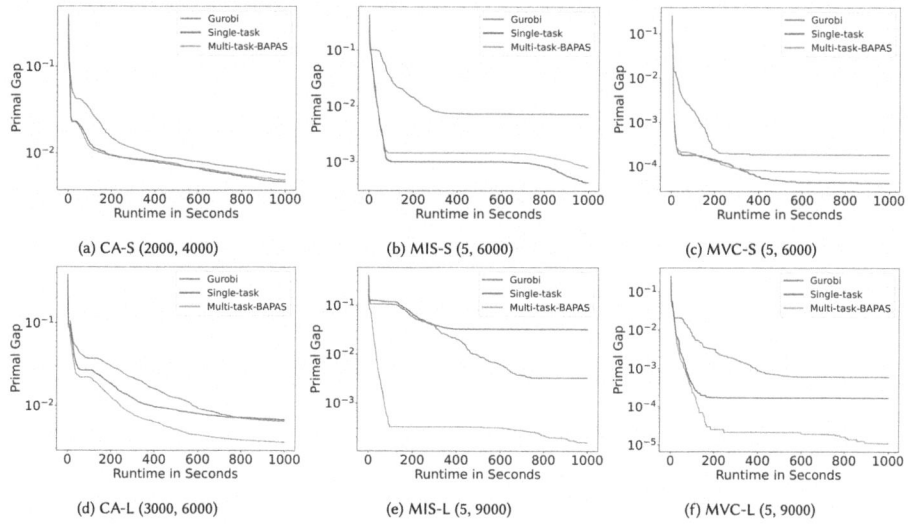

Fig. 2. Same-Task Performance: The Primal Gap (the lower, the better) as a runtime function averaged over 100 test instances on PAS and each benchmark. We compare the performance of *Gurobi* (green line), *Single-task* on PAS (blue line), and *Multi-task-BAPAS* (orange line). (Color figure online)

- How does *Multi-task* compare to *Single-task* and default solvers in performance on the same task? Additionally, how well does it generalize to larger instance sizes?
- Can the model, with its shared MILP embeddings, effectively fine-tune task-specific layers to handle new tasks, and how does its performance compare to both *Single-task* and default solvers?
- Does *Multi-task* consistently outperform *Single-task* in terms of generalization, as demonstrated through cross-task evaluations?

5.2 Results

Same-Task Performance:
To evaluate the same-task performance of a multi-task model, we test it on the tasks on which it is trained. We evaluate *Gurobi*, *Single-task*, and *Multi-task-BAPAS* on BACKDOOR and PAS using both S and L instances from CA, MIS and MVC. The detailed results are shown in Table 2 and Fig. 2. For 100 S instances, both *Single-task* and *Multi-task-BAPAS* outperform *Gurobi* consistently, demonstrating competitive performance. For BACKDOOR, *Multi-task-BAPAS* achieves a slightly better average runtime than *Single-task*, with improvements of 4.04%, 6.05%, and 1.56% in CA-S, MIS-S, and MVC-S, respectively. However, *Single-task* outperforms *Multi-task-BAPAS* in the number of instances, with 39 vs. 32 wins in MIS-S and 53 vs. 40 in MVC-S. For PAS, while *Single-task* consistently

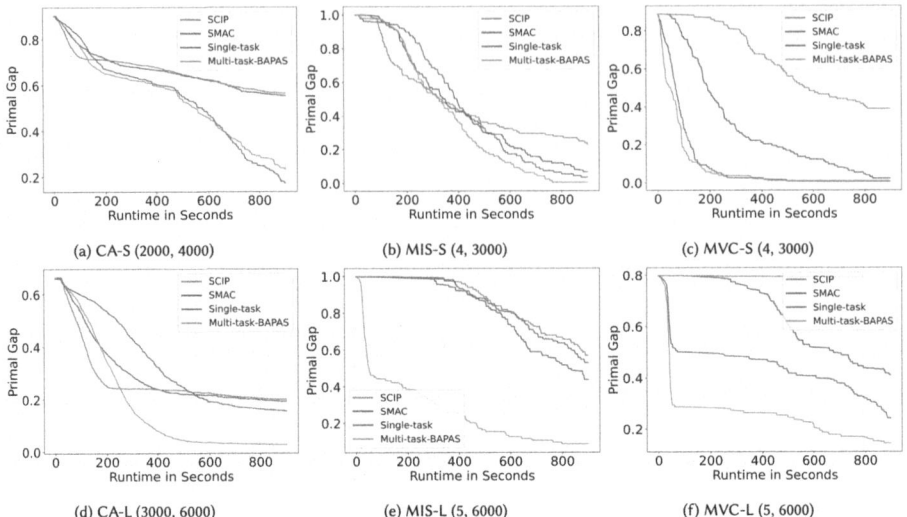

Fig. 3. New-task Performance: The Primal Gap (the lower, the better) as a runtime function averaged over 100 test instances on CONFIGURATION and each benchmark. We compare the performance of *SCIP* (green line), *SMAC* (red line), *Single-task* on CONFIGURATION (blue line), and *Multi-task-BAPAS* (orange line). (Color figure online)

achieves a better primal gap at the end of the time cutoff, *Multi-task-BAPAS* demonstrates slightly better performance in terms of the primal integral, securing 2 and 14 more wins in CA-S and MVC-S, respectively.

The results differ significantly when evaluating generalization performance on 100 L instances (directly tested without training). The multi-task model substantially outperforms both *Single-task* and *Gurobi*. For BACKDOOR, the multi-task model achieves the best average runtime across all benchmarks, showing an approximately 15% improvement over *Gurobi*, compared to a 9% improvement by *Single-task*. The multi-task model also secures more than 40 wins per benchmark. Additionally, a shift in the distribution of wins is observed, with *Gurobi* achieving a more significant number of wins and *Single-task* showing a reduced number of wins compared to the multi-task model. For PAS, *Multi-task-BAPAS* demonstrates a significantly better primal gap across all benchmarks at the end of the runtime cutoff. While the *Single-task* initially performs better than *Gurobi* on CA-L, it is eventually surpassed by *Gurobi* at the end of the runtime. Additionally, *Single-task* performs poorly on MIS-L, even underperforming compared to *Gurobi*. In contrast, *Multi-task-BAPAS*, trained on the same data, avoids this issue entirely. Regarding the primal integral, *Multi-task-BAPAS* achieves improvements of 49.42%, 92.75%, and 71.51% over *Gurobi* and secures wins in most instances, further highlighting its superior performance.

This result aligns with the understanding that *Single-task* are highly specialized for the specific distribution of the instances they are trained on, leading to optimal performance within the same data distribution. However, our *Multi-task-*

Table 3. Cross-task Evaluations: Solve time (seconds) for BACKDOOR and Primal Integral for PAS and CONFIGURATION averaged over 100 test instances for each benchmark. We compare the performance of *Gurobi*, *Single-task*, and correspond *Multi-task*. Results include the mean and the number of instances each approach wins. The number of Wins does not add up to 100 for CONFIGURATION because *SMAC* wins in some instances. The reforming entries are highlighted in bold for clarity.

Tasks	Approaches	CA-L		MIS-L		MVC-L	
		Mean	Wins	Mean	Wins	Mean	Wins
BACKDOOR Solve Time	*Gurobi*	793.10	17	530.41	25	427.86	20
	Single-task	749.08	28	470.73	31	390.04	24
	Multi-task-PASCONFIG	**696.39**	**55**	**446.61**	**44**	**354.35**	**56**
PAS Primal Integral	*Gurobi*	19.97	8	37.95	0	3.84	0
	Single-task	14.11	32	56.11	0	1.49	10
	Multi-task-BACONFIG	**12.12**	**60**	**3.13**	**100**	**1.04**	**90**
CONFIGURATION Primal Integral	*SCIP*	247.97	6	782.65	1	714.79	0
	Single-task	306.77	4	768.61	4	398.38	21
	Multi-task-BAPAS	**162.21**	**77**	**234.45**	**90**	**231.92**	**71**

BAPAS demonstrate competitive performance across most benchmarks, with only slight underperformance in some instances. When it comes to generalization, the benefits of multi-task training become apparent. By learning a generalized embedding that avoids overfitting to a single distribution, *Multi-task-BAPAS* achieves significantly better performance, leading to superior performance on unseen instances where *Single-task* struggles.

New-Task Performance:
We compare *SCIP*, *SMAC*, *Single-task*, and *Multi-task-BAPAS* on CONFIGURATION using S and L instances from CA, MIS and MVC. Results are presented in Fig. 3 and the table with the primal integral is available in the Appendix. First, in our new task of learning to generate configurations, *Single-task* outperforms in the primal gap over baselines on CA-S and MVC-S, and also achieving 18.27% and 66.97% improvements in the primal integral over *SMAC*. While it initially generates worse solutions than the baselines on MIS-S, it surpasses them by the end of the runtime cutoff. This demonstrates that our new task can effectively learn to generate configurations that enhance solving performance. Now focusing on *Multi-task-BAPAS*, we fine-tune only the task-specific layer using the same 200 configuration instances as the *Single-task*, while keeping the rest of the network architecture trained on BACKDOOR and PAS. Surprisingly, the multi-task model performs similarly to, or even better than, the *Single-task* on S instances. *Multi-task-BAPAS* achieves better solution progress within the first 600 s on CA-S and MVC-S and consistently outperforms the *Single-task* throughout the runtime on MIS-S. For primal integral, *Multi-task-BAPAS* reaches an average of 11.3% over *Single-task* across three S benchmarks.

Considering generalization performance on L instances, *Single-task* struggles, performing consistently worse than *SMAC* on CA-L and MIS-L, possibly indicating that this task is harder to generalize than others. In contrast, *Multi-task-BAPAS* performs much better than other approaches, achieving the best primal gap and primal integral, particularly excelling on CA-L and MIS-L, where the *Single-task* model underperforms in the primal gap. For the primal integral, *Multi-task-BAPAS* shows substantial improvements over *Single-task* with gains of 47.13%, 69.49%, and 41.78% on CA-L, MIS-L, and MVC-L, respectively. Additionally, the win rate of *Multi-task-BAPAS* increases significantly when transitioning from S to L instances, rising from 44 to 77 on CA, 23 to 90 on MIS, and 59 to 71 on MVC. These experiments highlight the effectiveness of our multi-task learning framework. By fine-tuning only the task-specific layer while leveraging embeddings from two prior tasks, *Multi-task-BAPAS* matches or outperforms *Single-task*, especially excelling in generalization to larger instances.

Cross-Task Evaluations:
To further validate the robustness of our approach, we performed cross-task evaluations on the learned MILP embeddings against *Single-task*. Specifically, we evaluated *Multi-task-PASCONFIG* on BACKDOOR, *Multi-task-BACONFIG* on PAS, and *Multi task-BAPAS* on CONFIGURATION, assessing their generalization performance on 100 large instances from each benchmark. Table 3 highlights consistent improvements of the multi-task models over the solver and *Single-task* across all benchmarks. These experiments demonstrate that the success of our multi-task training framework is not limited to a single pair of tasks but is broadly effective. Additional tables and figures for S and L instances and comparisons among different multi-task models on the same task are included in the Appendix.

6 Conclusion

This paper introduced a multi-task learning framework for MILP, unifying diverse tasks through shared representations and task-specific fine-tuning. Our approach demonstrated competitive performance with specialized models and significantly improved generalization across problem sizes and functions. Future work includes integrating tasks with dynamic features and extending the framework to different problem domains, advancing toward general foundation models for MILP optimization.

Acknowledgments. We would like to thank the anonymous reviewers for their constructive feedback to improve this paper. The National Science Foundation (NSF) supported the research under grant number 2112533: "NSF Artificial Intelligence Research Institute for Advances in Optimization (AI4OPT)".

References

1. Achterberg, T., Berthold, T., Hendel, G.: Rounding and propagation heuristics for mixed integer programming. In: Operations Research Proceedings 2011: Selected Papers of the International Conference on Operations Research (OR 2011), 30August–2 September 2011, Zurich, Switzerland, pp. 71–76. Springer (2012)
2. Albert, R., Barabási, A.L.: Statistical mechanics of complex networks. Rev. Mod. Phys. **74**(1), 47 (2002)
3. Alvarez, A.M., Louveaux, Q., Wehenkel, L.: A machine learning-based approximation of strong branching. INFORMS J. Comput. **29**(1), 185–195 (2017)
4. Berthold, T.: Primal heuristics for mixed integer programs. Ph.D. thesis, Zuse Institute Berlin (ZIB) (2006)
5. Bolusani, S., et al.: The SCIP optimization suite 9.0. arXiv preprint arXiv:2402.17702 (2024)
6. Brody, S., Alon, U., Yahav, E.: How attentive are graph attention networks? arXiv preprint arXiv:2105.14491 (2021)
7. Cai, J., Huang, T., Dilkina, B.: Learning backdoors for mixed integer programs with contrastive learning. arXiv preprint arXiv:2401.10467 (2024)
8. Cai, J., Kadioglu, S., Dilkina, B.: Balans: multi-armed bandits-based adaptive large neighborhood search for mixed-integer programming problem. arXiv preprint arXiv:2412.14382 (2024)
9. Cai, J., et al.: Getting away with more network pruning: from sparsity to geometry and linear regions. In: International Conference on Integration of Constraint Programming, Artificial Intelligence, and Operations Research, pp. 200–218. Springer (2023)
10. Chen, J., et al.: Minigpt-v2: large language model as a unified interface for vision-language multi-task learning. arXiv preprint arXiv:2310.09478 (2023)
11. Chmiela, A., Khalil, E., Gleixner, A., Lodi, A., Pokutta, S.: Learning to schedule heuristics in branch and bound. Adv. Neural. Inf. Process. Syst. **34**, 24235–24246 (2021)
12. Dilkina, B., Gomes, C.P.: Solving connected subgraph problems in wildlife conservation. In: Lodi, A., Milano, M., Toth, P. (eds.) CPAIOR 2010. LNCS, vol. 6140, pp. 102–116. Springer, Heidelberg (2010). https://doi.org/10.1007/978-3-642-13520-0_14
13. Dilkina, B., Gomes, C.P., Malitsky, Y., Sabharwal, A., Sellmann, M.: Backdoors to combinatorial optimization: feasibility and optimality. In: van Hoeve, W.-J., Hooker, J.N. (eds.) CPAIOR 2009. LNCS, vol. 5547, pp. 56–70. Springer, Heidelberg (2009). https://doi.org/10.1007/978-3-642-01929-6_6
14. Ding, J.Y., et al.: Accelerating primal solution findings for mixed integer programs based on solution prediction. In: Proceedings of the AAAI Conference on Artificial Intelligence, vol. 34, pp. 1452–1459 (2020)
15. Dinur, I., Safra, S.: On the hardness of approximating minimum vertex cover. Ann. Math. 439–485 (2005)
16. Drakulic, D., Michel, S., Andreoli, J.M.: Goal: a generalist combinatorial optimization agent learner. arXiv e-prints pp. arXiv-2406 (2024)
17. Ferber, A., Song, J., Dilkina, B., Yue, Y.: Learning pseudo-backdoors for mixed integer programs. In: International Conference on Integration of Constraint Programming, Artificial Intelligence, and Operations Research, pp. 91–102. Springer (2022)

18. Fischetti, M., Monaci, M.: Backdoor branching. In: International Conference on Integer Programming and Combinatorial Optimization, pp. 183–191. Springer (2011)
19. Floudas, C.A., Lin, X.: Mixed integer linear programming in process scheduling: modeling, algorithms, and applications. Ann. Oper. Res. **139**, 131–162 (2005)
20. Gasse, M., et al.: The machine learning for combinatorial optimization competition (ML4CO): results and insights. In: NeurIPS 2021 Competitions and Demonstrations Track, pp. 220–231. PMLR (2022)
21. Gasse, M., Chételat, D., Ferroni, N., Charlin, L., Lodi, A.: Exact combinatorial optimization with graph convolutional neural networks. In: Advances in Neural Information Processing Systems, vol. 32 (2019)
22. Gurobi Optimization, LLC: Gurobi Optimizer Reference Manual (2024). https://www.gurobi.com
23. Han, Q., et al.: A GNN-guided predict-and-search framework for mixed-integer linear programming. arXiv preprint arXiv:2302.05636 (2023)
24. He, H., Daume III, H., Eisner, J.M.: Learning to search in branch and bound algorithms. In: Advances in Neural Information Processing Systems, vol. 27 (2014)
25. Hendel, G.: Adaptive large neighborhood search for mixed integer programming. Math. Program. Comput. **14**, 185–221 (2022)
26. Hosny, A., Reda, S.: Automatic MILP solver configuration by learning problem similarities. Ann. Oper. Res. **339**(1), 909–936 (2024)
27. Huang, T., Dilkina, B.: Enhancing seismic resilience of water pipe networks. In: Proceedings of the 3rd ACM SIGCAS Conference on Computing and Sustainable Societies, pp. 44–52 (2020)
28. Huang, T., Ferber, A.M., Tian, Y., Dilkina, B., Steiner, B.: Searching large neighborhoods for integer linear programs with contrastive learning. In: International Conference on Machine Learning, pp. 13869–13890. PMLR (2023)
29. Huang, T., Ferber, A.M., Zharmagambetov, A., Tian, Y., Dilkina, B.: Contrastive predict-and-search for mixed integer linear programs. In: International Conference on Machine Learning. PMLR (2024)
30. Huang, W., Huang, T., Ferber, A.M., Dilkina, B.: Distributional MIPLIB: a multi-domain library for advancing ml-guided milp methods. arXiv preprint arXiv:2406.06954 (2024)
31. Kadioglu, S., Malitsky, Y., Sellmann, M., Tierney, K.: ISAC–instance-specific algorithm configuration. In: ECAI 2010, pp. 751–756. IOS Press (2010)
32. Khalil, E., Le Bodic, P., Song, L., Nemhauser, G., Dilkina, B.: Learning to branch in mixed integer programming. In: Proceedings of the AAAI Conference on Artificial Intelligence, vol. 30 (2016)
33. Khalil, E.B., Dilkina, B., Nemhauser, G.L., Ahmed, S., Shao, Y.: Learning to run heuristics in tree search. In: IJCAI, pp. 659–666 (2017)
34. Khalil, E.B., Morris, C., Lodi, A.: MIP-GNN: a data-driven framework for guiding combinatorial solvers. In: Proceedings of the AAAI Conference on Artificial Intelligence, vol. 36, pp. 10219–10227 (2022)
35. Khalil, E.B., Vaezipoor, P., Dilkina, B.: Finding backdoors to integer programs: a Monte Carlo tree search framework. In: Proceedings of the AAAI Conference on Artificial Intelligence, vol. 36, pp. 3786–3795 (2022)
36. Kingma, D.P., Ba, J.: Adam: a method for stochastic optimization. arXiv preprint arXiv:1412.6980 (2014)
37. Labassi, A.G., Chételat, D., Lodi, A.: Learning to compare nodes in branch and bound with graph neural networks. Adv. Neural. Inf. Process. Syst. **35**, 32000–32010 (2022)

38. Land, A.H., Doig, A.G.: An automatic method for solving discrete programming problems. Springer (2010)
39. Leyton-Brown, K., Pearson, M., Shoham, Y.: Towards a universal test suite for combinatorial auction algorithms. In: Proceedings of the 2nd ACM Conference on Electronic Commerce, pp. 66–76 (2000)
40. Li, S., Kulkarni, J., Menache, I., Wu, C., Li, B.: Towards foundation models for mixed integer linear programming. arXiv preprint arXiv:2410.08288 (2024)
41. Lin, J., Xu, M., Xiong, Z., Wang, H.: Cambranch: contrastive learning with augmented MILPs for branching. arXiv preprint arXiv:2402.03647 (2024)
42. Lindauer, M., et al.: Smac3: a versatile Bayesian optimization package for hyperparameter optimization. J. Mach. Learn. Res. **23**(54), 1–9 (2022)
43. Liu, S., Johns, E., Davison, A.J.: End-to-end multi-task learning with attention. In: Proceedings of the IEEE/CVF Conference on Computer Vision and Pattern Recognition, pp. 1871–1880 (2019)
44. Lodi, A., Zarpellon, G.: On learning and branching: a survey. TOP **25**, 207–236 (2017)
45. Nair, V., et al.: Solving mixed integer programs using neural networks. arXiv preprint arXiv:2012.13349 (2020)
46. Oord, A.V.D., Li, Y., Vinyals, O.: Representation learning with contrastive predictive coding. arXiv preprint arXiv:1807.03748 (2018)
47. Paulus, M.B., Zarpellon, G., Krause, A., Charlin, L., Maddison, C.: Learning to cut by looking ahead: cutting plane selection via imitation learning. In: International Conference on Machine Learning, pp. 17584–17600. PMLR (2022)
48. Pohl, I.: Heuristic search viewed as path finding in a graph. Artif. Intell. **1**(3–4), 193–204 (1970)
49. Scavuzzo, L., Aardal, K., Lodi, A., Yorke-Smith, N.: Machine learning augmented branch and bound for mixed integer linear programming. Math. Program. 1–44 (2024)
50. Song, J., Lanka, R., Yue, Y., Dilkina, B.: A general large neighborhood search framework for solving integer programs. In: Annual Conference on Neural Information Processing Systems (NeurIPS) (2020)
51. Song, J., Lanka, R., Zhao, A., Bhatnagar, A., Yue, Y., Ono, M.: Learning to search via retrospective imitation. arXiv preprint arXiv:1804.00846 (2018)
52. Tang, Y., Agrawal, S., Faenza, Y.: Reinforcement learning for integer programming: learning to cut. In: International Conference on Machine Learning, pp. 9367–9376. PMLR (2020)
53. Tarjan, R.E., Trojanowski, A.E.: Finding a maximum independent set. SIAM J. Comput. **6**, 537–546 (1977)
54. Tong, J., Cai, J., Serra, T.: Optimization over trained neural networks: taking a relaxing walk. In: International Conference on the Integration of Constraint Programming, Artificial Intelligence, and Operations Research, pp. 221–233. Springer (2024)
55. Valentin, R., Ferrari, C., Scheurer, J., Amrollahi, A., Wendler, C., Paulus, M.B.: Instance-wise algorithm configuration with graph neural networks. arXiv preprint arXiv:2202.04910 (2022)
56. Williams, R., Gomes, C.P., Selman, B.: Backdoors to typical case complexity. In: IJCAI, vol. 3, pp. 1173–1178 (2003)
57. Xu, L., Hutter, F., Hoos, H.H., Leyton-Brown, K.: Hydra-MIP: automated algorithm configuration and selection for mixed integer programming. In: RCRA Workshop on Experimental Evaluation of Algorithms for Solving Problems with Combinatorial Explosion at the International Joint Conference on Artificial Intelligence (IJCAI), pp. 16–30 (2011)

Breaking the Symmetries of Indistinguishable Objects

Özgür Akgün[1], Mun See Chang[1]($^{\boxtimes}$), Ian P. Gent[1], and Christopher Jefferson[2]

[1] School of Computer Science, University of St Andrews, St Andrews, UK
{ozgur.akgun,msc2,ian.gent}@st-andrews.ac.uk
[2] School of Science and Engineering, University of Dundee, Dundee, UK
cjefferson001@dundee.ac.uk

Abstract. Indistinguishable objects often occur when modelling problems in constraint programming, as well as in other related paradigms. They occur when objects can be viewed as being drawn from a set of unlabelled objects, and the only operation allowed on them is equality testing. For example, the golfers in the social golfer problem are indistinguishable. If we do label the golfers, then any relabelling of the golfers in one solution gives another valid solution. In this paper, we show how we can break the symmetries resulting from indistinguishable objects. We show how these symmetries induce symmetries of types built from indistinguishable objects, for example in a matrix indexed by indistinguishable objects. We then show how the resulting symmetries can be broken correctly and completely. As the method can be prohibitively expensive, we also study methods for breaking the symmetry only partially. In ESSENCE, a high-level modelling language, indistinguishable objects are encapsulated in 'unnamed types'. We provide an implementation to automatically break symmetries of unnamed types.

Keywords: Symmetries · Modelling · Constraint programming · Automated model transformations

1 Introduction

Symmetries have long been understood to be both widely occurring and a source of inefficiency for solving technologies. As a result, this has been an exceptionally well-studied topic in constraint programming [13], Boolean satisfiability [25], and mixed-integer programming [17]. A particularly important case of symmetries is where the problem has indistinguishable objects. These are objects which, when interchanged, give us essentially the same situation. For example, two machines of the same model are equivalent in a factory scheduling problem, and any valid schedule will give an equivalent schedule when two such machines are interchanged. Further complications are introduced when we have multiple sets of indistinguishable objects, and we are not allowed to interchange objects of the different sets.

When modelling problems with symmetries, due to the limited choices of representations, one tends to *introduce* symmetries that are not in the original problem. These symmetries must then be broken, e.g. by adding symmetry breaking constraints. High-level modelling languages such as ESSENCE allow one to specify the problems in a more abstract way, and symmetries introduced by modelling can then be handled automatically by CONJURE, an automatic model rewriting tool. In ESSENCE, 'unnamed types' were introduced to capture the notion of indistinguishable objects [5], to express sets of objects whose labels are not important. Compound types can be constructed in terms of these unnamed types, for example, we can have sets of tuples of indistinguishable objects. However, while unnamed types were present in the first version of ESSENCE, previously CONJURE ignored the symmetry of unnamed types and simply transformed them into integers. Handling these symmetries is significantly more difficult than other symmetries already managed by CONJURE. We show how these symmetries can be broken *automatically*, without requiring expertise in symmetry breaking.

To do so, we extend ESSENCE to support permutations of types. These permutations allow us to represent the symmetries of indistinguishable objects and provide a base for a general and extensible framework for breaking the symmetries introduced by these indistinguishable objects. By showing how we implement permutations, we show how other technologies that want to deal with the symmetries of indistinguishable objects can adopt a similar approach. As the types of ESSENCE variables can be arbitrary nested, and the list of objects we want to handle might be extended in the future, we give the semantics of ESSENCE types recursively, in terms of a much smaller set of mathematical objects. We use this semantics to define how symmetries of indistinguishable objects induce symmetries of objects constructed from them. We also show how a well-defined total ordering on a compound type can be built up in terms of the ordering of its constituent types.

These ingredients let us generalise the lex-leader symmetry breaking method to compound types. We will illustrate our technique in the context of CONJURE, but we will also discuss how other modelling languages can take the same approach to remove symmetries due to indistinguishable objects. Often we do not want to break all of the symmetries in a model since that encoding would require too many constraints and so will be detrimental to performance. In fact, checking if an assignment satisfies lex-leader symmetry breaking constraints is NP-hard in general [7]. For this reason, we also explore weaker, partial forms of symmetry breaking, offering a modelling choice between fast and complete symmetry breaking. We show, using well-known constraint models containing unnamed types, how our symmetry breaking encodings can be applied to them using this abstraction. As an example, we show that the commonly used 'double-lex' method [9] naturally arises from our methods.

In Sect. 2, we give a brief overview of symmetry breaking in constraint programming and the ESSENCE language, and then define the symmetries of unnamed types in Sect. 3. We then define the symmetries induced by unnamed types and see how we can break them, completely or incompletely, using a newly defined total order of the values of any type. In Sect. 4, we describe the method implemented in CONJURE and in Sect. 5 give some case studies.

Motivating Example: Social Golfers Problem. The problem asks for a schedule for p people playing golf over w weeks in g groups per week [14]. To attain maximum socialisation, no two different golfers play in the same group as each other in two different weeks. Notice that the problem only cares whether or not two golfers are the same person or different people. That is, the golfers are all *indistinguishable*. Similarly, the groups are also indistinguishable, and so are the weeks. For example, the constraints take no account of weeks being consecutive or otherwise: two weeks are either identical or distinct. This means that given any solution to the problem, we can obtain another solution through any permutation of the golfers. Similarly, we can also permute the groups and the weeks freely. Furthermore, we can apply any permutation of the golfers, groups and weeks concurrently to get another solution. With so many symmetries in hand, manually breaking them may require some modelling expertise, whether the aim is to break all symmetries completely or to do some form of partial symmetry breaking. The approach we take in this paper allows this to be done automatically. We will do this in the context of CONJURE, working on the *unnamed types* allowed by ESSENCE, in this case for golfers, weeks and groups. Our approach is general and could be applied in other modelling languages.

2 Background

A *constraint satisfaction problem* (CSP) $\mathcal{P}$ with n variables is a triple (V, D, C), where $V = \{V_1, V_2, \ldots, V_n\}$ is the set of *variables*, D consists of sets $\text{Dom}(V_i)$, called the *domain* of V_i, for each $1 \leqslant i \leqslant n$, and $C = \{C_1, C_2, \ldots, C_k\}$ is the set of *constraints*, where each C_i is a subset of the Cartesian product $\times_{1 \leqslant i \leqslant n} \text{Dom}(V_i)$. An *assignment* of the variables in V is an n-tuple $(a_1, a_2, \ldots, a_n)$, where each $a_i \in \text{Dom}(V_i)$. An assignment is a *solution* to $\mathcal{P}$ if it is in the intersection $\bigcap_{1 \leqslant i \leqslant k} C_i$. The *solution set* to $\mathcal{P}$ is the set of all solutions to $\mathcal{P}$.

A *permutation* of a set Ω is a bijection from Ω to itself. We typically denote permutations using the cycle notation. That is, a permutation $\sigma := (a_{11}, a_{12}, \ldots, a_{1k_1})(a_{21}, a_{22}, \ldots, a_{2k_2}) \ldots (a_{r1}, a_{r2}, \ldots, a_{1k_r})$ means that, for all i, we have $a_{ij} \mapsto a_{i(j+1)}$ for $j < k_i$ and $a_{ik_i} \mapsto a_{i1}$. The composition of two permutations σ_1 and σ_2 is denoted by $\sigma_1 \sigma_2$, the inverse of a permutation σ is denoted by σ^{-1}. For $\omega \in \Omega$ and a permutation σ over Ω, we denote the image of ω under σ by ω^σ. A *permutation group* G over a set Ω is a set of permutations over Ω that is closed under compositions and inverses. Such G necessarily contains the identity 1_G, the permutation that fixes all points in Ω. The set of all permutations over Ω is called the *symmetric group* of Ω, denoted by $\text{Sym}(\Omega)$.

A *group action* of a group G on a set Ω is a map $\phi : G \times \Omega \to \Omega$ such that $\phi(1_G, \omega) = \omega$ for all $\omega \in \Omega$ and $\phi(gh, \omega) = \phi(h, \phi(g, \omega))$. When such a group action exists, we say that G *acts on* Ω and G is a *symmetry group* of Ω. Further, we write ω^g for $\phi(g, \omega)$, which aligns with the notation of permutation application because $\text{Sym}(\Omega)$ naturally acts on Ω. Group actions formalise what we intuitively understand as symmetries: the identity permutation should fix the object in question, and applying two permutations consecutively should be the same as applying the composition of the two permutations.

When we have a symmetry group G acting on the domains of the decision variables V, we have an equivalence relation on the domain set D (and hence the set of all solutions), where two assignments a and a' are equivalent if there exists a permutation $g \in G$ such that $a^g = a'$. A *symmetry breaking constraint* is a constraint which, when added, removes or reduces symmetric values from consideration. A *sound* (resp. *complete*) symmetry breaking is one where at least one (resp. exactly one) solution from each equivalence class is preserved. Many symmetry breaking strategies use some ordering. For a total ordering $\leqslant$ on set A, the *lexicographical ordering* $\leqslant_{lex}$ *over* $\leqslant$ is the total ordering on the tuples/matrices over A such that $(t_1, t_2, \ldots, t_k) \leqslant_{lex} (t'_1, t'_2, \ldots, t'_k)$ if and only if either $t_i = t'_i$ for all i, or there is an i such that $t_i < t'_i$ and $t_j = t'_j$ for all $j < i$.

Symmetry breaking for constraint programming is very well studied (see [13] for an overview). Two closely related concepts are interchangeability of values [10] and intensional permutability of variables [23], but they only concern value and variable symmetries respectively. Our work differs from these as we take a type-directed view of indistinguishable objects, which means that types with interchangeable values can be used to build higher-level types. Depending on how we build these higher-level types, the symmetries can be variable or value symmetries, or indeed both or neither, as we shall see. This suggests that a new, more general, method of reasoning with symmetries is needed.

Symmetries are also introduced by CONJURE when the abstract types in ESSENCE are refined to lower-level types in ESSENCE PRIME (see Sect. 2.1). Currently all but the symmetries arising from unnamed types are automatically broken by CONJURE (see [2] for more details). Symmetries in the constraint modelling language MINIZINC is also extensively studied (see, for example, [6,18]), but our work here differs in that we consider higher-level abstract types. Outwith constraint programming, the symmetries of indistinguishable objects are exploited in the SAT solver SYMCHAFF, which requires a symmetry description as input [24]. It is shown that such a description can be generated from annotating PDDL models, but we operate on a much higher level language. The field of lifted inference started by Poole [21] also uses the symmetries of indistinguishable objects to improve efficiency, but the focus is on probabilistic reasoning and model counting.

2.1 ESSENCE as a Modelling Language

ESSENCE and ESSENCE PRIME are both constraint specification languages. The domains of decision variables in an ESSENCE or ESSENCE PRIME problem specification are defined by adding attributes and/or bounds to built-in types. For example, in ESSENCE, we can have a variable of domain set (size 3) of matrix indexed by int(1..5) of bool. In this paper, matrices indexed by $[I_1, I_2, \ldots, I_k]$ refers to k-dimensional matrices where the values are accessed by values of $I_1 \times I_2 \times \ldots \times I_k$, so an entry of such a matrix m is $m[i_1, i_2, \ldots, i_k]$, where each i_j is in I_j.

Types in ESSENCE and ESSENCE PRIME are divided into two kinds: atomic and compound. The atomic types of ESSENCE PRIME are Booleans and integers, while ESSENCE further supports enumerated types and unnamed types. Compound types, such as matrices, are defined using atomic types and can be arbitrarily nested. ESSENCE PRIME only support the matrix compound domain, while ESSENCE also supports tuples, records and variants. ESSENCE further supports abstract decision variables of set, multiset, sequence, function, relation and partition. Non-abstract domains are also called *concrete* domains. For more details, see [2] or the documentation at https://conjure.readthedocs.io.

Remark 1. For a type T, we denote its set of all possible values by $\mathrm{Val}(T)$, which is defined in terms of matrices, multisets and tuples: the values of bool, int and enum are what one would expect; the values of a tuple, record, variant and sequence can be naturally defined as tuples; the values of a matrix are matrices, the values of a set, mset, partition can be naturally defined as (nested) multisets; for types τ_i, the values of a function $\tau_1 \to \tau_2$ are subsets of $\mathrm{Val}(\tau_1) \times \mathrm{Val}(\tau_2)$ such that there are no two elements with the same value in its first position; the values of a relation $(\tau_1 * \tau_2 * \ldots * \tau_k)$ are subsets of $\mathrm{Val}(\tau_1) \times \mathrm{Val}(\tau_2) \times \ldots \times \mathrm{Val}(\tau_k)$. Note that the representations presented here give abstract meaning to the types, carefully selected to simplify the theory, but may not adhere to the representation of the underlying implementations.

There are two advantages in defining $\mathrm{Val}(T)$ in terms of matrices, multisets and tuples. Firstly, this allows us to avoid large case splits over types, e.g. when defining the symmetries of compound objects built from unnamed types. Furthermore, this makes our method more general and applicable across other platforms. As long as we define the values of a type in a similar way, either for a new type in ESSENCE or a type in other systems, the method described in this paper still applies. Note also that we can also have all types to be defined as multisets and tuples only. This is because a matrix m indexed by $[\tau_1, \tau_2, \ldots, \tau_k]$ of τ can be represented in terms of multisets and tuples as $\{(i_1, i_2, \ldots, i_k, m[i_1, i_2, \ldots, i_k]) \mid i_j \in \tau_j \text{ for all } 1 \leq j \leq k\}$.

CONJURE transforms a problem specification (a *model*) in the ESSENCE language into a problem specification in the ESSENCE PRIME language, through a series of rewrites or transformations (see [2]). Concrete domains are represented directly in ESSENCE PRIME, possibly by separating into their components. These transformations are straightforward and do not introduce any symmetries.

The abstract types are removed in a series of rewrites called *refinements*. For each abstract domain, there is a choice as to how it can be translated into concrete domains. Such a choice is what we call a *representation* of the abstract domain. The constraints involving the abstract domains are rewritten according to the selected representation. In some cases, an abstract type is represented as a concrete type that satisfies certain constraints (e.g. sets as lists with all different elements). Such constraints are called *structural constraints*. We will not detail representations used in CONJURE here, but will describe the ones which we use in examples. We direct interested readers to [2, Table 5] for a summary of the representations in CONJURE. *Modelling symmetries* are introduced if we translate abstract decision variables to a variable with a bigger domain, which may result in the increase of solution number. In CONJURE, modelling symmetries without unnamed types are *always* broken completely. The complete symmetry breaking constraint for the refinement of each abstract type in ESSENCE is well studied. Please refer to [2] for details.

3 Unnamed Types and How to Break Them

To model indistinguishable objects, we use the concept from ESSENCE of 'unnamed types'. While ESSENCE provides unnamed types as a built-in type and we implement our techniques in CONJURE, our work applies generally. For any modelling situation to which unnamed types apply, the techniques we propose can be used whether or not the modelling language used provides unnamed types. The advantage of having unnamed types built in, as in ESSENCE, is simply that no additional work is necessary to recognise the existence of indistinguishable objects. In this section we show the value in modelling with unnamed types, discuss the symmetries inherent in unnamed types and how to break them.

3.1 Modelling with Unnamed Types

We shall briefly show some examples of how one would use unnamed types in modelling, to illustrate their usefulness for high-level modelling and the difficulties in breaking their symmetries. Note that we can model these problems using more abstract ESSENCE types, but we choose to avoid these abstract types in this paper in hope to better illustrate the potential use of unnamed types.

Recall the **social golfer problem** from Sect. 1. A model may have, as decision variable, `matrix indexed by [int(1..w), int(1..p)] of int(1..g)`, together with constraints for maximal socialisation and to make sure that the group sizes are as expected, which we omit here. As we have seen, the golfers, weeks and groups can be permuted while still giving us valid schedules. The labels for golfers, weeks and groups, encoded as integers here, do not matter – permutations of them, when done consistently, will give us equivalent solutions. So we can define `golfers`, `weeks` and `groups` as unnamed types of size p, w and g respectively, using the syntax `letting golfers be new type of size p`, and similarly for `weeks` and `groups`. Then we can take the decision variable to be `matrix indexed by [weeks, golfers] of groups`, and let CONJURE handle the symmetries of unnamed types automatically.

The **template design problem** [27] arises in a printing factory that is asked to print $c_1, c_2, \ldots, c_k$ copies of designs $d_1, d_2, \ldots, d_k$ respectively. Designs are printed on large sheets of paper and each sheet can hold at most s designs.

A *template* is defined by the designs to be printed on a sheet (at most s of them, can be repeated, order does not matter). Given a number n, we want to find n templates $t_1, t_2, \ldots, t_n$, and the number of copies for each of them to satisfy the printing order, while minimising the total number of printing. One might model this problem with two decision variables: (i) `matrix indexed by [int(1..n)] of int`, called M_1, to encode the number of copies needed for each template; and (ii) `matrix indexed by [int(1..n), int(1..k)] of int(1..s)`, called M_2, to encode the number of copies of each design in each template. As before, the labels of the templates, currently integers 1 to n, do not matter – permuting the labels gives equivalent solutions. Since the templates are used as indices in two decision variables, any symmetry handling of one must be consistent with the other. For example, enforcing that M_1 must be sorted and at the same time enforcing that rows of M_2 are sorted may give us a wrong result. An alternative would be to replace `int(1..n)` in the indices for both M_1 and M_2 with an unnamed type of size n. CONJURE shall then automatically and consistently break their symmetries.

The **set-theoretic Yang-Baxter problem** asks for a special class of solutions to the infamous Yang-Baxter equations, which gives insights to various subfields of algebra and combinatorics (see [3] for references). A special class of the set-theoretic solutions of the Yang-Baxter equation can be modelled as a mapping $\varphi : X \times X \to X$ which satisfies certain constraints (see [3] for details), for a set X. As is common in mathematics, the elements X are unlabelled and interchangeable. In this case, if X is realised as a concrete set of $\{x_1, x_2, \ldots, x_n\}$, then any permutation of the elements of X in the definition of φ gives another mapping that is essentially the same (or equivalent). As, again, is common in mathematics, we want to count the number of solutions up to equivalences, which means that we want to remove the symmetries due to the interchangeability of the elements of X. Such a map φ can therefore be modelled as a `matrix indexed by [T,T] of T`, where T is an unnamed type. In this problem, the same unnamed type is used both as indices (twice) and elements of a matrix, so swapping two values in T requires swapping two rows and two columns, and also all occurrences of the two values for all variables. So the symmetries of this matrix is neither a variable nor a value symmetry (see [13]), the two most well-studied families of symmetries in constraint programming, as we require the synchronisation of both. Expressing correct symmetry breaking constraints requires significant expertise in constraint modelling, which limits the ability of many constraint users to deal with the symmetries of their problems.

3.2 Symmetries of Unnamed Types

As we have seen above, when modelling, we often want to express that two items are equivalent or indistinguishable from each other. In ESSENCE, we model these using *unnamed types*, which are sets of known size with implicit symmetries: values of an unnamed type are unlabelled and hence interchangeable. The values of unnamed types are not ordered and the only operations allowed on unnamed types are equality and inequality. Unnamed types are atomic so they can be used to construct compound domains in many ways, including as members of a set, the domain or image of a function, or the indices of a matrix.

In order to define symmetries inhabited by variables constructed from unnamed types, we need to first define the set of possible values of unnamed types. This proves to be difficult since as soon as we enumerate its values, we have put a label on its elements and hence introduced symmetries. We therefore define unnamed types as an enumerated type that comes with a symmetry group:

Definition 1. *An* unnamed type T *of size* n *is an* ESSENCE *type with value set* $\mathrm{Val}(T) = \{1_T, 2_T, \ldots, n_T\}$, *together with a symmetry group* $\mathrm{Sym}(\mathrm{Val}(T))$ *consisting of all permutations of* $\mathrm{Val}(T)$.

For example, in the social golfer problem, we can represent three possible golfers as an unnamed type G of size 3. Then $\mathrm{Val}(G)$ is $\{1_G, 2_G, 3_G\}$. Note that a value of an unnamed type is unique to its type. For example, 1_T can only be a value of the unnamed type T, and not at the same time be a value of another unnamed type U. This means that the values of distinct unnamed types T and U are always disjoint. To simplify notations, we write $\mathrm{Sym}(T)$ for $\mathrm{Sym}(\mathrm{Val}(T))$.

As we are dealing with indistinguishable objects, the symmetry group in Definition 1 is a symmetric group. However, our method of breaking these symmetries completely never uses the properties of symmetric groups. So the method is generalisable to the case where we have an atomic type T with any permutation group G on its values as its symmetry group.

Induced Symmetries on Compound Types. When solving a problem with unnamed types, we only want to retain the solutions up to symmetries. In the social golfer example, a solution is equivalent to another assignment where we have swapped two of the golfers, and so we should only output one of them. However, a decision variable can be an arbitrarily-nested construction of various ESSENCE types. So next we define how the symmetries of unnamed types induce symmetries of the compound variables constructed from unnamed types.

Definition 2. *Let T be an unnamed type of size n and X be a compound type. Then $\mathrm{Sym}(T)$ is a symmetry group of $\mathrm{Val}(X)$, where the action is defined recursively by: for all $g \in \mathrm{Sym}(T)$, and $x \in \mathrm{Val}(X)$,*

1. *if x is a value of type T, then x^g is the image under the action on $\mathrm{Val}(T)$;*
2. *if x is atomic and x is not of type T, then $x^g = x$;*
3. *if x is a matrix indexed by $[I_1, I_2, \ldots, I_k]$ of E, then the image x^g is a matrix where its i-th element $x^g[i]$ is $(x[i^{(g^{-1})}])^g$, where $i^{(g^{-1})}$ denotes the preimage of i under g;*
4. *if x is a multiset $\{v_1, v_2, \ldots, v_k\}$, then $x^g = \{v_1^g, v_2^g, \ldots, v_k^g\}$;*
5. *if x is a tuple $(v_1, v_2, \ldots, v_k)$, then $x^g = (v_1^g, v_2^g, \ldots, v_k^g)$.*

One can check that this indeed gives a group action. While there may be other possible group actions, we chose the most natural one. Recall from Remark

1 that possible values of a non-atomic variable can be constructed from only matrices, multisets and tuples. So Definition 2 *does* in fact define the image of all possible types in ESSENCE, by deducing from Remark 1, and considering sets as a special case of multisets (with multiplicity one for each element). As noted earlier, as long as we define types in a similar way, we can obtain the action on any compound type of other modelling languages using Definition 2.

Example 1. Let T be an unnamed type of size 3, and let X be of type `function T → int(4..5)`. A possible value of X is $x = \{(1_T, 4), (2_T, 5), (3_T, 4)\}$, representing the function that maps 1_T and 3_T to 4, and 2_T to 5. Consider the permutation $g = (1_T, 2_T)$ swapping 1_T and 2_T and leaving 3_T fixed. Then the image of x under g is $x^g = \{(1_T, 4), (2_T, 5), (3_T, 4)\}^g = \{(1_T, 4)^g, (2_T, 5)^g, (3_T, 4)^g\} = \{(1_T^g, 4^g), (2_T^g, 5^g), (3_T^g, 4^g)\} = \{(1_T^g, 4), (2_T^g, 5), (3_T^g, 4)\} = \{(2_T, 4), (1_T, 5), (3_T, 4)\}$, representing a function that maps 2_T and 3_T to 4, and 1_T to 5, and where the subsequent expression rewriting uses Parts 4,5,2,1 of Definition 2 respectively.

One may find the presence of preimage in the matrix index of Definition 2 to be unintuitive, but it is needed so we have a group action. One can also check that this definition is consistent to if we represent matrices as sets of tuples.

Example 2. Consider a 1-dimensional matrix $m = [a, b, c]$ indexed by elements of unnamed types $1_T, 2_T, 3_T$. Let g be the permutation $(1_T, 2_T, 3_T)$ and h be $(1_T, 2_T)$. From Definition 2, we find that the image of m first under g and then under h is $(m^g)^h = [c, a, b]^h = [a, c, b]$. We get the same value if we take the image of m under the composition of g and h, as $gh = (2_T, 3_T)$ and $m^{gh} = [a, b, c]^{(2_T, 3_T)} = [a, c, b]$. However, if Definition 2 had defined $(m^g)[i]$ to be $(m[i^g])^g$ instead, $(m^g)^h = [b, c, a]^h = [c, b, a]$, which is not m^{gh}.

Symmetries of Multiple Unnamed Types. If a variable X is constructed from multiple unnamed types, say T and U, any combination of elements in $\mathrm{Sym}(T)$ and $\mathrm{Sym}(U)$ also permute $\mathrm{Val}(X)$. We say that the direct product $\mathrm{Sym}(T) \times \mathrm{Sym}(U)$ also acts on $\mathrm{Val}(X)$. In general, the *direct product* of groups $G_1, G_2, \ldots, G_k$ is the Cartesian product $G_1 \times G_2 \times \ldots \times G_k$ consisting of all k-tuple $(g_1, g_2, \ldots, g_k)$ where each $g_i \in G_i$. If each G_i acts on a set Ω_i and the Ω_i's are disjoint, the direct product $G_1 \times G_2 \times \ldots \times G_k$ acts on the disjoint union $\bigcup_{1 \leqslant i \leqslant k} \Omega_i$ by $\alpha^{(g_1, g_2, \ldots, g_k)} = \alpha^{g_i}$ if $\alpha \in \Omega_i$.

Definition 3. *Let $T_1, T_2, \ldots, T_m$ be distinct unnamed types. Then the direct product $D := \mathrm{Sym}(T_1) \times \mathrm{Sym}(T_2) \times \ldots \times \mathrm{Sym}(T_m)$ is a symmetry group of $\mathrm{Val}(X)$, where the action is defined by $v^{(g_1, g_2, \ldots, g_m)} = (\cdots ((v^{g_1})^{g_2}) \cdots)^{g_m}$, for each element $(g_1, g_2, \ldots, g_m)$ of D and $v \in \mathrm{Val}(X)$, and the application of each g_i is as defined in Definition 2.*

This gives a group action because each $\mathrm{Sym}(T_i)$ acts on $\mathrm{Val}(T_i)$ and distinct unnamed types are disjoint sets. Further, as the g_i's permute disjoint sets of points, they commute. That is, $g_i g_j = g_j g_i$ for all i, j. Then, since we have a group action, taking images under them is commutative since $(v^{g_i})^{g_j} = v^{(g_i g_j)} = v^{(g_j g_i)} = (v^{g_j})^{g_i}$ for all i, j. So the order in which we take the images when considering permutations of different unnamed types does not matter, hence the order of unnamed types in the direct product also does not matter.

Example 3. Let T and U be unnamed types of size 2 and 4 respectively, and let M be of type `matrix indexed by [T, int(1..3)] of U`. Then $D = \text{Sym}(T) \times \text{Sym}(U)$ acts on $\text{Val}(M)$. Consider $m = [[1_U, 2_U, 3_U], [2_U, 3_U, 4_U]] \in \text{Val}(M)$ where we write the elements of m in the order $1_T, 2_T$, and let $g = (1_T, 2_T)$ and $h = (1_U, 3_U)(2_U, 4_U)$. Here g swaps 1_T and 2_T, whereas h and swaps 1_U and 3_U and also 2_U and 4_U at the same time. Then $(g, h) \in D$ and by Definition 3, $m^{(g,h)} = ([[1_U, 2_U, 3_U], [2_U, 3_U, 4_U]]^g)^h$, which gives $[[2_U, 3_U, 4_U]^g, [1_U, 2_U, 3_U]^g]^h = [[2_U, 3_U, 4_U], [1_U, 2_U, 3_U]]^h$ as g permutes the indices and fixes values not from T. Now as h fixes the indices $1_T, 2_T$ and the indices $1, 2, 3$, this is just $[[2_U, 3_U, 4_U]^h, [1_U, 2_U, 3_U]^h] = [[(2_U)^h, (3_U)^h, (4_U)^h], [(1_U)^h, (2_U)^h, (3_U)^h]]$. Finally, h permutes values in $\text{Val}(U)$, so $m^g = [[4_U, 1_U, 2_U], [3_U, 4_U, 1_U]]$.

If X is our only decision variable, then we obtain an equivalence relation on the set of all solutions, where two solutions x and y in $\text{Val}(X)$ are equivalent if $x^g = y$ for some $g \in G$. If we have multiple decision variables $V_1, V_2, \ldots, V_d$, then we can reduce to the case where there is only one decision variable, which is the tuple $(V_1, V_2, \ldots, V_d)$. This is particularly important when we want to ensure the consistent application of permutations of unnamed types across multiple variables, such as in the template design problem from Sect. 3.1.

3.3 Breaking the Symmetries of Unnamed Types

A common and general way to break symmetries is to use the lex-leader constraints [7]. In general, for a group G acting on the domain $\text{Val}(X)$ of a variable X, the (value) lex-leader constraint $LL_\preceq(G, X)$ for X under G with respect to a total ordering $\preceq$ of $\text{Val}(X)$ is the constraint $\forall \sigma \in G. X \preceq X^\sigma$. This asserts that, in any G-induced equivalence class of $\text{Val}(X)$, only one value (the smallest one) can be assigned to X. Recall that we treat our problems as containing a single variable, which may be a tuple, so we need to only consider symmetry breaking for a single variable X. In this paper, we always base our symmetry breaking on a lex-leader constraint $LL_\preceq(G, X)$. In this situation, sound symmetry breaking constraints will be implied by $LL_\preceq(G, X)$, and complete symmetry breaking constraints will imply $LL_\preceq(G, X)$ as well.

To completely break the unnamed type symmetries of a variable X, we use $LL_\preceq(G, X)$, where G is the symmetry group acting on $\text{Val}(X)$ and $\preceq$ is a total ordering on $\text{Val}(X)$. So we can eliminate unnamed type symmetries in the following way, the proof of which follows from the correctness of lex-leader constraints eliminating all but one solution in each equivalence class.

Proposition 1. *Starting with an* ESSENCE *model M with unnamed types $T_1, T_2, \ldots, T_k$ of size $s^1, s^2, \ldots, s^k$ respectively, we can obtain an equivalent* model (in the sense that there is a bijection between the solution sets) without unnamed types in the following way. Letting $V_1, V_2, \ldots, V_n$ be the decision variables of M, first replace each unnamed type T_i by an enumerated type with values $1_{T_i}, 2_{T_i}, \ldots, s^i_{T_i}$. Then let X be a new decision variable representing the tuple $(V_1, V_2, \ldots, V_n)$. Finally, we add the lex-leader constraint $LL_\preceq(\text{Sym}(T_1) \times \text{Sym}(T_2) \times \ldots \times \text{Sym}(T_k), X)$, where $\preceq$ is a total ordering on $\text{Val}(X)$.

Therefore, we need a total ordering of $\text{Val}(T)$ for any type T that is not constructed from unnamed types. We shall define these orderings in the next subsection. This will then inform us on how we can refine constraints of the form $X \preceq Y$ when X and Y are of abstract types.

Before we move on, note that there are many papers which consider different methods of generating subsets of symmetry breaking constraints (see [13] for an overview). The most common technique is to replace $LL_\preceq(G, X)$ with $LL_\preceq(S, X)$ for some subset S of G, to obtain sound but incomplete symmetry breaking constraints. When G is a direct product $G_1 \times G_2$, we may use $LL_\preceq(G_1 \cup G_2, X)$ instead. When G is the symmetric group $\text{Sym}(\{x_1, x_2, \ldots, x_n\})$, we can take $LL_\preceq(S, X)$ where $S = \{(x_i, x_j) \mid 1 \leq i, j \leq n\}$ consists of all permutations in G that swap two elements. Alternatively, we can take S to be $\{(x_i, x_{i+1}) \mid 1 \leq i < n\}$, the set of all permutations in G that swaps any consecutive points. Examples of the constraints added can be found in the .trace files of our repository https://github.com/stacs-cp/CPAIOR2025-Symmetry.

Total Ordering for all Types. We shall describe a general approach to defining a total ordering $\preceq_T$ of $\text{Val}(T)$ for any given type T not constructed from unnamed types. To simplify notations, we drop the subscript T when doing so will not cause confusion. The actual ordering used does not matter for correctness, as long as it is a total order. We first define an order on multisets in terms of the ordering on its members' type. We then show how ordering on other types can be defined in terms of the multiset ordering. This is the ordering used in our implementation in CONJURE.

For the ordering on multiset, we used an ordering very similar to one in the literature [12,16]. Let M be a type consisting of multisets of elements of type S and $\preceq_S$ is an ordering of $\text{Val}(S)$. We say that $m_1 \preceq_M m_2$ if and only if one of the following is true: (i) $m_2 = \emptyset$; (ii) $m_1, m_2 \neq \emptyset$ and $\min(m_1) \prec_S \min(m_2)$; (iii) $m_1, m_2 \neq \emptyset$ and $\min(m_1) = \min(m_2)$ and $m_1 \setminus \{\min(m_1)\} \preceq_M m_2 \setminus \{\min(m_2)\}$. This ordering may be unintuitive, but it is chosen so that ordering on multisets is equivalent to lex-ordering of a natural representation (specifically the occurrence representation; see, for example, [16] for proof). Since multisets are abstract types, constraints $X \preceq_M Y$, when X and Y are multisets, will need to be refined, and we can do so using this occurrence representation. The ordering for all other types can be found in the following definition.

Definition 4. *Let T be an* ESSENCE *type not constructed from any unnamed types. We define a total ordering $\preceq_T$ for values $\text{Val}(T)$ of type T recursively by:*

1. *if* Val(T) *consists of integers, we take* $\preceq_T$ *to be* $\leqslant$ *on integers;*
2. *if* Val(T) *consists of Boolean, we use* `false` $\preceq_T$ `true`*;*
3. *if* Val(T) *consists of enumerated types, then* $x \preceq_T y$ *if* x *occurs before* y *in the definition of the enumerated type;*
4. *if* Val(T) *consists of matrices or tuples of an inner type* S*, then take* $\preceq_T$ *to be lexicographical order* $\preceq_{Slex}$ *over an order* $\preceq_S$ *for the inner type;*
5. *if* Val(T) *consists of multisets of type* S*, take* $\preceq_T$ *to be the* $\preceq_M$ *above.*

Note again that using Remark 1, this definition gives an ordering for *all* ESSENCE types. An ordering for all types in other modelling languages can be defined in a similar way, by defining compound types in terms of multisets, tuples and matrices, and defining a concrete total ordering on each atomic type.

Remark 2. We can therefore refine constraints of the form $X \preceq Y$ to: $X \leqslant Y$ for the appropriate atomic ordering $\leqslant$ if X and Y are atomic; $X \preceq_{lex} Y$ if Val(X) and Val(Y) are matrices or tuples, and to $[-\text{freq}(X,i)|i \in X] \preceq_{lex} [-\text{freq}(Y,i)|i \in Y]$, using an ordering of X, if they are (multi)sets, where freq(X,i) gives the number of occurrence of i in X. As the base cases in Definition 4 are ordering $\leqslant$ of atomic types, these $\preceq_{lex}$ will eventually be rewritten to $\leqslant_{lex}$.

4 Implementation in Conjure

We outline the implementation of the symmetry breaking method described in this paper within CONJURE. The complete implementation can be found in the CONJURE repository at https://github.com/conjure-cp/conjure.

Permutations. We introduce a new ESSENCE type `permutation`. We allow permutations of integers, enumerated types or unnamed types. Permutation is available as a domain constructor, using keywords `permutation of`, and takes either an integer, enumerated or unnamed type domain as argument. Permutation values and expressions are written in cycle notations. Each permutation has an attribute `NrMovedPoints`, which gives the number of points that are not fixed by the permutation.

Permutations can be naturally represented as bijective total functions, which in turn can be represented as 1-dimensional matrices, where the element at a certain index gives the image of the index. Each permutation is stored with its inverse. This is because it turns out we almost always need to use the inverse of a permutation during symmetry breaking, and storing both the permutation and its inverse was the most efficient option in practice. The operator `permInverse` gets the inverse of a permutation. It uses the fact that the inverse of the inverse of a permutation is the original permutation, so there is no need to calculate any further permutation applications when calling `permInverse` twice.

The operator `image` takes a permutation g on a type T and a value x of type T such that `image`(g,x) gives the image of x under g. The more general operator `transform`(g,X) represents the image of the induced action of g on the values of its second argument. If g is a permutation on a type T, and the type of X contains no reference to T, then $\mathtt{transform}(g, X)$ is rewritten to X.

Unnamed Types. Members of an unnamed type can only be used as operands of an equality expression with other values of the same unnamed type. Unnamed types domains, similar to enumerated types, are eventually converted to $\mathtt{int(1..s)}$ where s is the size of the unnamed type. During refinements, unnamed types are converted to *tagged integers*, which behaves like integers but remembers the name ($\mathtt{IntTag}$) of the unnamed type it comes from. The tags are important for correct permutation applications, and we are careful to only add valid symmetry breaking constraints on each type of tagged integer. Permutations on an unnamed type T are refined into permutations on integers tagged with T, and $\mathtt{image}$ and $\mathtt{transform}$ will only change integer values or variables with the appropriate tag. The refinement rules of CONJURE ensure that when tagged integers are refined, the tag is preserved. For example a set of integers tagged with T is refined into a matrix indexed by integers tagged with T of Booleans. As seen in Proposition 1, each declaration of an unnamed type T will be removed, and all other domains constructed using T are replaced with tagged (with T) integers.

Symmetry Breaking Constraints. As discussed in Sect. 3.3, we break the symmetries induced by unnamed types using lex-leader constraints. Our implementation allows sound but incomplete symmetry breaking by replacing the group in LL with a subset, which is determined by run-time flags. For a model with unnamed types $T_1, T_2, \ldots, T_k$ and decision variables $V_1, V_2, \ldots, V_n$, the first set of flags defines the subsets of the unnamed symmetries $\mathrm{Sym}(T_i)$ to be used in the lex-leader: for each $i \in \{1, 2, \ldots, k\}$, we take $S_i := \{(j_{T_i}, (j+1)_{T_i}) \mid 1 \leqslant j < |T_i|\}$ if with the $\mathtt{Consecutive}$ flag; $S_i := \{(t, u) \mid t, u \in T_i\}$ with the $\mathtt{AllPairs}$ flag; and $S_i := \mathrm{Sym}(T_i)$ with the $\mathtt{AllPermutations}$ flag. The second set of flags determines whether to take the product or the union of these S_i's: the constraint used is $\bigwedge_{1 \leqslant i \leqslant k} LL(S_i, (V_1, V_2, \ldots, V_n))$ if with the $\mathtt{Independently}$ flag; and is $LL(S_1 \times S_2 \times \ldots \times S_k, (V_1, V_2, \ldots, V_n))$ if with the $\mathtt{Altogether}$ flag. Each $LL(S, X)$ is expressed as the conjunction, over all possible permutations $g \in S$, of expressions of form $X \mathrel{.\mathtt{<=}} \mathtt{transform}(g, X)$. Here $\mathrel{.\mathtt{<=}}$ represents the total order $\preceq_T$ from Definition 4, where T can be any suitable type.

Permutation Application. Starting from $X \mathrel{.\mathtt{<=}} \mathtt{transform}(g, X)$ from above, if g is a list of permutations $[g_1, g_2, \ldots, g_r]$ representing an element of a direct product, we rewrite expressions of the form $X \mathrel{\mathtt{<=}} \mathtt{transform}([g_1, g_2, \ldots, g_r], X)$ to the conjunction of $X \mathrel{.\mathtt{<=}} x_i$ and $x_i = \mathtt{transform}(g_i, x_{i-1})$ for $1 \leqslant i \leqslant k$, where x_0 is X and the x_i's are new variables. The refinement rules for $\mathtt{transform}(g, x)$, when g is a permutation, follow from Definition 2. CONJURE's general design, which applies rewrite rules until all high-level types and operators are removed, can easily handle this new set of rules. E.g. for a matrix X, we rewrite each entry $\mathtt{transform}(g, X)[i]$ to $\mathtt{transform}(g, X[\mathtt{transform}(\mathtt{permInverse}(g), i)])$. These internal $\mathtt{transform}$ and $\mathtt{permInverse}$ are further refined, until all permutations have been removed.

Refining Ordering Constraints. Each constraint of the form X.<= Y is refined to symOrder(X).<= symOrder(Y). Here symOrder(X) signifies that we are to rewrite the expression using Remark 2. So X.<= Y will eventually be written to expressions of the form X' <=lex Y' or X' <= Y' for some X' and Y', where <= is the order of atomic types and <=lex the lexicographic ordering over <=. The lexicographic constraints will typically contain every variable, but can often be simplified, e.g. $[a, b, c, d] \leqslant_{lex} [a, d, b, d]$ can be simplified to $[b, c] \leqslant_{lex} [d, b]$. Rather than perform these simplifications while initially generating and refining the lexicographic ordering constraints, a set of general rules for simplifying lexicographic ordering constraints [30] is run after refinement is finished. Finally the model undergo further refinements until only types in ESSENCE PRIME remain.

Table 1. How unnamed types occur in some problems

Problem	Type
Lam's Problem [28]	matrix indexed by [T,T] of ?
Set-theoretic Yang-Baxter [3]	matrix indexed by [T,T] of T
Balanced Incomplete Block Design [22]	matrix indexed by [T_1, T_2] of ?
Social Golfers [14]	matrix indexed by [T_1, T_2] of T_3
Covering Array [26]	matrix indexed by [T_1, T_2] of T_3
Template Design [27]	matrix indexed by [T_1] of ? matrix indexed by [T_1, T_2] of ?
Rack Configuration [15]	function $T \to$?
Semigroups [8]	function $(T, T) \to$ T
Vellino's Problem [1]	function: $T_1 \to T_2$ function: $T_1 \to$ mset of T_3
Sports Tournament Scheduling [29]	relation of ($T_1 * T_2 *$ set of T_3)

5 Case Studies and Discussion

We give a few case studies to demonstrate our implementation of the method. Future work will include more in-depth experimentation. We consider the problems from Sect. 3.1, and three further problems with matrix types as decision variables, but with varying number of unnamed types occurring in various positions, as well as some examples where the decision variables are of different types. The types of the decision variables of these problems are summarised in Table 1, where the T and T_i's are all distinct unnamed types, and '?' denotes other types that are not constructed from any unnamed types. The models in ESSENCE, and the automatically generated ESSENCE PRIME models, for all combinations of flags, can be found at https://github.com/stacs-cp/CPAIOR2025-Symmetry.

The resulting models were manually inspected for correctness. In particular, the number of solutions for small instances of the set-theoretic Yang-Baxter equation and the semigroup problem are consistent with those in the literature [19,20].

The different methods of symmetry breaking provide an easy way of choosing between different trade-offs. `Altogether-AllPermutations` will break all symmetries, producing an exact list of symmetry-broken solutions, at the cost of a very large number of constraints. The fact that we need many constraints is not surprising, as the symmetries of unnamed types include several cases which have been proved theoretically difficult. Consider the solutions to a problem with a decision variable of type `matrix indexed by [T,T] of bool`, with no constraints. These solutions can be viewed as directed graphs, so completely breaking the unnamed type symmetries is equivalent to finding the canonical image of these graphs. Similarly, a `set of set (size 2) of T` can be viewed as the edges of an undirected graph. Checking if two solutions of these problems are equivalent is therefore as hard as the graph isomorphism problem. Further, a variable of type `matrix indexed by [T1,T2] of bool` has row and column symmetries, and efficient generation of complete symmetry breaking constraints is also at least as hard as the graph isomorphism problem [4].

If we consider a matrix indexed by two unnamed types, e.g. in the balanced incomplete block design problem, then `Independently-Consecutive` produces "double-lex", one of the most widely used symmetry breaking methods [9]. The constraints generated by `Independently-AllPairs` have also been used in practice [11], and can lead to faster solving. Deciding exactly what level of symmetry breaking is best for a particular class of problem is future work.

6 Conclusion and Future Work

In this paper, we show how the symmetries of indistinguishable objects can be broken completely together with an implementation in ESSENCE. We do so by introducing the new **permutation** type, as well as a total ordering for all possible types in ESSENCE, which allows us to express and automatically generate symmetry breaking constraints for unnamed types. We have also seen how we can soundly but incompletely break unnamed type symmetries by controlling the permutations used in the lex-leader constraints. Our abstract treatments of types and unnamed type symmetries make our method generalisable to any other solving paradigms with indistinguishable objects. This paper also serves as a theoretical background for further work in this area.

Much further work awaits. The symmetry breaking method here can be prohibitively expensive in some cases. We therefore will investigate how some relaxations of the ordering constraints can be used to give faster symmetry breaking method. Furthermore, the static ordering described in Sect. 3.3 can also be a source of inefficiency. This is because, when rewriting multisets to their occurrence representations in Remark 2, the elements in $[-\text{freq}(X,i)|i \in X]$ must be sorted according to the total ordering of the inner types. This can be particularly difficult when we have deeply nested types. In general, it is not possible to produce a single global ordering which can be refined to a simple and efficient set of constraints in all possible representations. We shall therefore investigate the use of representation-specific total orderings.

Note that the symmetry breaking techniques in this paper will work even when the symmetry groups on unnamed types are not necessarily symmetric groups. We shall explore how the new permutation type can be used by an expert user for more control on symmetry breaking, in allowing arbitrary permutation groups, and see how we can automatically handle the symmetries for commonly occurring permutation groups, such as the chessboard symmetry. Generally, we intend to perform a more extensive analysis of the symmetry breaking constraints produced from our methods.

Acknowledgments. This work was supported by the Engineering and Physical Sciences Research Council EP/Y000609/1.

Disclosure of Interests. The authors have no competing interests to declare that are relevant to the content of this article.

References

1. Akgün, Ö.: CSPLib problem 116: Vellino's problem. http://www.csplib.org/Problems/prob116
2. Akgün, Ö., Frisch, A.M., Gent, I.P., Jefferson, C., Miguel, I., Nightingale, P.: Conjure: automatic generation of constraint models from problem specifications. Artif. Intell. **310** (2022)
3. Akgün, Ö., Mereb, M., Vendramin, L.: Enumeration of set-theoretic solutions to the Yang-Baxter equation. Math. Comput. **91**(335), 1469–1481 (2022)
4. Anders, M., Brenner, S., Rattan, G.: The complexity of symmetry breaking beyond lex-leader. In: Principles and Practice of Constraint Programming (2024)
5. Bakewell, A., Frisch, A.M., Miguel, I.: Towards automatic modelling of constraint satisfaction problems: a system based on compositional refinement. In: International Workshop on Modelling and Reformulating CSPs (2003)
6. Baxter, N., Chu, G., Stuckey, P.J.: Symmetry declarations for minizinc. In: Proceedings of the Australasian Computer Science Week Multiconference, pp. 1–10 (2016)
7. Crawford, J., Ginsberg, M., Luks, E., Roy, A.: Symmetry-breaking predicates for search problems. In: Proceedings of the Fifth International Conference on Principles of Knowledge Representation and Reasoning, vol. 96, pp. 148–159 (1996)
8. Distler, A., Jefferson, C., Kelsey, T., Kotthoff, L.: The semigroups of order 10. In: Milano, M. (ed.) Principles and Practice of Constraint Programming, pp. 883–899. Springer, Heidelberg (2012)
9. Flener, P., et al.: Breaking row and column symmetries in matrix models. In: Principles and Practice of Constraint Programming, pp. 462–477. Springer (2002)
10. Freuder, E.C.: Eliminating interchangeable values in constraint satisfaction problems. In: Proceedings of the Ninth National Conference on Artificial Intelligence, pp. 227–233 (1991)
11. Frisch, A.M., Hnich, B., Kiziltan, Z., Miguel, I., Walsh, T.: Global constraints for lexicographic orderings. In: Principles and Practice of Constraint Programming, pp. 93–108. Springer (2002)

12. Frisch, A.M., Miguel, I., Kiziltan, Z., Hnich, B., Walsh, T.: Multiset ordering constraints. In: International Joint Conferences on Artificial Intelligence, vol. 3, pp. 221–226 (2003)
13. Gent, I.P., Petrie, K.E., Puget, J.F.: Symmetry in constraint programming. In: Rossi, F., van Beek, P., Walsh, T. (eds.) Handbook of Constraint Programming, Foundations of Artificial Intelligence, Chap. 10, vol. 2, pp. 329–376. Elsevier (2006). https://doi.org/10.1016/S1574-6526(06)80014-3
14. Harvey, W.: CSPLib problem 010: Social golfers problem. http://www.csplib.org/Problems/prob010
15. Kiziltan, Z., Hnich, B.: CSPLib problem 031: Rack configuration problem. http://www.csplib.org/Problems/prob031
16. Kiziltan, Z., Walsh, T.: Constraint programming with multisets. In: International Workshop on Symmetry in Constraint Satisfaction Problems, pp. 9–20 (2002)
17. Margot, F.: Symmetry in integer linear programming. In: 50 Years of Integer Programming 1958-2008: From the Early Years to the State-of-the-Art, pp. 647–686 (2009)
18. Mears, C., Niven, T., Jackson, M., Wallace, M.: Proving symmetries by model transformation. In: Principles and Practice of Constraint Programming, pp. 591–605. Springer (2011)
19. OEIS Foundation Inc.: The number of non-degenerate involutive set-theoretic solutions of the Yang-Baxter equation of order n up to isomorphism, Entry A290887 in The On-Line Encyclopedia of Integer Sequences (2025). http://oeis.org/A290887
20. OEIS Foundation Inc.: Number of nonisomorphic semigroups of order n, Entry A027851 in The On-Line Encyclopedia of Integer Sequences (2025). http://oeis.org/A027851
21. Poole, D.: First-order probabilistic inference. In: International Joint Conference on Artificial Intelligence, pp. 985–991. Morgan Kaufmann Publishers Inc. (2003)
22. Prestwich, S.: CSPLib problem 028: Balanced incomplete block designs. http://www.csplib.org/Problems/prob028
23. Roy, P., Pachet, F.: Using symmetry of global constraints to speed up the resolution of constraint satisfaction problems. In: Workshop on Non Binary Constraints (1998)
24. Sabharwal, A.: SymChaff: exploiting symmetry in a structure-aware satisfiability solver. Constraints **14**, 478–505 (2009)
25. Sakallah, K.A.: Symmetry and satisfiability. In: Biere, A., Heule, M., van Maaren, H. (eds.) Handbook of Satisfiability: Second Edition. Frontiers in Artificial Intelligence and Applications, Chap. 13. IOS Press (2021)
26. Selensky, E.: CSPLib problem 045: The covering array problem. http://www.csplib.org/Problems/prob045
27. Smith, B.: CSPLib problem 002: template design. http://www.csplib.org/Problems/prob002
28. Walsh, T.: CSPLib problem 025: Lam's problem. http://www.csplib.org/Problems/prob025
29. Walsh, T.: CSPLib problem 026: sports tournament scheduling. http://www.csplib.org/Problems/prob026
30. Öhrman, H.: Breaking symmetries in matrix models. MSc thesis, Department of Information Technology, Uppsala University (2005)

Breaking Symmetries from a Set-Covering Perspective

Michael Codish[1] and Mikoláš Janota[2]

[1] Department of Computer Science, Ben-Gurion University of the Negev, Beer-Sheva, Israel
mcodish@bgu.ac.il
[2] Czech Technical University in Prague, Prague, Czechia
mikolas.janota@gmail.com

Abstract. We formalize symmetry breaking as a set-covering problem. For the case of breaking symmetries on graphs, a permutation covers a graph if applying it to the graph yields a smaller graph in a given order. Canonical graphs are those that cannot be made smaller by any permutation. A complete symmetry break is then a set of permutations that covers all non-canonical graphs. A complete symmetry break with a minimal number of permutations can be obtained by solving an optimal set-covering problem. The challenge is in the sizes of the corresponding set-covering problems and in how these can be tamed. The set-covering perspective on symmetry breaking opens up a range of new opportunities deriving from decades of studies on both precise and approximate techniques for this problem. Application of our approach leads to optimal LexLeader symmetry breaks for graphs of order $n \leq 10$ as well as to partial symmetry breaks which improve on the state-of-the-art.

1 Introduction

Graph search problems are about finding simple graphs with desired structural properties. Such problems arise in many real-world applications and are fundamental in graph theory. Solving graph search problems is typically hard due to the enormous search space and the large number of symmetries in graph representation. For graph search problems, any graph obtained by permuting the vertices of a solution (or a non-solution) is also a solution (or a non-solution), which is isomorphic, or "symmetric". When solving graph search problems, the presence of an enormous number of symmetries typically causes redundant search effort by revisiting symmetric objects. To optimize the search we aim to restrict it to focus on one "canonical" graph from each isomorphism class.

The focus on symmetry has facilitated the solution of many open instances of combinatorial search problems and graph search problems in particular. For example, the proof that the Ramsey numbers $R(3,3,4)$ and $R(4,5)$ are equal to 30 [5] and to 25 [14] respectively, the solution for the Sudoku minimum number of clues problem [26], and the enumeration of all non-word representable graphs of order twelve [20].

One common approach to eliminate symmetries is to add *symmetry breaking constraints* which are satisfied by at least one member of each isomorphism class [8,32,34]. A symmetry breaking constraint is called *complete* if it is satisfied by exactly one member of each isomorphism class and *partial* otherwise.

In many cases, symmetry breaking constraints, complete or partial, are expressed in terms of conjunctions of "lex-constraints". Each constraint corresponds to one symmetry, σ, which is a permutation on vertices, and restricts the search space to consider assignments that are lexicographically smaller than their permuted form obtained according to σ. Similarly, a set of permutations is identified with the conjunction of the lex-constraints for the elements of the set. A complete symmetry break is a set that satisfies exactly the set of canonical graphs. Of course, if one considers the set of all permutations, then the corresponding symmetry break is complete but too large to be of practical use.

Codish et al. [7] introduce a partial symmetry break, which is equivalent to considering the quadratic number of permutations that swap a pair of vertices. Rintanen et al. [29] enhance this approach for directed graphs. Itzhakov and Codish [19] observe that a complete symmetry break can be defined in terms of a small number of lex-constraints. They compute compact complete symmetry breaking constraints for graphs with 10 or less vertices. Similar approach is taken by Dančo et al. in the context of finite models [10]. It is known that breaking symmetry by adding constraints to eliminate symmetric solutions is intractable in general [1,9]. So, we do not expect to find a complete symmetry break of polynomial size that identifies canonical graphs which are lex-leaders.

In general, previous works focus on sets of permutations. Given a set Π of permutations, one typically asks questions of the form: "Which symmetries are broken by Π", "Are all symmetries broken by Π?", "Can we add a permutation to Π and break a symmetry not yet broken?", or "Can we remove a permutation from Π and still break the same symmetries?".

In this paper we take a different view. We focus on individual permutations. We say that a permutation covers a graph if its application on the graph yields a smaller graph. Each permutation "covers" a set of graphs. A complete symmetry break is a set of permutations, the elements of which cover all non-canonical graphs. One essential question is: "Which permutations are *essential* because they alone cover some graph?" We call such permutations "backbones".

The set-cover problem is a classical problem in computer science and one of the 21 NP-complete problems presented in the seminal paper by Karp [22] in 1972. Given a set of elements, $\mathcal{U}$, called the universe, and a collection S of subsets of $\mathcal{U}$ whose union equals $\mathcal{U}$, the set-cover problem is to identify a smallest sub-collection of S whose union equals $\mathcal{U}$.

By viewing symmetry breaking as a set-cover problem we make available a wide range of techniques which have been studied for many decades and applied to find set-covers, both exact and approximate. For example, [3,15,18,23,24,30]. The main challenge derives from the fact that in the set-cover perspective of symmetry breaking: (a) the universe consists of all non-canonical graphs which is a set of the order 2^{n^2}, and (b) the number of subsets considered corresponds to the non-identity permutations which is a set of order $n!$.

In this paper we focus on the search for optimal complete symmetry breaks for graphs. These are derived as solutions to minimal set-cover problems. To tame the size of the corresponding set-cover problems we focus on three classic optimizations applied to set-cover problems [30]. In our context we call these: (1) graph dominance, (2) permutation dominance, and (3) identification of permutation backbones.

Heule also addresses the problem of computing optimal symmetry breaks for graphs [16,17]. Heule seeks an answer in terms of the number of clauses in a CNF representation of the corresponding symmetry breaking constraint. For up to $n = 5$ vertices, Heule computes CNF size-optimal compact and complete symmetry breaks. We aim to find symmetry breaks that are optimal in their number of lex-constraints. We show that this can be done for all of the cases in which there exist complete and compact symmetry breaks based on lex-constraints. Namely, for graphs of orders $n \leq 10$.

2 Preliminaries and Notation

Throughout this paper we consider simple graphs, i.e. undirected graphs with no self-loops. The adjacency matrix of a graph G is an $n \times n$ Boolean matrix. The element at row i and column j is *true* if and only if (i, j) is an edge. We denote by $vec(G)$ the sequence of length $\binom{n}{2}$ which is the concatenation of the rows of the upper triangle of G. In abuse of notation, we let G denote a graph in any of its representations. The set of simple graphs on n vertices is denoted $\mathcal{G}_n$. An *unknown graph* of order-n is represented as an $n \times n$ adjacency matrix of Boolean variables which is symmetric and has the values *false* (denoted by 0) on the diagonal. We consider the following lexicographic ordering on graphs.

Definition 1 (ordering graphs). *Let G_1, G_2 be known or unknown graphs with n vertices and let $s_1 = vec(G_1)$ and $s_2 = vec(G_2)$ be the strings obtained by concatenating the rows of the upper triangular parts of their corresponding adjacency matrices. Then, $G_1 \leq G_2$ if and only if $s_1 \leq_{lex} s_2$.*

When G_1 and G_2 are unknown graphs, then the lexicographic ordering, $G_1 \leq G_2$, can be viewed as specifying a *lexicographic order constraint* over the variables in $vec(G_1)$ and $vec(G_2)$.

The group of permutations on $\{1 \ldots n\}$ is denoted S_n. We represent a permutation $\pi \in S_n$ as a sequence of length n where the ith element indicates the value of $\pi(i)$. For example: the permutation $[2, 3, 1] \in S_3$ maps as follows: $\{1 \mapsto 2, 2 \mapsto 3, 3 \mapsto 1\}$. Permutations act on graphs and on unknown graphs in the natural way. For a graph $G \in \mathcal{G}_n$ and also for an unknown graph G, viewing G as an adjacency matrix, given a permutation $\pi \in S_n$, then $\pi(G)$ is the adjacency matrix obtained by mapping each element at position (i, j) to position $(\pi(i), \pi(j))$ (for $1 \leq i, j \leq n$). Alternatively, $\pi(G)$ is the adjacency matrix obtained by permuting both rows and columns of G using π. Two graphs $G, H \in \mathcal{G}_n$ are *isomorphic* if there exists a permutation $\pi \in S_n$ such that $G = \pi(H)$.

Example 1. The following depicts an unknown, order-4, graph G, its permutation $\pi(G)$, for $\pi = [1, 2, 4, 3]$, and their vector representations. The lex-constraint $G \leq \pi(G)$ can be simplified as described by Frisch et al. [13] to: $\langle x_2, x_4 \rangle \leq_{lex} \langle x_3, x_5 \rangle$.

$$\mathbf{G} = \begin{bmatrix} 0 & x_1 & x_2 & x_3 \\ x_1 & 0 & x_4 & x_5 \\ x_2 & x_4 & 0 & x_6 \\ x_3 & x_5 & x_6 & 0 \end{bmatrix} \quad \pi(\mathbf{G}) = \begin{bmatrix} 0 & x_1 & x_3 & x_2 \\ x_1 & 0 & x_5 & x_4 \\ x_3 & x_5 & 0 & x_6 \\ x_2 & x_4 & x_6 & 0 \end{bmatrix} \quad \begin{array}{l} vec(G) = \langle x_1, x_2, x_3, x_4, x_5, x_6 \rangle \\ vec(\pi(G)) = \langle x_1, x_3, x_2, x_5, x_4, x_6 \rangle \end{array}$$

A *graph search problem* is a predicate, $\varphi(G)$, on an unknown graph G, which is invariant under the names of vertices. In other words, it is invariant under isomorphism. A solution to $\varphi(G)$ is a satisfying assignment for the variables of G. Graph search problems include existence problems, where the goal is to determine whether a simple graph with certain graph properties exists, enumeration problems, which are about finding all solutions (modulo graph isomorphism), and extremal problems, where we seek the smallest/largest solution with respect to some target (such as the number of edges or vertices in a solution). Solving graph search problems is typically hard due to the enormous search space and the large number of symmetries.

A symmetry break for graph search problems is a predicate, $\psi(G)$, on a graph G, which is satisfied by at least one graph in each isomorphism class of graphs. If ψ is satisfied by exactly one graph in each isomorphism class then we say that ψ is a complete symmetry break. Otherwise it is partial. A classic complete symmetry break for graphs is the LexLeader constraint [28] defined as follows:

Definition 2 (LexLeader). *Let G be an unknown order-n graph, Then,*

$$\text{LexLeader}(n) = \bigwedge \{\, G \leq \pi(G) \,|\, \pi \in S_n \,\} \quad (1)$$

The LexLeader constraint is impractical as it is composed of a super-exponential number of constraints, one for each permutation of the vertices. A symmetry break which is equivalent to the LexLeader constraint is called a LexLeader symmetry break. In [19], the authors present a methodology to compute LexLeader symmetry breaks which are much smaller in size.

Definition 3 (canonizing set of permutations). *Let $\Pi \subseteq S_n$ be a set of permutations such that*

$$\forall G.\ \text{LexLeader}(n)(G) \Leftrightarrow \bigwedge_{\pi \in \Pi} G \leq \pi(G)$$

In this case we say that Π is a canonizing set of permutations.

Fig. 1. Optimal set-cover for $n = 4$

In a nutshell, the algorithm presented in [19] computes a canonizing set Π initialized to the empty set by incrementally performing $\Pi \leftarrow \{\pi'\} \cup \Pi$ as long as there exists a permutation π' and a graph G such that

$$\bigwedge_{\pi \in \Pi} G \leq \pi(G) \ \wedge \ \pi'(G) < G$$

There are additional subtleties related to the removal of "redundant" permutations from the resulting set Π (see [19]). The canonizing sets obtained are surprisingly small. Their sizes are detailed on the right.

order	3	4	5	6	7	8	9	10
size	2	3	7	13	37	135	842	7853

3 The Set-Covering Perspective

The set-cover perspective on symmetry breaking is based on the notion that a permutation covers the set of graphs that it makes smaller.

Definition 4 (cover). *Let $\pi \in S_n$. Then, π covers an order-n graph G if $\pi(G) < G$. We denote the set $cover(\pi) = \{ G \,|\, \pi(G) < G \}$. For a set of permutations $\Pi \subseteq S_n$, we denote the set $cover(\Pi) = \{ G \,|\, \pi \in \Pi, \pi(G) < G \}$.*

Let Π be a canonizing set of permutations. It follows from Definition 3 that a graph G is canonical if and only if the constraints $G \leq \pi(G)$ hold for all $\pi \in \Pi$. The contra-positive states that G is non-canonical, if and only if there exists $\pi \in \Pi$ such that $\pi(G) < G$. Hence, we can view a canonizing set Π as a set such that for every non-canonical graph G, Π contains a permutation π such that π covers G. We can state this as follows:

Observation 1. *A set of permutations is canonizing if and only if it covers all non-canonical graphs.*

While Observation 1 might appear trivial, the nature of the statement leads to a new and alternative view on symmetry breaking in terms of set-covering. The goal for complete symmetry breaking is to cover all non-canonical graphs by a set of permutations. Of course, the set of all permutations cover all non-canonical graphs. However we can seek a smaller set of permutations and in particular, a set of the smallest size.

Definition 5 (optimal lex-constraint symmetry break). *An optimal lex-constraint symmetry break is a canonizing set of permutations of minimal cardinality. The optimal lex-constraint symmetry break problem is that of finding an optimal lex-constraint symmetry break.*

Fig. 2. Optimal Symmetry break for $n = 4$ as a set-cover problem

Example 2. Consider the case for graphs of order-4. Each graph G is represented as the integer value corresponding to the six digit binary sequence $vec(G)$ (viewed lsb first). The 11 canonical graphs $\{0, 12, 30, 32, 44, 48, 52, 56, 60, 62, 63\}$ are not covered by any of the permutations. We detail below the sets of graphs covered by three of the permutations. Figure 1 illustrates that these three permutations cover all of the non-canonical graphs of order-4. This is an optimal cover. The dots on the bottom represent an enumeration of the non-canonical graphs.

```
[1,2,4,3]:{2,3,8,9,10,11,14,15,18,19,26,27,34,35,40,41,42,43,46,47,50,51,58,59}
[1,3,2,4]:{1,5,9,13,16,17,19,20,21,23,24,25,27,28,29,31,33,37,41,45,49,53,57,61}
[2,1,3,4]:{2,3,4,5,6,7,14,15,18,19,22,23,34,35,36,37,38,39,46,47,50,51,54,55}
```

The matrix depicted as Fig. 2(a) depicts the set-cover problem corresponding to the cover sets for all of the permutations (excluding the row for the identity permutation). The rows correspond to the permutations, the columns correspond to the graphs, a cell is colored black if the corresponding permutation covers the corresponding graph, and white otherwise.

Three optimizations apply to simplify a set-cover problem. We present these, adapted to the context of symmetry breaking, from the presentation in [30].

Optimization 1 (permutation dominance): If permutation π_1 covers a subset of the graphs covered by permutation π_2, then we say that π_2 dominates π_1. In this case we discard the row corresponding to permutation π_1. In Example 2, four permutations can be excluded because of optimization 1. Figure 2(b) depicts the matrix after removing the corresponding four rows.

$cover([3, 2, 1, 4]) \subseteq cover([3, 2, 4, 1])$ $\qquad cover([3, 4, 1, 2]) \subseteq cover([3, 4, 2, 1])$
$cover([4, 2, 3, 1]) \subseteq cover([4, 2, 1, 3])$ $\qquad cover([4, 3, 2, 1]) \subseteq cover([4, 3, 1, 2])$

Optimization 2 (graph dominance): If graph G_1 is covered by a (non-empty) subset of the permutations that cover graph G_2, then we say that G_1 dominates G_2. In this case we discard the column corresponding to permutation G_2. In Fig. 2(b), 50 of the 55 graphs are dominated and discarded. Figure 2(c) depicts the matrix after removing the corresponding 50 columns and Fig. 2(d) depicts the matrix after applying Optimization 1 one more time.

Optimization 3 (backbone permutation): If a column contains a single one, this corresponds to the case when a graph is covered by exactly one permutation π. In this case we say that π is a backbone permutation and must be in any optimal set-cover. In the terminology of [30] the row is "essential". In this case the row corresponding to π and all columns corresponding to graphs covered by π are removed. In Fig. 2(d), column 2 contains a single one at row 2, so row 2 and columns 1, 2 are removed. Column 5 contains a single one at row 5, so row 5 and columns 4, 5 are removed. The remaining matrix contains one column with two rows. The column contains a single one corresponding to a third backbone permutation. Application of the optimizations renders an empty matrix.

The progression in Fig. 2(a–d) illustrates that the minimal set-cover problem given as Example 2 is solved by repeated application of these three optimizations identifying three backbone permutations which in this example cover all of the non-canonical graphs.

The General Case: In the general case, after repeated application of the above mentioned optimizations, we may still need to solve the remaining set-cover problem. In our implementation, we formulate the problem as an *optimization pseudo-boolean* (OPB) problem and solve it using RoundingSat [12].

Experiment 1: Table 1 summarizes the computation of optimal lex-constraint symmetry breaks by reduction to the set-cover problem. For graphs of orders 4, 5, and 6 we compute the matrix representations of the corresponding set-cover problems and apply the three optimizations described above. In all three cases the final matrix is empty and the optimal cover is found.

The first column details the order of the graph. The second column details the size of the initial matrix (rows × columns). The third column details the size of the cover sets (number of 1's in each row). The

Table 1. Optimal Symmetry Breaks via Set-Cover

order	initial	cover sizes	opt	time	2016
4	23 × 53	24 − 30	3	4.62	3
5	119 × 990	448 − 510	6	0.67	7
6	719 × 32612	15360 − 16380	13	103.02	13

forth column (opt) details the size of the minimal set-cover. The fifth column details the computation time (in seconds). The sixth column (2016) details the size of the canonizing set reported in [19].

For graphs of order-7, we need to construct a matrix of size 5039 × 2096108 and manipulate its rows and columns. This task is beyond the capability of our implementation. In the following sections we discuss ways to implement the three optimizations mentioned above before the explicit construction of the matrix representation of the problem.

4 A Concise Representation of $cover(\pi)$

Coming back to Example 2: How do we compute the set of graphs covered by a permutation π? By Definition 4, $cover(\pi)$ is the set of solutions of the constraint $\pi(G) < G$ where G is an unknown graph. We note that many of these sets tend to contain about half of the non-canonical graphs.

Example 3. Recall the setting of Example 1. The set of graphs covered by the permutation $\pi = [1, 2, 4, 3]$ is the set of solutions to the lex-constraint $\pi(G) < G$:

$$\langle x_1, x_3, x_2, x_5, x_4, x_6 \rangle <_{lex} \langle x_1, x_2, x_3, x_4, x_5, x_6 \rangle$$

which simplifies to $\langle x_3, x_5 \rangle <_{lex} \langle x_2, x_4 \rangle$ and has the following 24 solutions (in decimal representation): 2, 3, 8, 9, 10, 11, 14, 15, 18, 19, 26, 27, 34, 35, 40, 41, 42, 43, 46, 47, 50, 51, 58, 59.

We introduce the following to specify that a sequence is lexicographic smaller than another at position i. The right-hand side in Eq. (2) is a set of equality constraints which in our case involve Boolean variables and constants.

Definition 6 (lexicographic smaller at position i). *Let* $\bar{a} = \langle a_1, \ldots, a_m \rangle$ *and* $\bar{b} = \langle b_1, \ldots, b_m \rangle$. *Then,*

$$\bar{a} <^i_{lex} \bar{b} \Leftrightarrow \{(a_1 = b_1), \ldots, (a_{i-1} = b_{i-1}), (a_i = 0), (b_i = 1)\} \qquad (2)$$

For a permutation π, we say that π makes a graph G smaller at position i if $vec(\pi(G)) <^i_{lex} vec(G)$.

Observation 2

$$\langle a_1, \ldots, a_m \rangle <_{lex} \langle b_1, \ldots, b_m \rangle \Leftrightarrow \bigvee_{i=1}^{m} \langle a_1, \ldots, a_m \rangle <^i_{lex} \langle b_1, \ldots, b_m \rangle \qquad (3)$$

We now demonstrate that each of the "$<^i_{lex}$" constraints in the disjunction on the right side of Eq. (3) is both straightforward to solve and its set of solutions has a concise representation.

Definition 7 (patterns). *Let π be a permutation, G be an unknown graph of order-n with $vec(G) = \langle x_1, \ldots, x_m \rangle$ and let $1 \leq i \leq m$. The pattern,* $\mathsf{pats}_i(\pi)$, *is the result of applying the most general unifier of the equations from the right side in Eq. (2) to $vec(G)$. If the equations have no solution, then* $\mathsf{pats}_i(\pi) = \bot$. *We denote the set of non-$\bot$ patterns corresponding to a permutation π by* $\mathsf{pats}(\pi)$. *The patterns corresponding to a set of permutations Π is denoted* $\mathsf{pats}(\Pi)$.

Example 4. Recall $\pi = [1, 2, 4, 3]$ and the constraint $\langle x_1, x_3, x_2, x_5, x_4, x_6 \rangle <_{lex} \langle x_1, x_2, x_3, x_4, x_5, x_6 \rangle$ from Example 3. For $i = 1, 3, 5, 6$, $\mathsf{pat}_i(\pi) = \bot$. For instance, when $i = 1$ the equations $\{x_1{=}0, x_1{=}1\}$ have no solution. For $i = 2$, applying the most general unifier of $\{x_1{=}x_1, x_3{=}0, x_2{=}1\}$ to $\langle x_1, x_2, x_3, x_4, x_5, x_6 \rangle$ results in the pattern $\mathsf{pat}_2(\pi) = \langle x_1, 1, 0, x_4, x_5, x_6 \rangle$. For $i = 4$, applying the most general unifier of $\{x_1{=}x_1, x_3{=}x_2, x_2{=}x_3, x_5{=}0, x_4{=}1\}$ to $\langle x_1, x_2, x_3, x_4, x_5, x_6 \rangle$ results in the pattern $\mathsf{pat}_4(\pi) = \langle x_1, x_2, x_2, 1, 0, x_6 \rangle$.

The solutions of a constraint $\pi(G) <_{lex} G$ are concisely represented by the corresponding patterns in $\mathsf{pat}(\pi)$.

Example 5. Recall the two patterns detailed in Example 4 for the permutation $\pi = [1, 2, 4, 3]$ (see Table 2). The solutions for the constraint $\pi(G) < G$ are obtained as the set of $2^4 + 2^3$ instances of these two patterns. These are exactly the 24 solutions specified in Example 3.

Table 2. Constraints and patterns for $\pi(G) < G$ from Examples 4 and 5

index	constraint	pattern	# sols.
$i = 2$	$\{x_1{=}x_1, x_3{=}0, x_2{=}1\}$	$\langle x_1, 1, 0, x_4, x_5, x_6 \rangle$	2^4
$i = 4$	$\{x_1{=}x_1, x_3{=}x_2, x_2{=}x_3, x_5{=}0, x_4{=}1\}$	$\langle x_1, x_2, x_2, 1, 0, x_6 \rangle$	2^3

The set of graphs covered by a permutation π may contain an exponential number of graphs. However, π uniquely yields a (small) set of patterns $\mathsf{pats}(\pi)$ representing the graphs that are covered by π. This means that questions about the sets of graphs covered by π can be addressed on these patterns. Phrased as a constraint problem (over the Boolean domain), the set of graphs of order-n covered by a permutation π is as follows where $m = \binom{n}{2}$:

$$cover(\pi) = \bigvee\nolimits_{\mathsf{pat} \in \mathsf{pats}(\pi)} \bigwedge\nolimits_{i \in 1..m} x_i = \mathsf{pat}[i]$$

A given graph G with $vec(G) = \langle x_1, \ldots, x_m \rangle$ is covered by π if it is an instance of one of the patterns in $\mathsf{pats}(\pi)$.

Observe also that the sets of graphs represented by the different patterns for a permutation π are disjoint: Each pattern denotes the set of graphs that get smaller "for the first time" at a specified index i. It follows that it is easy to compute the number of graphs covered by a given permutation. For example, by summing the numbers in the right column of Table 2.

5 A Symbolic Approach to Set-Cover Optimizations

The classic set-cover assumes an explicit representation of the problem. Here we operate on the elements symbolically by considering the dominance relation between permutations and identifying backbones. These two concepts interact, as exemplified by Fig. 3, where permutations are depicted as the sets of graphs that they cover. Permutation A is a backbone because it covers a graph not covered by other permutations. Permutation E is not strictly a backbone because it covers a graph covered also by F, but since F is dominated by E, it can be removed and E becomes a backbone. Permutations B and C present a more subtle scenario: Neither appears to be dispensable, *but* if we ignore the graphs already covered by A, they cover the same set of graphs (a single graph). Hence,

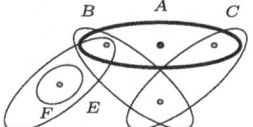 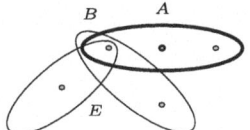 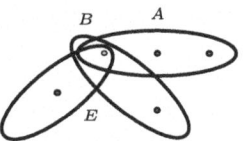

Fig. 3. Backbone and dominance interaction

Algorithm 1. Get all permutations in Π' dominated by Π

1: **procedure** GETDOMINATED(Π, Π')
2: $\quad R \leftarrow \{\}; \Gamma \leftarrow \mathsf{pats}(\Pi)$
3: $\quad$ **for all** $\pi \in \Pi'$ **do**
4: $\quad\quad$ **if** $\bigwedge_{\gamma \in \mathsf{pats}(\pi)}.unsat(\mathsf{notCovered}(\Gamma) \cup \mathsf{covered}(\gamma))$ **then**
5: $\quad\quad\quad R \leftarrow R \cup \{\pi\}$
6: $\quad$ **return** R

it is safe to remove either of them (non-deterministically). Once F and C are removed, E, B, and A are all backbones.

The key observation is that the dominance relation can be considered *modulo* existing backbones because these must be present in the cover. Hence, we first detect the set of backbones (see Sect. 6), then remove any permutation dominated by another one, modulo the backbones. This is repeated until a fixed point is found (because new backbones may appear after the dominance pruning).

Definition 8 (set of permutations dominates a permutation). *Let Π be a set of permutations and π a permutation. If $cover(\pi) \subseteq cover(\Pi)$, then we say that Π dominates π.*

Definition 9 (permutation dominates modulo β). *Let π_1, π_2 be permutations and β a set of (backbone) permutations. We say that π_1 dominates π_2 modulo β if $cover(\pi_2) \setminus cover(\beta) \subseteq cover(\pi_1) \setminus cover(\beta)$.*

Rather than encoding the dominance relation explicitly as a constraint problem, we develop an encoding via the pattern representation (Definition 7). This will enable a concise encoding amenable to incremental solving.

Stating that π_1 dominates π_2 modulo β is equivalent to stating that $\beta \cup \{\pi_1\}$ dominates π_2. Further, in terms of the pattern representation: to determine, whether a set of permutations Π dominates a permutation π, we only need to check that $\mathsf{pats}(\Pi)$ dominates each pattern of $\mathsf{pats}(\pi)$.

To determine, whether a set of patterns Γ dominates a pattern γ, we look for a graph that is *not* covered by Γ but is covered by γ. If such a graph exists, Γ does *not* dominate γ. Hence, for a graph G with $vec(G) = \langle x_1, \ldots x_m \rangle$, we wish to decide the satisfiability of the following sets of constraints $\mathsf{notCovered}(\Gamma) \cup \mathsf{covered}(\gamma)$ for each $\gamma \in \mathsf{pats}(\pi)$.

$$\mathsf{notCovered}(\Gamma) = \{\bigvee_{i \in 1..m} x_i \neq \gamma'[i] \mid \gamma' \in \Gamma \} \ \%G \text{ is not covered by } \Gamma \quad (4)$$
$$\mathsf{covered}(\gamma) = \quad \{\bigwedge_{i \in 1..m} x_i = \gamma[i]\} \quad \% \ G \text{ is covered by } \gamma \quad (5)$$

To encode the problem into SAT, fresh variables (Tseitin variables [33]) are introduced to represent the constraints of the form $x_i = x_j$, i.e. define fresh $e_{ij} \Leftrightarrow (x_i = x_j)$ as 4 clauses $\{\neg e_{ij} \vee \neg x_i \vee x_j, \neg e_{ij} \vee x_i \vee \neg x_j, e_{ij} \vee x_i \vee x_j, e_{ij} \vee \neg x_i \vee \neg x_j\}$. This encoding is concise and CNF-friendly. The set of graphs covered by a pattern corresponds to single a conjunction of e_{ij} literals. The set of graphs *not* covered by a pattern corresponds to a single clause of $\neg e_{ij}$ literals. The e_{ij} variables may be reused across multiple patterns and SAT calls.

Algorithm 2. Refinement of a set S of permutations modulo β

1: **procedure** REFINE(S,β)
2:     $S \leftarrow S \setminus \beta;\quad T \leftarrow S$ ▷ *permutations to test*
3:     **while** $T \neq \{\}$ **do**
4:        $\pi \leftarrow \mathsf{pick}(T);\quad T \leftarrow T \setminus \{\pi\}$ ▷ *pick arbitrary permutation from T*
5:        $R \leftarrow \text{GETDOMINATED}(\beta \cup \{\pi\}, S \setminus \{\pi\})$
6:        $S \leftarrow S \setminus R;\quad T \leftarrow T \setminus R$ ▷ *remove dominated permutations*
7:     **return** S

Algorithm 1 finds all permutations in a set Π' dominated by a set Π. It relies on the concise pattern representation to use incremental SAT [11] by loading the clauses for notCovered(pats(Π)) into the solver just once. Each pattern of a permutation π requires a separate SAT call. However, if there is a pattern $\gamma \in \mathsf{pats}(\pi)$ not dominated by Π, then further SAT calls are unnecessary. Also, satisfiable tests on Line 4 provide witness graphs, which can be used to prune the for all loop from other $\pi \in \Pi'$ for which it is also a witness.

The procedure GETDOMINATED is readily used in the procedure REFINE (Algorithm 2), to remove any permutation dominated by some other permutation in the set modulo the given set of backbones β. In Sect. 6 we will look at algorithms to calculate the set of backbones.

6 Backbones for Symmetry Breaking

The problem of finding backbones has the nature of one quantifier alternation, because we are asking whether there *exists* a permutation π that covers some graph g such that g is not covered *for all* other permutations $\pi' \neq \pi$. In this section we explore two approaches to finding backbones: by iterating over all graphs in Sect. 6.1; and by iterating over all permutations in Sect. 6.2. In the experimentation reported here we apply a combination of two of these (see Sect. 7).

6.1 Backbone Calculation via Iteration on Graphs

The basic algorithm to find permutation backbones is sketched as Algorithm 3. The set of backbones β is initialized to the empty set. The algorithm iterates over all graphs G in their $vec(G)$ representation, starting from the empty graph (all zeroes), and ending with the complete graph (all ones). For each graph, we test if it is a backbone graph and if so, add a permutation to β. At Line 5 there is a call to BACKBONESTATUS. This procedure is assumed to return the set S of permutations that cover G and is refined modulo β. First, the permutations that cover G are computed applying an allSAT encoding, and then procedure refine is applied to remove permutations dominated by others modulo β. The call at Line 8 increments $vec(G)$ viewing it as a binary number with least significant bit on the right.

Algorithm 3. find backbones – the fantasy of iterating over all graphs

1: **procedure** BACKBONES(n)
2:     initialize $\beta = \emptyset$
3:     initialize $G = \langle 0, 0, \ldots, 0, 0 \rangle$
4:     **while** $G \leq \langle 1, 1, \ldots, 1, 1 \rangle$ **do**
5:         $S \leftarrow$ BACKBONESTATUS(G, β)
6:         **if** $S == \{\pi\}$ **then**
7:             $\beta \leftarrow \beta \cup \{\pi\}$
8:         $G \leftarrow$ INCREMENT(G)
9:     **return** β

This approach might seem dubious, as there are $2^{\binom{n}{2}}$ graphs to consider along the way. We detail three optimizations that enable to "leap" forward in the iteration such that the algorithm is able to compute backbones at least for $n \leq 10$. The first two optimizations are "exact". The third is heuristic. It might miss some backbones but will not wrongly identify a backbone.

Is G Covered by a Backbone? In the while-loop at Line 4, before the expensive call at Line 5, we could test if the current graph G is already covered by one of the backbone permutations found so far. In this case we need not check the status of G. This test is efficient given the compact representation pat(β) of the graphs covered by β. Moreover, in this case, the call at line 8 can be replaced by a "leap" in the iteration. The idea is that if G is covered by β, then it is likely that many of the consecutive graphs are also covered for the same reason.

For example, let $w = 00110$, consider the order-6 graph $G = w.0^{10}$, and assume G is covered by a backbone with pattern pat $= \langle x_1, x_1, x_2, 1, 0, x_3, \ldots, x_{12} \rangle$. Observe that any graph with the prefix w is also an instance of pat as the last 10 elements in pat consist solely of free variables. Hence we increment w to $w' = 00111$ and iteration continues with $w'.0^{10}$ thus skipping 2^{10} iterations.

Is G a Canonical Graph? Consider the case where the call at Line 6 results in $S = \emptyset$. This implies that G is canonical as it is not covered by any permutation. Again, we can apply a "leap" in the iteration at Line 8. The idea is that if G is canonical and has a suffix of k zeros, then in may cases, changing a single zero from the suffix to a one does not make it easier to find a permutation that makes G smaller. We test this "per digit" and after k checks we can jump 2^k steps in the iteration.

For example, let $w = 000001100010101011111$. The order-8 graph $G = w.0^7$ is canonical and has a suffix with 7 zeros. With 7 checks, we determine that replacing a zero by a one in the suffix does not result in a graph that has a cover. Indeed all of the $2^7 - 1$ graphs succeeding G are canonical. So, we increment w to $w' = 000001100010101100000$ and iteration continues with $G = w'.0^7$ thus skipping 2^7 steps.

Does G have a "Huge" Number of Covering Permutations? At Line 5 in Algorithm 3, in the call to procedure BACKBONESTATUS, we compute a set S

using an allSAT encoding to find all permutations that cover the graph G. As a heuristic, we abort this allSAT computation if a fixed bound of B permutations are found. We only apply the REFINE algorithm if less than B permutations are found. Also in this case we can apply a "leap" in the iteration at Line 8. The idea is that when G has a prefix of k zeros, then changing even a single zero from the suffix to a one, often times, does not make it easier to find a permutation that makes G smaller—so, G will still have a large number of covering permutations.

For example, assume $B = 10$ and denote $w = 00000000000000000010$. The graph $G = w.00000000$ has 10 or more covering permutations. After testing that changing any one of the last 8 zeros to a one results in a graph that still has at least 10 covering permutations, we increment w to $w' = 00000000000000000011$ and iteration continues with $G = w'.00000000$ thus skipping 2^8 steps.

6.2 Backbone Calculation as a SAT Call

Here, we show that it is possible to decide whether a permutation is a backbone by a SAT call; then we iterate over all permutations to find all backbones. The intuition is as follows. All the permutations cover the whole set of the non-canonical graphs. Removing a backbone will necessarily diminish the set of covered graphs. In another words, a backbone $\pi \in S_n$ must *not* be dominated by some set of permutations $\Pi \subseteq S_n \setminus \{\pi\}$.

Hence, to test whether a given permutation is a backbone is the test $\pi \notin \text{GETDOMINATED}(S_n \setminus \{\pi\}, \{\pi\})$ (see Algorithm 1). Such a test needs to be issued for every permutation, which makes the procedure expensive. This can be mitigated by quickly eliminating some backbone-candidates by choosing arbitrarily some set $\Pi \subset S_n$ and marking all $\text{GETDOMINATED}(\Pi, S_n \setminus \Pi)$ as non-backbones.[1] Some more sophisticated approaches could be considered, such as in [2,25].

This approach immediately generalizes to an arbitrary set of permutations $\Pi \subset S_n$, i.e. if some permutations were already removed from S_n due to being dominated, the backbone test becomes $\pi \notin \text{GETDOMINATED}(\Pi \setminus \{\pi\}, \{\pi\})$.

We refer to the approach to find backbones described here as Algorithm 4.

7 Solving the Set-Cover Problems

This section puts all of the components together to compute optimal symmetry breaks. We experimented with the two algorithms described in Sects. 6.1 (Algorithm 3) and Sect. 6.2 (Algorithm 4) to detect backbones. In our trials, the configuration which works best in practice is to first apply Algorithm 4 and then, starting from this result to apply the approach from Algorithm 3. We alternate the search for backbones with refining the set of permutations that need to be considered for the set-cover problem.

Table 3 provides details of the computation. The first four columns are about the computation of backbones and refined permutations. Columns bb_1 and bb_2

---

[1] Backbone pruning is also used in simpler, classical SAT backbones [21].

Table 3. Constraints and patterns for $\pi(G) < G$ from Examples 4 and 5

order	bb_1	bb_2	rows	cols	sets	opt	2016	$time_1$	$time_2$	$time_3$	enc. size
6	9	13	0	0	0-0	13	13	89 s	0 s	0 s	463
7	18	25	44	148	3-72	35	37	123 s	1 s	0 s	956
8	30	112	19	33	3-10	121	135	929 s	55 s	0 s	1,925
9	90	709	207	344	2-42	765	842	289 m	75 m	58 s	5,190
10	131	6,920	694	3,481	2-550	7,181	7,853	613 m	137 h	68 m	31,193

detail the number of backbones found after applying Algorithm 3 and then repeatedly applying Algorithm 4 alternating with Algorithm 2. The final set of permutations that need be considered in the set-cover problems is detailed in column rows. These are the rows in the matrix representation.

The next four columns in Table 3 are about creating and solving the matrix representation for the set-cover problems. The column cols details the number of graphs covered by the permutations in the rows. The column sets details the size range of the sets of graphs covered by the permutations in the rows. The column opt details the size of the optimal set-cover found, and the column 2016 details the number of permutations in the canonizing sets reported in [19].

The next three columns are about times (rounded to seconds, minutes, or hours): $time_1$ to compute backbones bb_1; $time_2$ to compute bb_2 and rows; and $time_3$ to generate the matrix representation, apply Optimizations 1–3, and solve the set-cover problem. Interestingly, for $n < 10$ we never needed to solve the general set-cover because everything is solved by repeated applications of the described optimizations. For $n = 10$, the optimizations reduce the initial matrix to one with 199 rows and 197 columns. We formulate the remaining set-cover problem as an *optimization pseudo-boolean* (OPB) problem and solve it by RoundingSat [12]. This OPB problem is solved in 0.01 s.

The final column (enc. size) shows the number of clauses to encode the symmetry break; this is done by negating all the patterns (see Eq. (5)) in the optimal cover—the pattern encoding is 3 orders of magnitude smaller than the direct lex-leader encoding (see Definition 3).

All of the complete symmetry breaks found in this paper were verified, using GANAK [31], to have a number of solutions corresponding to the number of non-isomorphic graphs (sequence A000088 of the OEIS [27]).

8 Backbones as a Partial Symmetry Break

Codish et al. [6,7] introduce a polynomial sized partial symmetry breaking constraint for graphs defined in terms of transpositions (permutations that swap two vertices). In fact, any set of permutations Π can be viewed as a partial symmetry break obtained by replacing S_n with Π in Eq. (1). Heule defines the notion of *redundancy ratio* to measure the precision of symmetry breaks on graphs [17]. In our terminology, the redundancy ratio for a set of permutations Π, which we denote $\rho(\Pi)$, is the ratio between the number of graphs that are not covered by Π and the number of isomorphism classes. If Π is canonizing, then $\rho(\Pi)$ is 1. Table 4 details the numbers, trns and bb_1, of transpositions and backbones found using Algorithm 3, together with the redundancy ratios $\rho(\mathsf{trns})$ and $\rho(\mathsf{bb}_1)$. One can observe that the symmetry break using backbones involves a relatively small set of permutations, and provides a partial symmetry break which is more precise.

Table 4. Partial symmetry breaks: transpositions vs. backbones

order	trns	$\rho(trns)$	bb_1	$\rho(bb_1)$
6	15	1.76	9	1.19
7	21	3.02	18	1.36
8	28	5.39	30	1.87
9	36	9.42	90	1.99
10	45	15.34	131	2.99

9 Conclusion

This paper formalizes symmetry breaking as a set-covering problem and this is the main contribution of the paper. As demonstrated by the paper, this formalization opens up a range of new opportunities for complete and partial symmetry breaking deriving from decades of studies on both precise and approximate techniques for this problem. We focus primarily on precise solutions to provide complete symmetry breaks for graphs which are optimal in the number of lex-constraints. We achieve this for all cases in which small complete symmetry breaks based on lex-constraints have been computed in [19]. Namely, for graphs of order $n \leq 10$. Interestingly (see Table 3), the symmetry breaks computed in [19] are less than 10% larger than the optimal ones.

An important ingredient is the notion of *patterns* to provide a concise representation of the (possibly exponential) set of graphs covered by a permutation. Another important ingredient is in the notion of a *backbone*, which is a permutation that is the only one that covers some graph that is not already covered by other permutations. We apply two types of optimizations: *before* the construction of the matrix representation of the set-cover problem; these are symbolic and rely on SAT encodings, and *after* the construction. Both types identify permutations and graphs that can be ignored in the search for an optimal cover, and both types identify backbone permutations.

It is interesting to note that for graphs of order $n < 10$, our construction of the optimal symmetry breaks never needed to solve a general set-cover problem. Everything is solved by repeated application of the described optimizations (before and after the construction of the matrix representation). For the case $n = 10$, the original set-cover problem of size $(10! \times 2^{45})$ is finally reduced to a set-cover problem of size (199×197). In all cases, the repeated identification of backbones can be seen as "driving" the construction.

In this paper we consider set-covers defined in terms of the sets of graphs covered by permutations. It is interesting to investigate set-covers defined in terms of patterns (which are similar to the implications defined in [4]) and in terms of "isolaters" as applied in [17].

Encouraged by the results in Table 4, where a small set of backbone permutations is shown to provide a strong partial symmetry break, it is interesting to further investigate the application of set-cover techniques to construct small and precise partial symmetry breaks.

Acknowledgments. The results were supported by the Ministry of Education, Youth and Sports within the dedicated program ERC CZ under the project *POSTMAN* no. LL1902 and co-funded by the European Union under the project *ROBOPROX* (reg. no. CZ.02.01.01/00/22_008/0004590). This article is part of the *RICAIP* project that has received funding from the European Union's Horizon 2020 research and innovation programme under grant agreement No 857306.

Disclosure of Interests. The authors have no competing interests to declare that are relevant to the content of this article.

References

1. Babai, L., Luks, E.M.: Canonical labeling of graphs. In: Proceedings of the Fifteenth Annual ACM Symposium on Theory of Computing, pp. 171–183. ACM (1983)
2. Belov, A., Janota, M., Lynce, I., Marques-Silva, J.: Algorithms for computing minimal equivalent subformulas. Artif. Intell. **216**, 309–326 (2014). https://doi.org/10.1016/J.ARTINT.2014.07.011
3. Caprara, A., Toth, P., Fischetti, M.: Algorithms for the set covering problem. Ann. Oper. Res. **98**(1–4), 353–371 (2000). https://doi.org/10.1023/A:1019225027893
4. Codish, M., Ehlers, T., Gange, G., Itzhakov, A., Stuckey, P.J.: Breaking symmetries with lex implications. In: Gallagher, J.P., Sulzmann, M. (eds.) FLOPS 2018. LNCS, vol. 10818, pp. 182–197. Springer, Cham (2018). https://doi.org/10.1007/978-3-319-90686-7_12
5. Codish, M., Frank, M., Itzhakov, A., Miller, A.: Computing the Ramsey number R(4, 3, 3) using abstraction and symmetry breaking. Constraints Int. J. **21**(3), 375–393 (2016)
6. Codish, M., Miller, A., Prosser, P., Stuckey, P.J.: Breaking symmetries in graph representation. In: Rossi, F. (ed.) IJCAI 2013, Proceedings of the 23rd International Joint Conference on Artificial Intelligence, Beijing, China, 3–9 August 2013, pp. 510–516. IJCAI/AAAI (2013). http://ijcai.org/proceedings/2013

7. Codish, M., Miller, A., Prosser, P., Stuckey, P.J.: Constraints for symmetry breaking in graph representation. Constraints **24**(1), 1–24 (2018). https://doi.org/10.1007/s10601-018-9294-5
8. Crawford, J.M., Ginsberg, M.L., Luks, E.M., Roy, A.: Symmetry-breaking predicates for search problems. In: Aiello, L.C., Doyle, J., Shapiro, S.C. (eds.) Proceedings of the Fifth International Conference on Principles of Knowledge Representation and Reasoning (KR 1996), Cambridge, Massachusetts, USA, 5–8 November 1996, pp. 148–159. Morgan Kaufmann (1996)
9. Crawford, J.M., Ginsberg, M.L., Luks, E.M., Roy, A.: Symmetry-breaking predicates for search problems. In: Aiello, L.C., Doyle, J., Shapiro, S.C. (eds.) Proceedings of the Fifth International Conference on Principles of Knowledge Representation and Reasoning (KR 1996), pp. 148–159. Morgan Kaufmann (1996)
10. Dančo, M., Janota, M., Codish, M., Araújo, J.J.: Complete symmetry breaking for finite models. In: AAAI 2025 (2025). https://arxiv.org/abs/2502.10155
11. Eén, N., Sörensson, N.: An extensible SAT-solver. In: Giunchiglia, E., Tacchella, A. (eds.) SAT 2003. LNCS, vol. 2919, pp. 502–518. Springer, Heidelberg (2004). https://doi.org/10.1007/978-3-540-24605-3_37
12. Elffers, J., Nordström, J.: Divide and conquer: towards faster pseudo-boolean solving. In: Lang, J. (ed.) Proceedings of the Twenty-Seventh International Joint Conference on Artificial Intelligence, IJCAI 2018, 13–19 July 2018, Stockholm, Sweden, pp. 1291–1299. ijcai.org (2018). https://doi.org/10.24963/IJCAI.2018/180
13. Frisch, A.M., Harvey, W.: Constraints for breaking all row and column symmetries in a three-by-two matrix. In: Proceedings of SymCon 2003 (2003)
14. Gauthier, T., Brown, C.E.: A formal proof of r(4, 5)=25. In: Bertot, Y., Kutsia, T., Norrish, M. (eds.) 15th International Conference on Interactive Theorem Proving, ITP 2024, 9–14 September 2024, Tbilisi, Georgia. LIPIcs, vol. 309, pp. 16:1–16:18. Schloss Dagstuhl - Leibniz-Zentrum für Informatik (2024). https://doi.org/10.4230/LIPICS.ITP.2024.16
15. Gupta, A., Lee, E., Li, J.: A local search-based approach for set covering. In: Kavitha, T., Mehlhorn, K. (eds.) 2023 Symposium on Simplicity in Algorithms, SOSA 2023, Florence, Italy, 23–25 January 2023, pp. 1–11. SIAM (2023). https://doi.org/10.1137/1.9781611977585.CH1
16. Heule, M.J.H.: The quest for perfect and compact symmetry breaking for graph problems. In: Davenport, J.H., et al. (eds.) 18th International Symposium on Symbolic and Numeric Algorithms for Scientific Computing, SYNASC 2016, Timisoara, Romania, 24–27 September 2016, pp. 149–156. IEEE Computer Society (2016). http://ieeexplore.ieee.org/xpl/mostRecentIssue.jsp?punumber=7827704
17. Heule, M.J.H.: Optimal symmetry breaking for graph problems. Math. Comput. Sci. **13**, 533–548 (2019)
18. Hoffman, K., Padberg, M.: Set covering, packing and partitioning problems, pp. 2348–2352. Springer, Boston (2001)
19. Itzhakov, A., Codish, M.: Breaking symmetries in graph search with canonizing sets. Constraints **21**(3), 357–374 (2016). https://doi.org/10.1007/s10601-016-9244-z
20. Itzhakov, A., Codish, M.: Incremental symmetry breaking constraints for graph search problems. In: Proceedings of the AAAI Conference on Artificial Intelligence, vol. 34, no. 02, pp. 1536–1543 (2020). https://doi.org/10.1609/aaai.v34i02.5513
21. Janota, M., Lynce, I., Marques-Silva, J.: Algorithms for computing backbones of propositional formulae. AI Commun. **28**(2), 161–177 (2015). https://doi.org/10.3233/AIC-140640

22. Karp, R.M.: Reducibility among combinatorial problems. In: Miller, R.E., Thatcher, J.W. (eds.) Proceedings of a symposium on the Complexity of Computer Computations, held 20–22 March 1972, at the IBM Thomas J. Watson Research Center, Yorktown Heights, New York, USA, pp. 85–103. The IBM Research Symposia Series. Plenum Press, New York (1972). https://doi.org/10.1007/978-1-4684-2001-2_9
23. Lei, Z., Cai, S.: Solving set cover and dominating set via maximum satisfiability. In: The Thirty-Fourth AAAI Conference on Artificial Intelligence, AAAI 2020, The Thirty-Second Innovative Applications of Artificial Intelligence Conference, IAAI 2020, The Tenth AAAI Symposium on Educational Advances in Artificial Intelligence, EAAI 2020, New York, NY, USA, 7–12 February 2020, pp. 1569–1576. AAAI Press (2020). https://doi.org/10.1609/AAAI.V34I02.5517
24. Liu, X., Fang, Y., Chen, J., Su, Z., Li, C., Lü, Z.: Effective approaches to solve p-center problem via set covering and sat. IEEE Access **8**, 161232–161244 (2020). https://doi.org/10.1109/ACCESS.2020.3018618
25. Marques-Silva, J., Janota, M., Belov, A.: Minimal sets over monotone predicates in boolean formulae. In: Sharygina, N., Veith, H. (eds.) Computer Aided Verification 25th International Conference, CAV 2013, Saint Petersburg, Russia, 13–19 July 2013. Proceedings. Lecture Notes in Computer Science, vol. 8044, pp. 592–607. Springer (2013)
26. McGuire, G., Tugemann, B., Civario, G.: There is no 16-clue Sudoku: solving the Sudoku minimum number of clues problem. CoRR abs/1201.0749 (2012). http://arxiv.org/abs/1201.0749
27. The on-line encyclopedia of integer sequences (2010). http://oeis.org
28. Read, R.C.: Every one a winner or how to avoid isomorphism search when cataloguing combinatorial configurations. Ann. Discrete Math. **2**, 107–120 (1978)
29. Rintanen, J., Rankooh, M.F.: Symmetry-breaking constraints for directed graphs. In: Endriss, U., et al. (eds.) ECAI 2024 - 27th European Conference on Artificial Intelligence, 19–24 October 2024, Santiago de Compostela, Spain - Including 13th Conference on Prestigious Applications of Intelligent Systems (PAIS 2024). Frontiers in Artificial Intelligence and Applications, vol. 392, pp. 4248–4253. IOS Press (2024). https://doi.org/10.3233/FAIA240998
30. Roth, R.: Computer solutions to minimum-cover problems. Oper. Res. **17**(3), 455–465 (1969). https://doi.org/10.1287/OPRE.17.3.455
31. Sharma, S., Roy, S., Soos, M., Meel, K.S.: GANAK: a scalable probabilistic exact model counter. In: Kraus, S. (ed.) Proceedings of the 28th International Joint Conference on Artificial Intelligence (IJCAI 2019), pp. 1169–1176. IJCAI, Macao, China (2019)
32. Shlyakhter, I.: Generating effective symmetry-breaking predicates for search problems. Discret. Appl. Math. **155**(12), 1539–1548 (2007)

33. Tseitin, G.S.: On the complexity of derivations in the propositional calculus. Studies in Constructive Mathematics and Mathematical Logic (1968)
34. Walsh, T.: General symmetry breaking constraints. In: Principles and Practice of Constraint Programming - CP 2006, 12th International Conference, CP 2006, Nantes, France, 25–29 September 2006, Proceedings, pp. 650–664 (2006)

Modeling and Solving the Generalized Test Laboratory Scheduling Problem

Philipp Danzinger[✉], Tobias Geibinger, Florian Mischek,
and Nysret Musliu

Christian Doppler Laboratory for Artificial Intelligence and Optimization for Planning and Scheduling, DBAI, Institute for Logic and Computation, TU Wien, Favoritenstraße 9-11, 1040 Vienna, Austria
{philipp.danzinger,tobias.geibinger,
florian.mischek,nysret.musliu}@tuwien.ac.at

Abstract. The Test Laboratory Scheduling Problem (TLSP) is an NP-hard scheduling problem based on the real-world scheduling requirements of an industrial test laboratory. TLSP requires the solver to find a grouping of tasks into jobs, and to schedule those jobs, assigning resources of different types (employees, workbenches, and equipment) and optimizing different soft constraints. Over time, new real-world scheduling requirements have emerged that necessitate a more flexible description of resources. To deal with such situations, in this paper, we propose Generalized TLSP (G-TLSP), a new problem extension of TLSP which unifies different resource types.

To solve G-TLSP, we propose a new Constraint Programming (CP) model and solve instances with exact CP solvers as well as with a Very Large Neighborhood Search (VLNS) algorithm. Our approaches are evaluated on existing instances as well as two new real-world instances. We achieve competitive performance with existing specialized solvers on converted TLSP instances and find high-quality solutions for the new real-world instances.

Keywords: Scheduling · Combinatorial Optimization · Modeling · Constraint Programming · Very Large Neighborhood Search

1 Introduction

Efficient scheduling of tasks and resources is crucial for many industrial applications. In particular, test laboratories face complex scheduling challenges due to their need to handle multiple projects simultaneously while efficiently utilizing expensive equipment and skilled employees. The Test Laboratory Scheduling Problem (TLSP) was introduced in [6,7] to capture these real-world requirements. A solver must find a grouping of tasks into jobs and schedule those jobs, assigning one of multiple modes and resources from different types, while optimizing different objectives. Jobs are the sole unit of scheduling and inherit most of their properties from their tasks.

© The Author(s), under exclusive license to Springer Nature Switzerland AG 2025
G. Tack (Ed.): CPAIOR 2025, LNCS 15762, pp. 188–204, 2025.
https://doi.org/10.1007/978-3-031-95973-8_12

A core feature of TLSP is its handling of three distinct types of resources: employees, workbenches, and equipment, each having different properties. For instance, the number of employees that must be assigned to a job is based only on the chosen mode, while the required number of other resource types depends on the tasks making up the job. In addition, some constraints apply only to employees, like the soft constraint of minimizing the number of different employees working on jobs of every single project. Furthermore, equipment is subdivided into equipment groups, and jobs can require a different number of units from each one. This system [3] successfully captured the requirements of several industrial laboratories, including the original industrial application, and successfully accommodated a number of additional requirements over time. However, new scheduling requirements have emerged from the existing industrial partner as well as other test laboratories that cannot be properly expressed within this framework. This includes extending the employee-related constraints to other types of resources or relaxing the requirement that the number of required resources to schedule a job depends only on the job's mode for employees, and only on its tasks for workbenches and equipment.

To expand the applicability of TLSP to a broader set of laboratory environments, we propose the Generalized TLSP (G-TLSP), a new problem variant that generalizes the resource types to a single concept. This generalization allows for more flexible resource modeling while maintaining the ability to solve existing TLSP instances by converting them.

Our main contributions are:

- A formal problem description of G-TLSP that generalizes TLSP's resource model.
- A new CP model for solving G-TLSP.
- An extension of an existing Very Large Neighborhood Search (VLNS) algorithm for TLSP with a new initialization mechanism that finds feasible solutions even for the largest test instances.
- Two new real-world problem instances.
- An evaluation of our new G-TLSP solution methods on the original TLSP benchmark instances, showing competitive performance compared to existing specialized TLSP solvers.

The rest of this paper is structured as follows: Sect. 2 reviews related work and provides further background on the TLSP. Section 3 presents our formal problem description for G-TLSP. Sections 4 and 5 detail our CP model and VLNS algorithm, respectively. Section 6 presents our experimental evaluation, and Sect. 7 concludes the paper.

2 Related Work

The Test Laboratory Scheduling Problem was first introduced in [6,7] to model the real-world scheduling requirements of industrial test laboratories. TLSP is an extension of the well-known Resource-Constrained Project Scheduling problem

(RCPSP) with several additional features and constraints, as well as multiple optimization objectives.

Some features of TLSP are similar to other problems from the literature, for instance multiple modes like in [5], or multiple projects like in the Resource Constrained Multi-Project Scheduling Problem (RCMPSP). A formulation by Wauters et al. [12] combines multiple modes with multiple projects.

One fairly unique feature of TLSP is its separation into a grouping and scheduling stage. An instance is composed of tasks, which first need to be grouped together into jobs. These jobs then become the only unit of scheduling and derive most of their scheduling requirements from their tasks. Batch scheduling [11] is a related concept, although to the best of our knowledge no other work uses the same grouping formalism as the TLSP. In contrast to typical RCPSP variants, the TLSP also features heterogeneous resources, where only a subset of employees, workbenches, or devices in an equipment group may be compatible with any given task. Similar restrictions on suitable resource assignments appear e.g. in the Multi-Skill Project Scheduling Problem (MSPSP) [1].

TLSP was first solved in [4] with a CP model and VLNS algorithm. Later works introduced additional approaches focused on Local Search [9] and hyper-heuristics [8]. The existing VLNS is usually competitive with other solution approaches in cases where it finds a feasible solution. However, it relies on a CP model to find a feasible initial solution, which causes it to time out on some large instances that other approaches can solve.

3 Problem Description

This Section provides a formal problem description for G-TLSP. The parts not related to resources are based on the original TLSP formulation from [6,7], while the parts related to resources are a novel contribution of this work.

A G-TLSP instance consists of a set of *projects* to schedule and an *environment*. Projects are subdivided into task families, which each contain a set of tasks. Each task has resource requirements and other attributes such as precedence relations. The environment specifies the length of the scheduling horizon, the available resources, and potentially information about an existing schedule. A solution consists of a grouping of tasks into jobs, as well as time slot and resource assignments for these jobs.

3.1 Environment

The scheduling horizon consists of a set of time slots $t \in T = \{0, \ldots, |T|-1\}$.

Next, there is a set of modes $m \in M = \{1, \ldots, |M|\}$. In a solution, each job needs to be assigned exactly one mode. Modes allow for trade-offs between a job's duration and resource consumption. To model the change in job duration, each mode $m \in M$ has an associated time factor $v_m \in \mathbb{R}^+$ that acts as a multiplier for the duration of tasks performed in that mode.

Additionally, the environment specifies what resources are available for scheduling. Resources in G-TLSP are partitioned into groups. The set of all resource groups is denoted $R^* = \{R_1, \ldots, R_{|R^*|}\}$. Each resource group with index i consists of individual resource units $R_i = \{1, \ldots, |R_i|\}$. Tasks may require any number of resources from any resource group.

Furthermore, $R^{linked} \subseteq R^*$ represents the set of resource groups for which the *linked tasks* constraint is active. As explained later in Sect. 3.2, tasks in an instance can be linked. The linked tasks constraint then forces the solver to assign the same resource units to jobs containing linked tasks, but only for resources in groups from R^{linked}. Likewise, $R^{min} \subseteq R^*$ is the set of resource groups for which the *number of resources* soft constraint is active, minimizing the number of distinct resource units used for each project.

When converting an existing TLSP instance to G-TLSP, R^* contains one resource group for employees, one for workbenches, and one for each equipment group in the original instance.

3.2 Projects and Tasks

A G-TLSP instance contains a set of projects $p \in P$, each with its own set of tasks $a \in A_p$. For notational convenience, p_a refers to the project that the task a belongs to. The set of all tasks is denoted $A^* := \bigcup_{p \in P} A_p$.

Furthermore, the tasks A_p in each project p are partitioned into task families $F_{p,i} \subseteq A_p$ indexed by the project p and an additional index i. The family of a task a is denoted by f_a, the set of families for project p is F_p, and the set of all families across projects is F^*. The set of tasks belonging to family f is denoted A_f. Task families serve a two-fold purpose: firstly, only tasks from the same family may be grouped together into a job. Secondly, each family has a setup time $s_f \in \mathbb{R}_0^+$ associated with it. When a job is formed from tasks in a family f, the setup time s_f gets added to the sum of the durations of its tasks to calculate the total duration of the job. The setup time is also subject to the mode time factor v_m discussed earlier in Sect. 3.1.

Finally, each project p contains a set L_p of sets of linked task sets. Formally, L_p is an equivalence relation over A_p. Jobs that each contain a task from the same linked set need to be assigned the same resource units for resource groups in R^{linked}, which was defined in Sect. 3.1.

Each task $a \in A^*$ has various properties associated with it:

- Each task has a release date $\alpha_a \in T$, which is the earliest possible start date. It also has a soft due date $\bar{\omega}_a \in T$, whose violation incurs a penalty, and a hard deadline $\omega_a \in T$, by which it must be completed.
- Each task has a set $M_a \subseteq M$ of available modes.
- Each task has a duration $d_a \in \mathbb{R}_0^+$.
- Each task has a set of predecessor tasks from the same project $\mathcal{P}_a \subseteq A_p$ where $p = p_a$. A task must be performed after every of its predecessor tasks, or grouped into the same job.

- From each resource group $R_g \in R^*$, task a requires a number $r_{a,g,m} \in \mathbb{N}_0$ of units that may also depend on the mode $m \in M$. It also has sets of available resources $R_{a,g} \subseteq R_g$, and a set of preferred resources $R^{Pr}_{a,g} \subseteq R_g$. Only available resources may be assigned to a job containing a, and assigning non-preferred resources incurs a penalty.

3.3 Initial Schedule

To support using G-TLSP for day-to-day scheduling operations, each instance can also contain information about an existing schedule. Every instance can contain a set of base jobs J^0. Formally, each base job $j \in J^0$ is a set of tasks, so $j \subseteq A_f$ for some task family $f \in F^*$. All tasks included in the same base job must be part of the same job in a solution. The tasks $a \in j$ are called *fixed tasks*. Note that a solution may still add additional tasks to the job. Additionally, as a subset of the base jobs, there are started jobs $J^{0S} \subseteq J^0$. Started jobs represent jobs that are already in progress at the start of the scheduling horizon. As such, in a solution, any job formed from a started job must have a start time of 0 and incurs no setup time. In practice, available resources for started jobs are usually restricted to match the existing assignments in preprocessing to prevent the solver from changing them.

3.4 Solution Description

A solution to a G-TLSP instance consists of:

- A partition J that groups tasks into jobs. Each job $j \in J$ is a set of tasks. Only tasks from the same task family inside the same project may be grouped together.
- An assigned mode for each job. The mode chosen for a job must be available for all of its tasks.
- A start time and completion time for all jobs. For every job, the start time must be bigger than or equal to the release time of all of its tasks and its completion time must be smaller than or equal to the deadline of its tasks. The difference between completion and start time must equal the job's duration. The duration of a job j containing tasks from family f and performed in mode m is given by $\lceil v_m \cdot (s_f + \sum_{a \in j} d_a) \rceil$. In the special case where $\exists j^0 \in J^{0S} \mid j \cap j^0 \neq \emptyset$, the job is considered a started job, and s_f in the formula is set to 0.
- Resource assignments for each job. For each resource group, the assigned resources must exactly cover the highest demand of any task in the job, and each assigned resource must be available for all tasks. Every individual resource unit may only be used by at most one job at the same time.

What follows is a formal description of the feasibility conditions a G-TLSP solution must satisfy, and of the preferences that serve as optimization targets.

Solution Variables. A solution consists of assignments to the following variables:

$$J \subseteq 2^{A^*} \qquad \ldots \text{Set of jobs}$$
$$\dot{s}_j \in T \qquad j \in J \qquad \ldots \text{Assigned start time}$$
$$\dot{n}_j \in T \qquad j \in J \qquad \ldots \text{Assigned completion time}$$
$$\dot{m}_j \in M \qquad j \in J \qquad \ldots \text{Assigned mode}$$
$$\dot{a}_{j,g} \subseteq R_g \qquad j \in J, R_g \in R^* \qquad \ldots \text{Job Resource Assignment}$$

For notational convenience, let j_a refer to the job j containing $a \in A^*$. Its uniqueness is ensured by the feasibility conditions. Similarly, let J_p refer to the set of jobs that include tasks from project p.

Feasibility Conditions

$$\exists! j \in J : a \in J \qquad \forall a \in A^* \qquad (H1)$$
$$p_{a_1} = p_{a_2} \wedge f_{a_1} = f_{a_2} \qquad \forall j \in J, a_1 \in j, a_2 \in j \qquad (H2)$$
$$\exists j \in J : j^0 \subseteq j \qquad \forall j^0 \in J^0 \qquad (H3)$$

The *Job Assignment* condition (H1) ensures every task is part of exactly one job. The *Job Grouping* condition (H2) enforces that all tasks within a job belong to the same project and family. The *Fixed Tasks* condition (H3) requires all tasks in a fixed job to be part of the same job in the solution.

$$\dot{n}_j - \dot{s}_j = \begin{cases} \left[\left(\sum_{a \in j} d_a \right) \cdot v_{\dot{m}_j} \right] & \text{if } \exists j^s \in J^{0S} : j^s \subseteq j \\ \left[\left(s_{f_j} + \sum_{a \in j} d_a \right) \cdot v_{\dot{m}_j} \right] & \text{otherwise} \end{cases} \qquad \forall j \in J \quad (H4)$$

$$\dot{s}_j \geq \max_{a \in j} \alpha_a \wedge \dot{n}_j \leq \min_{a \in j} \omega_a \qquad \forall j \in J \qquad (H5)$$
$$j_{a'} \neq j_a \implies \dot{n}_{j_{a'}} \leq \dot{s}_{j_a} \qquad \forall a \in A^*, a' \in \mathcal{P}_a \qquad (H6)$$
$$(\exists j^s \in J^{0S} : j^s \subseteq j) \implies \dot{s}_j = 0 \qquad \forall j \in J \qquad (H7)$$

The *Job Duration* condition (H4) requires the interval from start to finish of a job to align with the job's calculated duration, which is based on its task, chosen mode, and setup time (provided the job does not contain tasks from a started job). The *Time Window* condition (H5) specifies that each job must respect the release date and deadline of all of its tasks. The *Task Precedence* condition (H6) asserts that a job may only start after the completion of all its prerequisite jobs. Jobs inherit the precedence relation from their tasks. Tasks with a precedence relation may also be performed as part of the same job. The *Started Jobs* condition (H7) ensures that a job incorporating fixed tasks from a started job in the base schedule must begin at time slot 0.

$$|\{j \in J : \dot{s}_j \leq t \wedge \dot{n}_j > t \wedge r \in \dot{a}_{j,g}\}| \leq 1 \quad \forall R_g \in R^*, r \in R_g, t \in T \quad \text{(H8)}$$

$$|\dot{a}_{j,g}| = \max_{a \in j}(r_{a,g,\dot{m}_j}) \quad \forall j \in J, R_g \in R^* \quad \text{(H9)}$$

$$\dot{a}_{j,g} \subseteq \bigcap_{a \in j}(R_{a,g}) \quad \forall j \in J, R_g \in R^* \quad \text{(H10a)}$$

$$\dot{m}_{\xi(a)} \in M_a \quad \forall a \in A^* \quad \text{(H10b)}$$

$$\dot{a}_{j_{a_1},g} = \dot{a}_{j_{a_2},g} \quad \forall R_g \in R^{linked}, p \in P, L \in L_p, a_1, a_2 \in L \quad \text{(H11)}$$

The *Single Assignment* condition (H8) ensures at any time slot t, each resource unit in every resource group R_g is assigned to at most one job. The *Resource Requirement* condition (H9) specifies that for each job j and resource group R_g, the number of assigned resources exactly matches the highest requirement of any task in j, considering the job's mode. The *Resource Suitability* condition (H10a) guarantees that, for each resource Group R_g, resources assigned to a job are compatible with all tasks in the job. Likewise, (H10b) enforces compatibility of assigned modes. Finally, the *Linked Tasks* condition (H11) ensures that whenever two jobs contain tasks from the same linked task set, they are assigned the same resource units for all resource groups in R^{linked}.

Preferences. Similarly to TLSP, G-TLSP contains 5 preferences. Each one corresponds to one optimization objective and has an associated penalty $s_i, i \in \{1, 2, 3, 4, 5\}$.

$$s_1 = |J| \quad \text{(S1)}$$

$$s_2 = \sum_{j \in J} \sum_{R_g \in R^*} |\dot{a}_{j,g} \setminus \bigcap_{a \in j}(R_{a,g}^{Pr})| \quad \text{(S2)}$$

$$s_3 = \sum_{p \in P} \sum_{R_g \in R^{min}} |\bigcup_{j \in J_p} \dot{a}_{j,g}| \quad \text{(S3)}$$

$$s_4 = \sum_{j \in J} \max(\dot{n}_j - \min_{a \in j}(\bar{\omega}_a), 0) \quad \text{(S4)}$$

$$s_5 = \sum_{p \in P} \left(\max_{j \in J_p} \dot{n}_j - \min_{j \in J_p} \dot{s}_j \right) \quad \text{(S5)}$$

The *Number of Jobs* preference (S1) minimizes the total number of jobs. The *Preferred Resource* preference (S2) penalizes assignments of resources to jobs that are not preferred by all of the job's tasks. Next, the *Number of Resources* preference (S3) minimizes the number of different resources assigned to jobs

in every project. It is applied to the resource groups specified in R^{min}. The
Due Date preference (S4) penalizes jobs that end after their (soft) due date $\bar{\omega}_j$.
Finally, the *Project Completion Time* preference (S5) seeks to minimize the total
completion time for each project. That is, the time from the start of the first job
to the end of its last one.

4 CP Model

This Section proposes a CP model to solve G-TLSP. The structure of the model
follows the constraints described in Sect. 3, with several adaptions to make it
suitable and performant for a wide variety of CP solvers.

For one, since support for set variables is limited across solvers, the model
encodes the partitioning of tasks into jobs J by treating each task as a potential job. One set of scheduling variables is created for each task. To decide on
the grouping, an additional array of decision variables $\dot{\xi}(a) \in A^* \quad \forall a \in A^*$
allows each task to point at one task. Tasks pointing at themselves are called
representative tasks and are interpreted as a job. Tasks pointing at other tasks
are interpreted as being merged into a job represented by a different task, and
have their scheduling variables ignored. Constraints ensure that these pointers
actually represent a partition.

Next, since many solvers lack support for floating-point variables, task durations $d_a, a \in A^*$ and setup times $s_f, f \in F^*$ are multiplied by a scale factor $\mathcal{M}$
and rounded up to an integer in preprocessing. Finally, since many solvers only
support a single optimization objective, the model contains a single optimization
objective which is a weighted sum of soft constraint violations.

The model was implemented in the solver-independent constraint programming language MiniZinc [10], but we expect the formulation to easily carry over
to other languages and solvers.

4.1 Decision Variables

To build a solution, the solver needs to assign the following variables:

$\dot{\xi}(a) \in A^*$	$a \in A^*$	... Assigned representative task
$\dot{m}_a \in M$	$a \in A^*$	... Assigned mode
$\dot{s}_a \in T$	$a \in A^*$	... Assigned start time
$\dot{n}_a \in T$	$a \in A^*$	... Assigned completion time
$\dot{d}_a \in T$	$a \in A^*$	... Assigned duration
$\dot{a}_{a,r} \in \{0,1\}$	$a \in A^*, R_g \in R^*, r \in R_g$	... Job Resource Assignment

The decision variables for completion time and duration are redundant, but
simplify the model description and improve performance in practice. For each
task $a \in A^*$, the model instantiates scheduling variables $\dot{m}_a, \dot{s}_a, \dot{n}_a, \dot{d}_a$, and

$\dot{u}_{a,r}$ $\forall R_g \in R^*, r \in R_g$. If the solver decides $\dot{\xi}(a) = a$, a is interpreted as representing a job, and the assignments to its scheduling variables correspond to a scheduled job in the solution. Otherwise, its scheduling variables are ignored by all constraints, hold no significance, and are set to 0 for symmetry-breaking.

4.2 Hard Constraints

The first set of constraints describes the task grouping.

$$\dot{\xi}(\dot{\xi}(a)) = \dot{\xi}(a) \qquad a \in A^* \qquad \text{(CP-H1)}$$

$$p_a = p_{\dot{\xi}(a)} \wedge f_a = f_{\dot{\xi}(a)} \qquad a \in A^* \qquad \text{(CP-H2)}$$

$$\texttt{all\_equal}(\{\dot{\xi}(a) \mid a \in j^0\}) \qquad j^0 \in J^0 \qquad \text{(CP-H3)}$$

$$\dot{\xi}(a) \leq a \qquad a \in A^* \qquad \text{(CP-H1-SYMM)}$$

$$\dot{\xi}(a) \neq a \implies (\dot{m}_a = 1 \wedge \dot{s}_a = \dot{n}_a = \dot{d}_a = 0$$
$$\wedge \forall_{R_g \in R^*, r \in R_g} \dot{u}_{a,r} = 0) \qquad a \in A^* \qquad \text{(CP-SYMM)}$$

Constraint (CP-H1) forces representative tasks to point at themselves, which implies that $\dot{\xi}(a)$ partitions tasks into jobs. (CP-H2) enforces that tasks may only point at a representative task from the same project and family, which implies that only tasks from the same project and family can form a job. (CP-H3) enforces that tasks from fixed jobs are grouped together. Together, constraints (CP-H1–CP-H3) already describe a legal grouping. (CP-H1-SYMM) breaks the symmetry of choosing a representative task by forcing the task with the smallest index in any job to be the representative. (CP-SYMM) breaks the symmetry of making arbitrary assignments to the unused scheduling variables of tasks that are not selected as representative tasks.

$$\dot{s}_{\dot{\xi}(a)} \geq \alpha_a \wedge \dot{n}_{\dot{\xi}(a)} \leq \omega_a \qquad a \in A^* \qquad \text{(CP-H4)}$$

$$\dot{\xi}(a) = a \implies \dot{d}_a = \dot{n}_a - \dot{s}_a \qquad a \in A^* \qquad \text{(CP-DUR)}$$

$$Setup(a) = \begin{cases} 0 \text{ if } \exists j^s \in J^{0S}, a_2 \in j^s \mid \dot{\xi}(a_2) = a \\ \lceil s_{f_a} \cdot v_{\dot{m}_a} \cdot \mathcal{M} \rceil \text{ else} \end{cases} \qquad \text{(CP-SETUP)}$$

$$\dot{\xi}(a) = a \implies \dot{d}_a \cdot \mathcal{M} \geq Setup(a) + \sum_{a_2 \in A^* \mid \dot{\xi}(a_2) = a} \lceil d_a \cdot v_{\dot{m}_a} \cdot \mathcal{M} \rceil \qquad a \in A^*$$

$$\dot{\xi}(a) = a \implies (\dot{d}_a - 1) \cdot \mathcal{M} < Setup(a) + \sum_{a_2 \in A^* \mid \dot{\xi}(a_2) = a} \lceil d_a \cdot v_{\dot{m}_a} \cdot \mathcal{M} \rceil \qquad a \in A^*$$

$$\text{(CP-H5)}$$

Constraints (CP-H4) through (CP-H5) concern job durations. (CP-H4) enforces that the start date and deadline of every task inside a job must be respected. Constraint (CP-DUR) connects $\dot{d}_a$ to $\dot{s}_a$ and $\dot{n}_a$. Formula (CP-SETUP) defines a shorthand notation for the setup time of a job. Then, constraint (CP-H5) calculates and rounds up the duration of jobs.

$$\dot{\xi}(a) = \dot{\xi}(a_2) \vee \dot{n}_{\dot{\xi}(a_2)} \leq \dot{s}_{\dot{\xi}(a)} \qquad a \in A^*, a_2 \in \mathcal{P}_a \quad \text{(CP-H6)}$$

$$\dot{s}_{\dot{\xi}(a)} = 0 \qquad j^s \in J^{0S}, a \in j^s \quad \text{(CP-H7)}$$

$$\texttt{cumulative}((\dot{s}_a)_{a \in A^*}, (\dot{d}_a)_{a \in A^*}, (\dot{a}_{a,r})_{a \in A^*}, 1) \qquad R_g \in R^*, r \in R_g \quad \text{(CP-H8)}$$

Constraint (CP-H6) enforces task precedences and (CP-H7) forces jobs with started tasks to start at the beginning. Constraint (CP-H8) ensures that no resource is used by multiple jobs at the same time. In MiniZinc, `cumulative(s, d, a, c)` is a special scheduling constraint that enforces that a list of jobs with start times `s`, durations `d`, and demands `a` for a single replenishing resource never exceed its capacity `c`.

$$\dot{\xi}(a) = a \implies \sum_{r \in R_g} \dot{a}_{a,r} = \max_{a_2 \in A^* | \dot{\xi}(a_2) = a} r_{a_2, g, \dot{m}_a} \qquad a \in A^*, R_g \in R^* \quad \text{(CP-H9)}$$

$$\dot{a}_{\dot{\xi}(a), r} > 0 \implies r \in R_{a,g} \qquad a \in A^*, R_g \in R^*, r \in R_g \quad \text{(CP-H10a)}$$

$$\dot{m}_{\dot{\xi}(a)} \in M_a \qquad a \in A^* \quad \text{(CP-H10b)}$$

Constraint (CP-H9) ensures that each job is assigned exactly the correct number of resources from each resource group. Given the mode assigned to a job, for each resource group, the assignment exactly covers the highest demand from any of its tasks for that group. Constraint (CP-H10a) ensures that resources assigned to a job are available to all tasks that are part of the job. Constraint (CP-H10b) enforces that the mode of a job is available to all of its tasks.

$$\dot{a}_{\dot{\xi}(a), r} = \dot{a}_{\dot{\xi}(a_2), r} \qquad p \in P, (a, a_2) \in L_p, R_g \in R^{linked}, r \in R_g \quad \text{(CP-H11)}$$

Finally, constraint (CP-H11) ensures that for resource groups $R_g \in R^{linked}$, the jobs containing linked tasks have the same resource units assigned.

4.3 Soft Constraints

Since MiniZinc and many solvers only support single-objective optimization, the different soft constraints as a weighted sum of the individual constraint penalties. Publicly available test instances usually use uniform weights.

The weights are given by $w_i \in \mathbb{R}_0^+, i \in \{1, 2, 3, 4, 5\}$. The objective function is given by $\sum_{i \in \{1,2,3,4,5\}} (s_i \cdot w_i)$, where s_i is given by:

$$s_1 = \sum_{a \in A^* | \dot{\xi}(a)=a} 1 \tag{CP-S1}$$

$$s_2 = \sum_{a \in A^* | \dot{\xi}(a)=a} \sum_{R_g \in R^*} \sum_{r \in (R_g \setminus (\bigcap_{a' \in A^* | \dot{\xi}(a')=a} R_{a',g}^{Pr}))} \dot{a}_{a,r} \tag{CP-S2}$$

$$s_3 = \sum_{p \in P} \sum_{R_g \in R^{min}} \sum_{r \in R_g} ((\sum_{a \in A_p} \dot{a}_{a,r}) > 0) \tag{CP-S3}$$

$$s_4 = \sum_{a \in A^* | \dot{\xi}(a)=a} \max(0, \dot{n}_a - \min_{a' \in A^* \text{ s.t. } \dot{\xi}(a')=a} (\bar{\omega}_{a'})) \tag{CP-S4}$$

$$s_5 = \sum_{p \in P} (\max_{a \in A_p}(\dot{n}_a) - \min_{a \in A_p \text{ s.t. } \dot{\xi}(a)=a} (\dot{s}_a)) \tag{CP-S5}$$

Soft constraint (CP-S1) minimizes the number of created jobs. (CP-S2) minimizes the number of non-preferred assigned resources and (CP-S3) minimizes the number of different resources used for each project. It can apply to any set of resource groups, as specified by R^{min}. Next, (CP-S4) minimizes the violations of tasks' due dates. Finally, (CP-S5) minimizes the duration of projects.

4.4 Optimizations

In G-TLSP, the number of resources from a group R_g required to perform a task a is given by $r_{a,g,m}$, which also depends on the mode m. The model encodes this with constraint (CP-H9). This is a flexible way to model the requirements but can result in an excessive number of constraints when this flexibility is not needed. To solve this, we created an optimized variant of (CP-H9) that detects cases where the number of required resources is effectively independent of the task or the mode. It can then use a more efficient specialized variant of the constraint.

$$R^{\text{task-dependent}} := \{R_g \in R^* \mid \forall_{a \in A^*} \forall_{m_1, m_2 \in M_a} \quad r_{a,g,m_1} = r_{a,g,m_2}\}$$
$$R^{\text{mode-dependent}} := \{R_g \in R^* \mid \forall_{m \in M} \forall_{a,a' \in A^*} ($$
$$(m \in M_a \land m \in M_{a'}) \implies r_{a,g,m} = r_{a',g,m})\}$$

We define $R^{\text{task-dependent}}$ and $R^{\text{mode-dependent}}$ to be the sets of resource groups where the required number effectively depends only on the task or only on the mode, respectively. This can easily be computed during the model's compilation.

$$\dot{\xi}(a) = a \implies \sum_{r \in R_g} \dot{a}_{a,r} = \begin{cases} r_{a,g,\dot{m}_a} & \text{if } R_g \in R^{\text{mode-dependent}} \\ \max_{a_2 \in A^* | \dot{\xi}(a_2)=a} r_{a_2,g,1} & \text{if } R_g \in R^{\text{task-dependent}} \\ \max_{a_2 \in A^* | \dot{\xi}(a_2)=a} r_{a_2,g,\dot{m}_a} & \text{otherwise} \end{cases}$$
$$a \in A^*, R_g \in R^* \tag{CP-H9-OPT}$$

The optimized constraint (CP-H9-OPT) replaces (CP-H9). In the first case, where the requirement only depends on the mode, the inner max expression over potential other tasks in the same job becomes superfluous, since other tasks share the same required amounts. In the second case, the number depends only on the tasks and the inner expression does not change based on the mode, so an arbitrary mode (in this case the mode with index 1) can be used.

5 Very Large Neighborhood Search

To find good solutions for large instances, we employ a Very Large Neighborhood Search (VLNS) algorithm, which is an improved version of the algorithm from [4]. The original algorithm starts from an initial solution and then iteratively improves it. At each step, it destroys the schedule of a small number of projects and repairs it by solving a small CP instance. For a more detailed description of the algorithm, we refer to [4].

The original algorithm starts from a feasible initial solution that is found with CP. This causes it to time out on some large instances where CP is unable to find a feasible solution within the time limit. Our new version addresses this by instead starting from a (usually infeasible) solution obtained by a greedy construction heuristic, and extending the algorithm's main loop to enable it to deal with hard conflicts as well.

The greedy heuristic builds a schedule that satisfies most hard constraints, violating only the *linked tasks* and *resource conflict* constraints. It also locally reduces the number of violations of these constraints while building the schedule.

The heuristic starts by grouping tasks into jobs that are as large as possible and assigning the mode with the shortest duration. We have found this to work best empirically for our instances, both artificial and real-world. It then schedules projects in ascending order of their earliest deadline. When scheduling a project, it topologically sorts the project's jobs based on the task predecessors $\mathcal{P}_a, a \in A^*$. Based on this topological order, it computes the latest possible start time for every job that still allows scheduling all its successor jobs before their deadline. It then schedules the jobs in ascending topological order. To schedule a job, the algorithm tries out all possible start times, keeping track of the best achievable constraint violations. Since the *linked tasks* and *resource conflict* constraint violations are independent for each assigned resource unit, an optimal resource assignment for a given start time can be found by first scoring each resource individually, and then assigning the required number of resources for each group based on the best scores.

Once this initial solution has been constructed, the new VLNS algorithm iteratively tries to remove hard constraint violations by following a loop similar to the improvement loop from the original algorithm. Since the schedule may violate the *linked tasks* and *resource conflict* constraints, the generated subinstances for the CP solver are further preprocessed to remove conflicts between and within fixed projects. For instance, when overlapping jobs from fixed projects have the same resource r assigned, preprocessing removes the requirement and assignment for r from all these jobs. Then, a new blocker task is created. Its requirements are set to force it to occupy resource r for the combined time slots of the conflicting jobs. *Linked tasks* violations can be removed by simply deleting the linked task sets L_p for fixed projects p.

With these improvements, the new VLNS algorithm can find solutions for instances where the CP model, and thus the existing VLNS, times out.

6 Evaluation

To evaluate the performance of our new CP model for G-TLSP, we compare its performance to the state-of-the-art specialized TLSP model. We evaluate the approaches on 35 instances in total. This includes 33 existing TLSP instances (30 artificial and 3 real-world) from [6]. In addition, we provide 2 new real-world instances taken from the daily scheduling operations of a partnered test laboratory. The existing instances range from 5 projects and 13 tasks to 90 projects with 1573 tasks. Existing real-world instances are on the higher end of this range, containing 59 to 74 projects and 660 to 856 tasks. For a more complete description, we refer to [6]. In this work, we introduce two new real-world instances: Lab4 is available for TLSP and G-TLSP and contains 76 projects and 924 tasks. Lab5 is only available for G-TLSP and encompasses 89 projects and 879 tasks. The new instances are publicly available at https://dbai.tuwien.ac.at/staff/fmischek/TLSP/.

We perform all evaluations on a Linux machine with a 12-core AMD Ryzen 9 5900X CPU and 64GB of DDR4-3200 RAM. Since all solvers are single-threaded, we perform 6 runs in parallel to save time. MiniZinc [10] version 2.8.7 and Chuffed [2] version 0.10.4 are used for all tests. For every configuration, we perform three runs with different random seeds. Each run has a timeout of 30 min. For the TLSP results, the original solver from [4] is used and re-run to ensure a consistent hardware configuration.

To ensure a fair comparison of the CP models, we apply our extended VLNS algorithm from Sect. 5 to both the TLSP and G-TLSP evaluations. The previous algorithm would time out on all instances where CP times out, since it uses CP to find an initial feasible solution. We use the same VLNS hyper-parameters as [4], with the timeouts for solving sub-instances reduced by a factor of 4, which roughly accounts for the hardware differences according to benchmark results.

6.1 Results

The full results are presented in Table 1 and summarized in Fig. 1. Overall, the performance of our new G-TLSP solvers is similar to the existing specialized TLSP solvers. In both cases, the pure CP solver manages to prove optimality only for very small instances, quickly drops off in performance on larger instances, and fails to solve some of the largest ones to feasibility. In contrast, our improved VLNS algorithm is able to solve even the largest instances and provides much

Table 1. Detailed evaluation results for all solvers and instances. The G-TLSP solvers use the new CP model proposed in this work, while the TLSP solvers are specialized for the TLSP and taken from the literature. The best and average results out of three runs are reported. When only some runs found a feasible solution, the number of feasible runs is given instead of the average penalty. Solver timeouts are indicated with —. The best result for each instance is bold. * indicates that the solver could prove optimality. Instance Lab5 has no results for the TLSP solvers since the instance is only available for G-TLSP.

#	CP				VLNS			
	TLSP		G-TLSP		TLSP		G-TLSP	
	Best	Avg	Best	Avg	Best	Avg	Best	Avg
1	**57***	57	**57***	57	**57**	57	**57**	57
2	**71***	71	**71***	71	**71**	71	**71**	71
3	142	142	142	142	**141**	142	**141**	142
4	120	120	119	119	103	103	**102**	102
5	244	244	287	287	**240**	248	**240**	240
6	180	180	188	188	**140**	140	**140**	140
7	359	360	411	415	284	285	**283**	283
8	311	311	311	311	284	285	**282**	284
9	689	689	729	730	**415**	423	428	428
10	956	956	1029	1029	**502**	508	503	507
11	1053	1053	1055	1055	**812**	818	815	821
12	737	737	804	804	644	647	**643**	648
13	336	336	364	364	316	316	**313**	315
14	453	453	492	492	**410**	411	413	414
15	1723	1723	1753	1753	901	927	**887**	904
16	1574	1575	1588	1588	**1111**	1115	1113	1114
17	1424	1425	1516	1516	1045	1053	**1040**	1045
18	1847	1847	1803	1803	1369	1382	**1353**	1355
19	2715	2715	2805	2805	**1821**	1863	1964	2011
20	3076	3076	2857	2857	2142	2155	**2118**	2151
21	949	949	1137	1137	**569**	571	571	572
22	980	980	1053	1054	**713**	732	721	729
23	—	—	—	—	**1602**	1656	1644	1677
24	—	—	—	—	1747	1762	**1737**	1742
25	4340	4340	3694	3694	**1884**	1907	1940	1970
26	3823	3823	3812	3812	**2537**	2600	2579	2602
27	3743	3743	3378	3378	**1762**	1780	1781	1793
28	3147	3147	2647	2647	**2236**	2242	2239	2244
29	—	—	—	—	**2897**	2965	3062	3083
30	6543	6543	6295	6295	**4465**	4571	4585	4633
Lab1	4990	4990	4709	4709	**3449**	3471	3742	(2/3)
Lab2	3407	3407	3285	3285	2653	2696	**2642**	2700
Lab3	2971	2973	3013	3013	2598	2609	**2580**	2588
Lab4	—	—	—	—	4581	4733	**4491**	4551
Lab5			—	—			**5210**	(2/3)

smaller penalties. Compared to VLNS based on TLSP, the VLNS based on the new G-TLSP seems to fall behind on large artificial test instances, but usually finds better solutions for the real-world instances despite being more general.

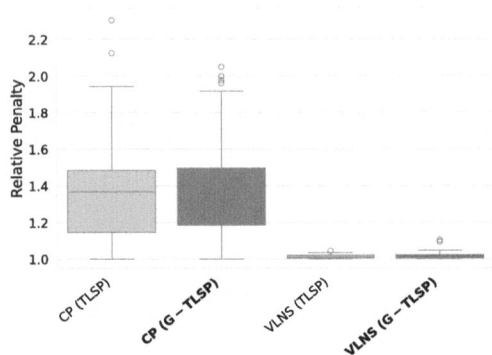

Fig. 1. Evaluation results overall. Penalties for each instance are normalized so that 1 corresponds to the lowest penalty for that instance across solvers. The G-TLSP solvers are based on the new CP model proposed in this work, while the TLSP existing specialized solvers for TLSP. Instances #23, #24, #29, Lab4, and Lab5 were excluded from the analysis because not all solvers were able to solve them within the time limit.

6.2 Deployment

The problem formulation and solution methods described in this paper have been successfully deployed in industrial practice. Our G-TLSP model, in particular via the VLNS algorithm, is currently being used for daily scheduling operations at a partnered industrial test laboratory. In addition, a second test laboratory is currently working on deploying the G-TLSP solvers for their environment.

7 Conclusion

In this paper, we introduced the Generalized TLSP (G-TLSP), a new variant of the Test Laboratory Scheduling Problem (TLSP) that unifies different resource types into a single framework. This generalization allows G-TLSP to capture additional real-world requirements and better carry over to other test laboratories. It also enables a more straight-forward integration of new constraints like resource mutual exclusion. In addition, TLSP instances can be converted and solved with a G-TLSP solver.

To solve G-TLSP, we proposed a new CP model and evaluated it both directly and as part of a VLNS algorithm. Our experimental results on 35 instances, including 2 new real-world instances, show that despite the increased flexibility of G-TLSP, our new model achieves competitive performance compared to specialized TLSP solvers from the literature. The CP-based exact solver could prove optimality for the smallest instances and found feasible solutions for most of the test instances. Our extended VLNS algorithm together with our new model is able to solve even the largest instances and performs particularly well on real-world instances, finding better solutions than the TLSP-based VLNS in several cases. The approaches presented in this paper, particularly the VLNS-based G-TLSP solver, are actively being used in practice.

Future work could investigate additional resource constraints that arise in practice. In addition, other solution approaches could be explored, like local search, parallelized VLNS, or learning-based methods.

Acknowledgments. The financial support by the Austrian Federal Ministry for Labour and Economy and the National Foundation for Research, Technology and Development and the Christian Doppler Research Association is gratefully acknowledged. In addition, this research was funded in part by the Austrian Science Fund (FWF) [10.55776/COE12]. Furthermore, Tobias Geibinger is a recipient of a DOC Fellowship of the Austrian Academy of Sciences at the Institute of Logic and Computation at the TU Wien.

Disclosure of Interests. The authors have no competing interests to declare that are relevant to the content of this article.

References

1. Bellenguez, O., Néron, E.: Lower bounds for the multi-skill project scheduling problem with hierarchical levels of skills. In: Burke, E., Trick, M. (eds.) PATAT 2004. LNCS, vol. 3616, pp. 229–243. Springer, Heidelberg (2005). https://doi.org/10.1007/11593577_14
2. Chu, G.: Improving combinatorial optimization. Ph.D. thesis, University of Melbourne, Australia (2011). http://hdl.handle.net/11343/36679
3. Danzinger, P., Geibinger, T., Janneau, D., Mischek, F., Musliu, N., Poschalko, C.: A system for automated industrial test laboratory scheduling. ACM Trans. Intell. Syst. Technol. **14**(1) (2023). https://doi.org/10.1145/3546871
4. Danzinger, P., Geibinger, T., Mischek, F., Musliu, N.: Solving the test laboratory scheduling problem with variable task grouping. In: Proceedings of the International Conference on Automated Planning and Scheduling, vol. 30, no. 1, pp. 357–365 (2020). https://doi.org/10.1609/icaps.v30i1.6681
5. Elmaghraby, S.E.: Activity networks: project planning and control by network models (1977). https://api.semanticscholar.org/CorpusID:60546726
6. Mischek, F., Musliu, N.: A local search framework for industrial test laboratory scheduling. In: Proceedings of the 12th International Conference on the Practice and Theory of Automated Timetabling (PATAT-2018), Vienna, Austria, 28–31 August 2018, pp. 465–467 (2018)
7. Mischek, F., Musliu, N.: A local search framework for industrial test laboratory scheduling. Ann. Oper. Res. **302**(2), 533–562 (2021). https://doi.org/10.1007/S10479-021-04007-1
8. Mischek, F., Musliu, N.: Leveraging problem-independent hyper-heuristics for real-world test laboratory scheduling. In: Proceedings of the Genetic and Evolutionary Computation Conference, GECCO '23, pp. 321–329. Association for Computing Machinery, New York (2023). https://doi.org/10.1145/3583131.3590354

9. Mischek, F., Musliu, N., Schaerf, A.: Local search approaches for the test laboratory scheduling problem with variable task grouping. J. Schedul. **26**(5), 457–477 (2023). https://doi.org/10.1007/s10951-021-00699-. https://ideas.repec.org/a/spr/jsched/v26y2023i5d10.1007_s10951-021-00699-2.html
10. Nethercote, N., Stuckey, P.J., Becket, R., Brand, S., Duck, G.J., Tack, G.: MiniZinc: towards a standard CP modelling language. In: Bessière, C. (ed.) CP 2007. LNCS, vol. 4741, pp. 529–543. Springer, Heidelberg (2007). https://doi.org/10.1007/978-3-540-74970-7_38
11. Potts, C.N., Kovalyov, M.Y.: Scheduling with batching: a review. Eur. J. Oper. Res. **120**(2), 228–249 (2000)
12. Wauters, T., Kinable, J., Smet, P., Vancroonenburg, W., Vanden Berghe, G., Verstichel, J.: The multi-mode resource-constrained multi-project scheduling problem. J. Sched. **19**(3), 271–283 (2014). https://doi.org/10.1007/s10951-014-0402-0

Parallelising Lazy Clause Generation with Trail Sharing

Toby O. Davies[1(✉)], Frédéric Didier[2], Laurent Perron[2], and Peter J. Stuckey[3,4]

[1] Google Research, Sydney, Australia
tobyodavies@google.com
[2] Google Research, Paris, France
{fdid,lperron}@google.com
[3] Monash University, Melbourne, Australia
peter.stuckey@monash.edu
[4] OPTIMA ARC Industry Training and Transformation Centre, Melbourne, Australia

Abstract. We investigate the effectiveness of splitting the search space in parallelising the state-of-the-art CP-SAT solver. One of the key barriers to effective search-space splitting in learning solvers is the generated sub-problems are not independent, leading to substantial communication-related overhead, substantial redundant work, or both. Our contributions attempt to mitigate this issue: job reassignment; and trail sharing. Jobs (sub-trees) are reassigned to new workers if the clause database of the currently assigned worker appears ill-suited to the region of the search-space, when doing so workers can share some of the state from their local trail. We argue a trail prefix can be viewed as a lossy compressed representation of much of the relevant information a worker has learnt about a job, this information can be exploited by the next worker assigned the same subtree to avoid some redundant work. We show these approaches complement standard portfolio and clause-sharing approaches, improving CP-SAT's performance on MiniZinc challenge benchmarks with a moderate number of worker threads. To enable these approaches, we also introduce "Buffered Work Stealing," which can be parameterised to emulate the two main existing approaches to search-space splitting in the literature: Work Stealing and Embarrassingly Parallel Search, as well as an intermediate configuration between these two extremes that slightly outperforms both.

1 Introduction

There have been many successful attempts at parallelising non-learning CP solvers, with many of them achieving near linear or even super-linear speedups [20]. One of the primary problems with parallelising non-learning CP solvers is load balancing. There are two main approaches, the first, Embarrassingly Parallel Search (EPS) [18], generates many more subproblems than workers, and relies on the law-of-large-numbers so the sum of the runtimes of these smaller subproblems assigned to one worker will be much closer to the mean than if the problem were split in fewer parts. The second approach is work-stealing, where threads which run out of work will steal part of the search space from another thread so they continue to have work to do [16]. There are also more delicate issues such as the interaction between branching heuristics and the work stealing/assignment scheme [8]. If the parallel solver ends up searching the search tree in an order which is significantly contrary to that specified by the branching heuristic in sequential search, the efficiency of the parallel solver may suffer. However, these issues have largely been addressed [15].

Effectively parallelising nogood learning CP solvers is significantly more difficult due to the fact that nogoods learned in one subtree may exponentially speed up closing a later subtree. Naively solving two subtrees in parallel can therefore result in doing almost twice as much work, leading to a tiny speedup from doubling the number of workers. This means the solver either needs to share clauses between workers, or do significant amounts of redundant work.

We should first mention that since nogood learning is adapted to CP solvers from SAT solvers, many of the problems occurring in the parallelisation of nogood learning CP solvers also occur in the parallelisation of clause learning SAT solvers, see the background of [19] for an overview of the approaches used in the SAT community.

Much of the work on parallelising nogood learning CP/SAT solvers has focused on other opportunities for improvement. For example, portfolio solvers which exploit that different solvers are good at different problem classes. Some parallel solvers run multiple threads individually with different random seeds or settings to exploit the heavy-tail behavior of some problem classes [11]. Objective probing (see e.g. [10]) can also speed up the search for good solutions. Whilst these improvements are useful, they will always provide strongly diminishing returns as more cores are added.

Pure portfolio solvers can only be as fast as the fastest solver no matter how many threads are used unless they share some information. Finding good solutions faster in optimisation problems cannot speed up the proof of optimality beyond knowing the best solution value from the beginning. In order to get scalable parallelism it is ultimately necessary to tackle search parallelism and somehow mitigate the issue of nogood learning tying the different parts of the search tree together.

The contributions of this paper are:

- An architecture for search space splitting, Buffered Work Stealing, that can be parameterised to emulate the two most popular search space splitting approaches for constraint programming, as well as mixes of the two.
- An architecture for job reassignment, where jobs that are judged to be outside the area of expertise of the current worker, are handed back in order to migrate jobs to a workers that are better suited to tackling them.
- The introduction of Trail Sharing, where a lossy representation of the search performed on a job is stored in shared tree, in order to avoid repeated work.

– Experiments showing that the new features allow faster closing of optimisation problems, and lead to better scalability as the number of workers grows. Overall the extensions substantially improve the performance of the worlds leading CP solver.

2 Background

2.1 Lazy Clause Generation

Lazy Clause Generation (LCG) [21] is a state-of-the-art technique for solving Constraint Programming problems by lazily encoding the problem into a Boolean satisfiability problem as conflicts are discovered during search. This allows the solver to take advantage of conflict analysis, clause learning, and back-jumping offered by Conflict-Directed Clause Learning SAT solvers without having to completely encode the problem into a large and likely largely redundant or irrelevant SAT model before starting to solve the model.

A constraint *program* $P = (V, D, C, o)$ consists of a set of (integer) variables V, a domain mapping each variable $v \in V$ to set of integers $D(v) \subset \mathbb{Z}$, a set of constraints over the variables V, and an objective term o defined over variables V to be minimized w.l.o.g. A *valuation* for P is a mapping θ from variables to value in their domain $\theta(v) \in D(v)$. A *solution* of P is a valuation θ which satisfies all constraints $\models \theta(c), c \in C$. An *optimal solution* of P is a solution θ such where there is no solution θ' where $\theta'(o) < \theta(o)$.

We treat D, C and P as logical formulae as follows $D \equiv \forall_{v \in V} v \in D(v)$, $C \equiv \forall_{c \in C} c$, $P \equiv D \wedge C$. We use notation $B \models F$ to mean in all solutions of B the formula F holds.

A CP solver reasons about *atomic constraints* [7] of the form $v = d$ and $v \geq d$, and their negations $v \neq d$ and $v < d$. An LCG solver lazily maps atomic constraints over variables in V to a Boolean representation. It uses Boolean variables $[\![v \geq d]\!], d \in D(v) \setminus \{min(D(v))\}$ to represent the integer variable v with domain $D(v)$. If b is a "Boolean" variable $D(b) = \{0, 1\}$, then we treat b as Boolean notation for $[\![b \geq 1]\!]$. LCG solvers usually also use Booleans of the form $atomx = d$, but these are not used by CP-SAT [17] (except in decomposing constraints) and hence omitted here.

When the LCG solver propagates a lower bound $v \geq d$, using constraint c it generates an explanation $E \Rightarrow v \geq d$ where E is set of atomic constraints where $D \wedge c \models (\bigwedge E) \to v \geq d$. This is mapped to the Boolean clause $L \to [\![v \geq d]\!]$ where $L = \bigwedge_{e \in E} rep(e)$, which is given to the SAT solver as an explanation for $[\![v \geq d]\!]$. Note that this clause is a consequence of the constraint program P. Similarly for propagation of upper bounds $v \leq d$ ($\neg[\![v \geq d+1]\!]$). Propagation failure is similarly explained.

Lazy clause generation uses the usual SAT mechanisms for conflict resolution: on propagation failure it starts conflict resolution with the clause explaining failure, eventually learning a nogood. It takes advantage of the knowledge of the meaning of integer literals to simplify the resulting nogood, e.g. $[\![x \geq 3]\!] \wedge [\![x \geq$

$6] \to \bot$ is simplified to $[\![x \geq 6]\!] \to \bot$. A significant advantage of LCG over solving the same problem with (Max)SAT is in most cases very few representation literals for each integer $v \in v$ need to be created. These literals and explanation clauses are created on demand and are not stored unless they occur in a nogood.

In abuse of notation we shall write $P \models F$ where F is a formula involving Boolean literals, to mean $P \models F'$ where F' is F with the atomic constraint bracketing removed. Since each explanation is a consequence of the program P, then by induction each nogood n generated by conflict resolution is also a consequences of P, i.e. $P \models n$.

To solve optimisation problems CP solvers use branch and bound: when they find a solution θ they add a constraint $o < \theta(o)$ to C to remove any solutions which are not better. In this optimisation approach all nogoods previously derived remain valid consequences of the updated program.

2.2 Parallelising Solvers

Discrete optimisation and Boolean satisfiability both implicitly require exploring a very large search tree of possible valuations for the variables involved, and/or alternatively building a succinct proof of optimality/unsatisfiability. Parallelising this exploration/proof is a well explored problem. There are two main approaches to this: *splitting* approaches try to split the tree into parts and assign sub-trees to workers to explore in parallel; while *portfolio* approaches have each parallel worker exploring the whole tree, but communicating to the other workers to help reduce the proof size.

Portfolio Methods. Portfolio methods are ostensibly the easiest form of parallelisation to implement. In the simplest case the individual workers need not communicate at all. Note that since tree exploration is a heavy tailed phenomena, even this simple approach can lead to super-linear speedups, especially for satisfiable problems [3]. But in a modern solver we can do more: CP portfolios always share the best incumbent solution value, so that we can use stronger branch and bound pruning in each worker. Once the solver learns nogoods a portfolio can become much stronger by sharing them, we explore this in Subsect. 2.3.

Portfolio methods are strongest when there is a significant difference between the searches performed by different workers in the portfolio. If the searches are too similar then the workers risk repeating work. Usually portfolio approaches lose much of their effectiveness above some number of workers, say 8 or 16, as the workers become too similar.

Search-Space Splitting. Search-space splitting approaches either statically or dynamically break up the tree to be explored into jobs, and share these jobs amongst the workers. In the simplest forms such as embarrassingly parallel search [18] (similar to *cube and conquer* [13]) the jobs are created at the start, and workers select the next job in a global queue once they finish their current task. In more complex forms such as *work stealing*, whenever a worker has no job it asks some other worker to create a new job for it, stealing part of an existing job. Search-space splitting can also share incumbent solutions and nogoods, just like portfolio searches. Search-space splitting ostensibly does not repeat work, but because of the nature of combinatorial problems, many of the different jobs they tackle will have strong similarities and once we have nogoods, much work they perform is repeated when compared to a single worker exploring the entire tree. Like portfolio methods work sharing methods decrease in benefit as the number of workers grow, but usually still gain some measurable advantage.

Auxiliary Methods. During parallel solving there are opportunities for workers to perform other activities than tree search in the hope of improving the overall performance of the parallelised solve. Examples of this are: solution finding methods, typically local search based approaches that look for solutions not using tree search; solution improving methods, which explore around solutions founds by other workers to see if they can find better solutions nearby, using either local search or large neighbourhood search; dedicated inference methods (*probing*), which try to derive binary clauses from the knowledge gained by multiple workers.

2.3 Clause Sharing

Once the solvers support nogoods then almost all parallel approaches will share nogoods; this is a powerful improvement since even portfolio workers can have their work reduced by reusing nogoods discovered by other workers. But too many nogoods slow down propagation, so sharing all nogoods learned by all workers is ineffective, and indeed workers produce nogoods at a prodigious rate. To combat this clause-sharing SAT solvers implement some policy for which and how many nogoods to share within a limited time, and which workers to share them with [3,19].

Parallel SAT solvers were the first to make use of clause sharing. The question was always which clauses to share with which other workers? Initial methods concentrated on short clauses [12], some even restricting to only unit clauses [4]. Later is was found that *literal block distance (LBD)* [2] was a useful indicator of the reusability of clauses, and that was used to control sharing [1]. Modern approaches [3,19] to clause sharing SAT solvers recognise that sharing changes over time so adapt limits to maintain a regular amount of clause sharing over the lifetime of a solve.

While good clause sharing strategies certainly improve parallel SAT solvers, the eventual scalability of parallel clause sharing solvers seems to be limited due to the sequential structure of resolution proofs [14].

3 Parallel CP-SAT

The CP-SAT solver (also called CP-SAT-LP) [17] is a state-of-the-art Lazy Clause Generation (LCG)-based solver [21]. In it's parallel configuration, it implements a very effective parallel portfolio which has won the parallel track of the Minizinc Challenge [22] every year since 2018. The portfolio workers have very distinct search behaviour (and inference strength), which kinds of worker are enabled in the default portfolio vary with the number of worker threads.

With 8 threads these include a core guided solver [10]; activity-based search with 3 different levels of "linearization", which vary how much effort is dedicated to the redundant linear-programming propagator; a "fixed search" worker, which uses the user-specified search strategy and the intermediate "default" linearization; a "quick restart" worker which uses a randomised search strategy and restarts every 10 conflicts; and two "incomplete" workers that perform large neighbourhood search and local search [9].

CP-SAT's clause-sharing implementation shares all unit and binary clauses, plus up to 4KiB/s in total of additional clauses using an adaptive LBD threshold, sharing clauses up to LBD 5 and length 32. All of these clauses are shared to all workers.

One of the key principles in CP-SAT's synchronisation of its parallel workers is for workers to synchronise their own state with shared state only at the root (i.e. immediately after restart, or a conflict triggers a backjump to decision level 0). This ensures that the critical loop of decision making and propagation need only read thread-local data. Typically important information (like learned unit or binary clauses, new incumbent solutions, and new objective bounds) are written to some shared buffer immediately, and other workers will read from this buffer the next time they backtrack to the root. Our implementation of Search-Space Splitting workers aims to follow this design principle.

3.1 Buffered Work Stealing

The first of our contributions is a simple novel approach to Search-Space Splitting we call "Buffered Work Stealing" (BWS) that aims to maintain the most desirable properties of Work Stealing and EPS. With appropriate parameters BWS can emulate either of these algorithms.

BWS workers communicate via a **shared tree**, whose leaves are jobs assigned to BWS workers. This tree stores at each node, (in addition to parent and child pointers): the **decision** made at that node (except at the root); the best **objective bound** at the node; if it is **closed**; and if it is **implied** (a node is implied iff it's sibling is closed). The shared tree keeps a buffer of any leaves that are not assigned to any worker. Whenever any node is closed, all its children are closed, and its parent will also be closed if the node was implied. An example part of a shared tree is shown in Fig. 1 (where the decision is the label of the parent edge).

BWS has two key parameters: L, the number of open leaves per BWS worker; and N, the maximum total number of nodes per BWS worker. BWS with W workers maintains a tree with at most NW total nodes (including closed nodes), workers may split their assigned leaf when there are fewer than LW open leaves. Workers with no assigned leaf are assigned a leaf from the buffer when they reach level 0 if an unassigned leaf is available.

If the buffer is empty (e.g. if insufficient jobs, have been generated so far, or if we have reached the NW total node limit), then the worker will search from the root, effectively operating as a portfolio worker for 1 restart and then try again to acquire a job from the buffer.

When $L = 1$ and N is a large number (10000 in our experiments), BWS emulates Work Stealing; when $L \gg 1$ (64 in our experiments) and $N = 2L$, BWS emulates EPS. If $L = 2$ and $N = 10000$, BWS has the best properties of Work Stealing and EPS—the tree is mostly split later during search, when more information is available to workers; but workers very rarely have to wait to be assigned a new a job, as there is a significant buffer of jobs that can be immediately assigned.

Workers only sync information about their assigned leaf with the shared tree when they reach decision level 0, (typically on restart). This means no expensive synchronisation is required in the hot propagation loop.

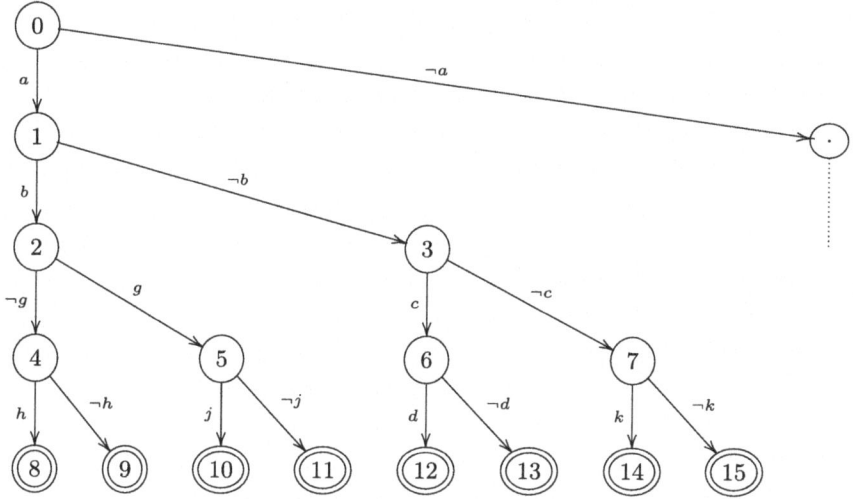

Fig. 1. Part of a shared tree, with leaf nodes in doubled circles.

Each worker takes decisions in the same order as the decisions in the path to their assigned leaf from the root in the shared tree. If a worker reaches a node that is marked as implied in the shared tree, and its decision literal d_i is not already propagated true, the worker propagates d_i with the reason $d_1 \wedge \cdots \wedge d_{i-1} \to d_i$ where $\{d_1, \ldots, d_{i-1}\}$ are the previous decisions.

For example, if node 11 of Fig. 1 is closed, then the branch at node 5 is no longer truly a branch, so a worker assigned node 10 should not treat j as a decision, but rather a consequence of earlier decisions, $a \wedge b \wedge g \to j$, is a correct explanation, if not necessarily the strongest. We weaken the explanation strength to avoid the need to share the nogoods that closed node 11.

If a worker reaches a decision d_i which is already negated (by local nogoods) then it learns a conflict $d_1 \wedge \cdots d_i \to \bot$ and the corresponding node in the shared tree (and all its descendants) are marked as closed, and the worker restarts and acquires a new leaf from the shared tree. For example, if the worker assigned node 15 ($[a, \neg b, \neg c, \neg k]$) propagates k after the first three decisions, it performs conflict analysis on the nogood $a \wedge b \wedge \neg c \wedge \neg k \to \bot$, and marks the node as closed.

When the worker has fixed all of the decisions it has been assigned, it begins branching normally using CP-SAT's default VSIDS-based branching strategy. Once propagation has finished at the next decision level after the end of the assigned job, workers propose to split their leaf using the decision they took immediately after their assigned job. So if a worker were assigned node 8 $[a, b, \neg g, h]$ (in Fig. 1) and then branched on d, once propagation has finished, they would propose to split their job into $[a, b, \neg g, h, d]$ (which the worker would keep), and $[a, b, \neg g, h, \neg d]$, which would be immediately placed in the unassigned buffer; updating the shared tree appropriately, i.e. adding new leaves $[a, b, \neg g, h, d]$ and $[a, b, \neg g, h, \neg d]$ under node 8.

The shared tree will reject a split if it would cause too many total nodes (NW) or open leaves (LW) in the shared tree. Additionally, to avoid creating too imbalanced a search tree, a split will only be accepted by the shared tree if the worker's assigned leaf is at a depth at most $log_2(LW) + 1$, excluding any implied nodes.

If a BWS worker has no assigned job and the buffer is empty, the worker performs a single restart as a portfolio worker. Initially all BWS workers race to split the root node.

3.2 Job Reassignment

In a parallel LCG or SAT solver, a critical consideration is which subset of nogoods to share with which other workers. Given that each worker can produce 1000 s of nogoods per second, the communication of all nogoods to all workers is infeasible. Even communicating a single nogood of length l to all other workers involves $O(Wl)$ work. Even more significant than the communication overhead, some of these clauses may be irrelevant or redundant given another worker's clause database, adding such redundant clauses slows down propagation for all workers until potentially expensive analysis detects the redundancy and removes the clause.

An alternate approach is to consider moving work to the worker that has the most relevant clause database. A job is defined by the $j \leq log_2(NW)$ decisions taken from the root of the search tree to the assigned leaf. So moving a job to a worker that is better suited is only $O(log_2(W))$ work (since N is a constant), yet can significantly reduce the time required to close a subtree compared to assigning to an arbitrary worker. Importantly moving jobs does not add any irrelevant clauses to any workers' clause databases, avoiding the slowdown that can be caused by over-sharing of clauses.

The main challenge of this approach is choosing when to move work from one thread to another. Our approach to this problem is inspired by Glucose's restart policy [2]—workers are initially assigned an arbitrary job, then periodically compare the average LBD of clauses learned while working on this job to a moving average of the LBD of clauses learned in the last 8 previously assigned jobs. When workers return their currently assigned tree to the unassigned buffer, or choose to keep their current tree, they add the average LBD of clauses learned on this job to the moving average. If their current leaf has a higher LBD than the moving average, the leaf is returned to the "unassigned" buffer, and a different leaf is assigned. Whenever a new tree is assigned the worker's restart strategy is reset (so e.g. the luby sequence starts again, amongst other things).

Workers have two phases when assigned a new tree—they initially do a single "trial" restart (in CP-SAT this is exactly 50 conflicts for the first restart after the restart policy is reset). If the average LBD among conflicts on this tree is no more than the threshold, they enter the second phase. In this phase workers check the average LBD of learned clauses (since the subtree was initially assigned) against the threshold at level 0 at most once per second, so each time a worker decided a tree is "good enough", it will be kept for a non-trivial length of time. When it does so, it adds the average LBD of clauses learned to the moving average. This is similar to the "one round per second" clause-sharing scheme used by HordeSat [3], and serves to limit the communication overhead.

Example 1. A leaf is assigned to a worker, the worker performs one restart (50 conflicts), on average the LBD of those conflicts is 7.6. Initially the worker's LBD moving average is 0 so the first subtree is returned after the first restart, a new leaf is assigned and we add 7.6 to the moving average. The worker then does one 50-conflict restart on the new leaf, these conflicts have an average LBD of 5.4, this is less than the moving average which is currently 7.6, so the tree is kept and 5.4 is added to the moving average (now $(7.6 + 5.4)/2 = 6.5$) After 1 s of work, the average LBD of all clauses since this tree was assigned is 5, less than the moving average of 6.5, so it is kept again, and 5 is added to the moving average (now $(7.6 + 5.4 + 5)/3 = 6$). In 1 more second, the average LBD of clauses since the restart heuristic was reset is 8, more than the threshold of 6, so the tree is returned, the restart heuristic reset, and 8 is added to the moving average (now $(7.6 + 5.4 + 6 + 8)/4 = 6.5$). This continues until there are over 8 samples in the moving average, then the oldest is dropped, so if the next 5 averages were 6, 5, 7, 9, and 5.6, the threshold would be $(5.4 + 6 + 8 + 6 + 5 + 7 + 9 + 5.6)/8 = 6.5$.

Note that it is theoretically possible for a worker to get stuck in "trial" mode restarts, however the small fixed-length moving average makes this extremely unlikely - each tree assigned only for 1 restart can only increase the LBD threshold, so eventually some subtree should become "good enough". The only realistic way for a worker to get stuck in this mode is if the worker is closing many subtrees in one restart, in which case we are happy for them to pull subtrees from the buffer at a greatly increased rate.

4 Trail Sharing

A job is defined by the sequence of decision literals taken to reach the leaf node in the shared search tree. But once we are moving jobs from one worker to another, simply making the same decisions as the previously assigned worker and resuming search will lead to significant redundant work, especially if the previous worker had performed significant work on the job. To avoid this repeated work without adding (significantly) to the communication cost, we also store a set of all literals that propagate at each open, non-implied node in the shared tree. By storing this information, each path from root to leaf effectively stores a copy of the trail up to that decision level. Workers copy this trail from the shared tree whenever they reach level 0, so rather than *clause sharing* they are *trail sharing*, and store any literals they propagated in the shared tree. The trail provides a compact, if lossy, representation of (potentially) a large amount of search by the previous worker(s).

But the trail generated by one worker for a sequence of decisions will not be the same for a different worker which has a different set of nogoods; many of the propagated literals in the trail will be caused by propagation of nogoods that are not present in the new worker. These must have propagated by some other worker, either one that gave up this job, or a worker that was working on another job that shares some prefix of decisions with this one.

To handle this Trail Sharing workers use the following behaviour: the new worker traverses the trail of literals T defining its new job from top to bottom as follows, initially with $depth = 0$, whenever propagation reaches fixed point:

- A decision literal l which is already true is marked as implied (and the sibling node is closed), all literals implied at $depth + 1$ are now implied at $depth$.
- A literal l which is implied at $depth$ and already true is ignored.
- A decision literal l which is unassigned is treated as a new decision. The variable $depth$ is incremented and we set $d_{depth} = l$ then all constraints propagate given this new decision.
- An implied literal l which is unassigned, is enqueued for propagation with the explanation $d_1 \wedge \cdots \wedge d_{depth} \to l$ and propagation is resumed. This simple explanation is valid, since it must be a consequence of all previous decisions on the trail, since it was a valid propagation for the earlier worker, all of whose nogoods are globally valid (See Theorem 1 below).
- A non-decision literal l which is false, creates a conflict with explanation $d_1 \wedge \cdots \wedge d_{depth} \to \bot$. The correctness of the explanation holds using the same reasoning above.

Example 2. Worker 1 is assigned job 12 $[a, \neg b, c, d]$ of Fig. 1; in processing the job it uses (local) nogoods to propagate $\neg b \to e$ and $e \wedge c \to f$. Suppose it gives up the job after its first trial restart with trail $T \equiv [a, \neg b, e^*, c, f^*, d]$ where propagated consequences are marked *. The consequence e will be added to node 3, and f to node 6. Next suppose worker 2 is assigned the job 12 it will see a trail, T, defining job 12. Suppose in traversing the trail it uses (local) nogood $a \to \neg b$

to remove the $\neg b$ decision (marking node 3 as implied, and node 2 as closed in the shared tree). But locally it does not propagate e; it thus adds e to its trail with explanation $a \to e$ (clearly a correct consequence of the globally true nogoods $a \to \neg b$ and $\neg b \to e$). Suppose local nogood $a \wedge e \to k$ then propagates. It then makes the decision c which in this case also propagates f since worker 2 shares the clause $e \wedge c \to f$. Finally it makes the decision d and begins new exploration with starting trail $[a, (\neg b)^*, e^*, k^*, c, f^*, d]$. At its next restart, this worker will mark node 2 as closed, adding consequences $\neg b$, e and k to node 1.

Overall the shared tree effectively stores all consequences that are discovered in the decisions encoded in the shared tree, even when the nogoods defining them are larger than the clause sharing policy will share amongst workers. The weakened explanations generated in the trail resulting from this policy are guaranteed to be correct, and are usually small, even if the 1UIP nogoods that initially propagated these implications are large, and thus may not be shared by CP-SAT's regular clause sharing policy.

Theorem 1. *The explanation for an unpropagated implied literal l appearing in trail T defined by $d_1 \wedge \cdots \wedge d_{depth} \to l$ is valid for constraint program P.*

Proof. Given the trail T was created by some worker p with a set of nogoods N_p, let $d'_1, \ldots, d'_m$ be the sequence of previous decision literals in the trail T before l appears. Now l is clearly a consequence of $d'_1, \ldots, d'_m$ and the constraint program P and N_p since it is generated by unit propagation. Clearly its also a consequence of $d'_1, \ldots, d'_m$ and the constraint program P, since every nogood in N_p is a globally valid consequence of P. i.e. $P \models d'_1 \wedge \cdots \wedge d'_m \to l$.

Now $d_1, \ldots, d_{depth}$ is a sub-sequence of $d'_1, \ldots, d'_m$ by construction. Suppose d'_i is the first decision literal omitted when $depth = k$, then it must have been propagated before reached in the new worker n using its nogoods N_n. Again since each nogood in N_n is a globally valid consequence of P, this means that $P \models d_1 \wedge \cdots \wedge d_k \to d'_i$. We can continue this argument for all omitted decision to show that $P \models d_1 \wedge \cdots \wedge d_{depth} \to d'_1 \wedge \cdots \wedge d'_m$. By transitivity the result holds. □

Additionally, we store polarity information on leaves in the shared tree when they are returned to further reduce repeated work. CP-SAT's decision heuristic computes and uses a "target phase" partial assignment, periodically resetting the branching polarity of literals in this partial assignment each "stable phase" [5,6]. This target phase is the longest conflict-free partial assignment found since the last stable phase. When replacing a job, BWS workers store up to the first 256 Boolean literals from their target phase at each leaf in the shared tree. The next worker assigned to this job sets the polarity of these literals to match the values taken by the last worker.

This cross-worker phase-saving allows workers to avoid re-exploring the same search space deeper than the assigned prefix. We expect this to primarily be useful in finding new solutions, more than closing jobs faster, as it should allow the next worker to skip some variable assignments that the previous worker has proven impossible, but if a variable has no valid value because the job is infeasible, the new worker will still likely have to try both assignments.

Table 1. Comparison of search space splitting algorithms and the pure portfolio baseline with 24 workers. The best value in each column is in **bold**.

	Time	Points	Closed	Speedup	Best
Trail Sharing	**21.69**	**3720**	**85.17%**	**1.19**	91.03%
Reassignment	21.74	3490	**85.17%**	1.16	90.95%
BWS	21.91	3219	**85.17%**	1.16	91.18%
Work Stealing	22.20	2766	85.02%	1.12	91.48%
EPS	22.59	2856	84.79%	1.12	90.72%
Baseline	23.15	3311	84.33%	–	**92.32%**

5 Experiments

We evaluated the performance of our algorithms on all 1315 MiniZinc Challenge instances [22] up to 2024 with a 10 min time limit. Since portfolio approaches typically achieve super-linear speedups with small numbers of workers but see strongly diminishing returns at larger numbers, we investigate the performance of *adding* search-space splitting workers to CP-SAT's existing portfolio. To test this, we replace $(N-8)/2$ portfolio workers with BWS workers. We chose an offset of 8 as this is the minimum recommended number of workers in CP-SAT's default configuration, and to take only half of workers above this threshold as portfolio workers do still scale above 8 workers, and more "incomplete" LS and LNS workers in particular are very useful in finding high-quality solutions to the largest problems.

We want to test two hypotheses: first to see if Reassignment, Trail Sharing, and BWS outperform existing approaches to Search Space Splitting, and also CP-SAT's state-of-the-art baseline portfolio; second, we want to show that these workers scale better than the portfolio approach as more worker threads are added.

To test our first hypothesis we compare adding the different search-space splitting algorithms to CP-SAT, to one another, and to the baseline (CP-SAT's default portfolio), over all 1315 instances from all minizinc challenges up to 2024. This experiment uses 24 cores per solve, in this configuration CP-SAT uses either: 8 BWS workers, 11 "complete" portfolio workers, and 5 "incomplete" workers; or 15 "complete" portfolio workers, and 9 "incomplete" workers. This number of workers was chosen as an arbitrary small number where we have a comparable number of portfolio, incomplete and BWS workers, allowing us to demonstrate the effectiveness of these workers at a scale that is usable on many machines.

Where we report results for Embarrassingly Parallel Search, and Work Stealing, we configured BWS workers to emulate these algorithms as described in Sect. 3.1, where we report results for BWS, we use $L = 2$ and $N = 10000$.

To test our second hypothesis, we compare the performance of the Baseline and Trail Sharing configurations with 32 and 56 workers each over the same minizinc challenge instances. We expect to see a larger gap between the performance of the two Trail Sharing configurations compared to the baseline.

Experiments were performed on 128-core Google Compute Engine instances with 2 AMD EPYC 7B13 CPUs each and 240 GB of RAM.

6 Results

In Table 1, we show the results of our first experiment. The "Time" column shows the 10-s shifted geometric mean solve time over all instances in seconds. "Points" shows a simplified version of MiniZinc challenge-style scoring, explained below. "Closed" shows the fraction of the instances closed. "Speedup" shows the mean speedup vs the baseline (i.e. baseline solve time divided by this solver's solve time, averaged over all instances). "Best" shows the fraction of instances where the solver found the best solution among all solvers tested.

A solver scores a point for each other solver that it outperforms on each benchmark instance. A solver outperforms another if it proves optimality or infeasibility faster, or if neither prove optimality, it scores a point if it finds a better solution, or the same quality solution faster.

Trail sharing performs best on 4 of the 5 metrics. Interestingly the baseline is best at finding the best solutions, but closes the fewest instances and does so more slowly than any of the configurations including search-space splitting workers. This suggests that a larger offset before starting to add splitting workers may be preferable to the offset of 8 we tested with, if finding the best solution is more important than proving it optimal. We suspect this effect is largely due to having fewer LNS workers, which specifically are designed to find high quality solutions; but these don't help significantly to complete the search.

The "points" metric is intended to capture the trade-off between these two important qualities in solvers, and both Trail Sharing, and Reassignment outperform the baseline on this metric.

BWS largely lives up to its promise of being the best of EPS and Work Stealing, outperforming both on 4 of the 5 metrics, Work Stealing does find the best solution more often. This may be because Work Stealing workers effectively become portfolio workers regularly due to the unassigned buffer being empty every time a worker closes their job.

In Fig. 2, we see that there is still measurable benefit to adding additional trail sharing workers past 32, when there is negligible benefit to adding more portfolio workers. In Fig. 3, we plot the solve time of each instance with trail sharing vs the baseline with 56 cores. Each instance below the diagonal is faster with trail sharing enabled, the mean speedup vs the baseline is $1.27\times$ using trail sharing, though there is clearly some overhead for solves taking less than 1 s. Of the 720 instances solved by either configuration in more than 1 s, 445 of them were faster with trail sharing, and 275 were slower. With 24 cores, the mean speedup was $1.19\times$, so the relative speedup compared to the baseline increased

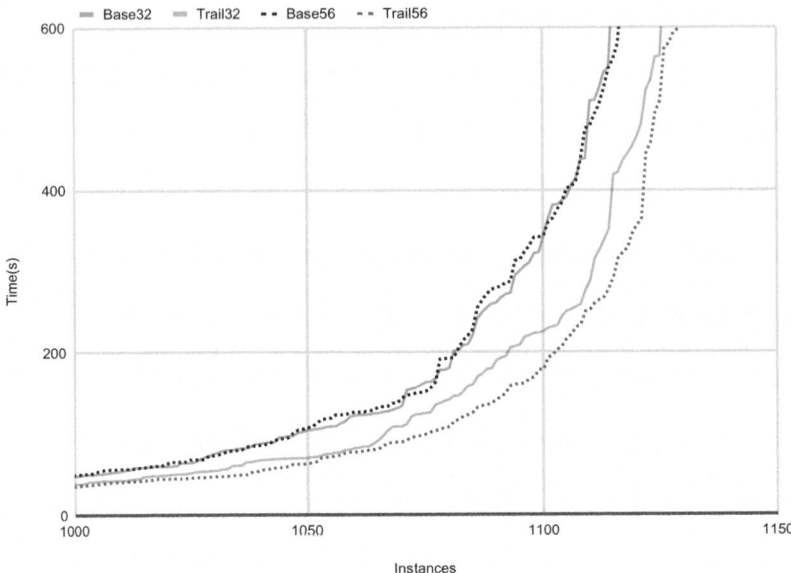

Fig. 2. Cactus plot of time to solve a number of instances, for baseline and trail sharing with 32 and 56 cores.

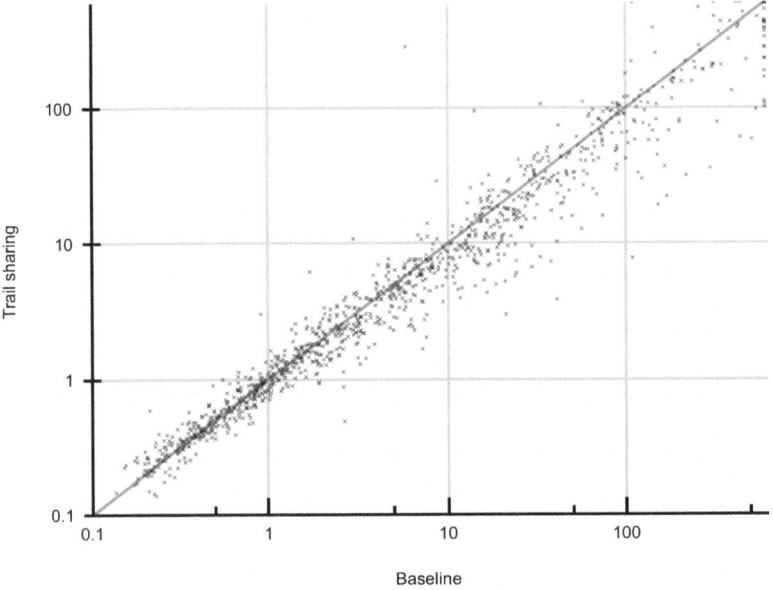

Fig. 3. Scatter plot of the solve time for each of the 1315 instances with 56 worker threads with and without trail sharing workers. Instances below the diagonal are faster with trail sharing enabled.

with more cores, again demonstrating that trail sharing scales better than CP-SAT's default clause-sharing portfolio.

7 Conclusions and Future Work

Search-space splitting techniques are complementary to more widely used clause-sharing portfolio approaches and combining them significantly improves the performance of CP-SAT on the MiniZinc challenge benchmarks when using at least 24 worker threads. These workers will be enabled by default in a future release of CP-SAT.

We introduce Trail Sharing, a lossy but very efficient way of sharing information that occurs near the root of the search. This is the information which has the potential to affect the shared search the most, and hence is most valuable to share, even imperfectly.

Interestingly while clause sharing portfolio solvers have been dominant in the SAT domain, search space splitting solvers appear to be especially useful for CP, where finding a good solution typically requires exploring more of the search tree, compared to finding a single solution or completing an unsatisfiability proof.

While our implementation uses shared memory to implement BWS and Trail Sharing, information is being copied between workers each restart, so BWS and Trail Sharing should also work in a distributed environment using explicit message passing. We suspect that scaling this approach may require some kind of distributed buffering where, each machine has a buffer that worker threads can pull from cheaply, and jobs are moved between machines periodically. Alternatively a nested approach may be effective, where each machine is treated like a single worker by a controlling process which would assign jobs to machines, and each machine could then further split those jobs among the worker threads on that machine.

BWS without reassignment could be applied to traditional non-learning CP or MIP solvers, where more independent subtrees should mean the benefit of splitting later is less significant, so we suspect BWS may more significantly outperform Work Stealing in such a solver.

The algorithms proposed in this paper also work for pure SAT, it would be interesting to integrate them into or compare them to state-of-the-art clause-sharing SAT solvers.

Finally, we believed the relative communication costs of job reassignment compared to nogood sharing are compelling. Our approach to job reassignment has a very simple notion of "expertise", we judge whether a job is "good" for a worker simply by looking at average LBD of nogoods produced. There is significant scope to explore different ways of assigning a jobs to the workers with the best "expertise" for the job that may substantially improve search space splitting approaches further.

Acknowledgments. The work of Peter Stuckey is partially supported by the Australian Research Council (OPTIMA ITTC IC200100009).

References

1. Audemard, G., Hoessen, B., Jabbour, S., Lagniez, J.-M., Piette, C.: Revisiting clause exchange in parallel SAT solving. In: Cimatti, A., Sebastiani, R. (eds.) SAT 2012. LNCS, vol. 7317, pp. 200–213. Springer, Heidelberg (2012). https://doi.org/10.1007/978-3-642-31612-8_16
2. Audemard, G., Simon, L.: Predicting learnt clauses quality in modern sat solvers. In: Proceedings of the 21st International Joint Conference on Artificial Intelligence, IJCAI'09, pp. 399–404. Morgan Kaufmann Publishers Inc., San Francisco (2009)
3. Balyo, T., Sanders, P., Sinz, C.: HordeSat: a massively parallel portfolio SAT solver. In: Heule, M., Weaver, S. (eds.) SAT 2015. LNCS, vol. 9340, pp. 156–172. Springer, Cham (2015). https://doi.org/10.1007/978-3-319-24318-4_12
4. Biere, A.: Lingeling and friends entering the SAT challenge 2012. In: Proceedinds of the SAT Challenge 2012: Solver and Benchmark Descriptions. Technical Report B-201202, University of Helsinki, pp. 33–34 (2012)
5. Biere, A., Fleury, M.: Chasing target phases. In: Workshop on the Pragmatics of SAT (2020)
6. Cai, S., Zhang, X., Fleury, M., Biere, A.: Better decision heuristics in CDCL through local search and target phases. J. Artif. Intell. Res. **74**, 1515–1563 (2022)
7. Choi, C., Lee, J., Stuckey, P.: Removing propagation redundant constraints in redundant modeling. ACM Trans. Comput. Logic **8**(4), 23 (2007). https://doi.org/10.1145/1276920.1276925
8. Chu, G., Schulte, C., Stuckey, P.J.: Confidence-based work stealing in parallel constraint programming. In: Gent, I.P. (ed.) CP 2009. LNCS, vol. 5732, pp. 226–241. Springer, Heidelberg (2009). https://doi.org/10.1007/978-3-642-04244-7_20
9. Davies, T.O., Didier, F., Perron, L.: Violationls: constraint-based local search in cp-sat. In: Dilkina, B. (ed.) Integration of Constraint Programming, Artificial Intelligence, and Operations Research, pp. 243–258. Springer, Cham (2024)
10. Ehlers, T., Stuckey, P.J.: Parallelizing constraint programming with learning. In: Quimper, C.-G. (ed.) CPAIOR 2016. LNCS, vol. 9676, pp. 142–158. Springer, Cham (2016). https://doi.org/10.1007/978-3-319-33954-2_11
11. Gomes, C.P., Selman, B., Crato, N., Kautz, H.: Heavy-tailed phenomena in satisfiability and constraint satisfaction problems. J. Autom. Reason. **24**(1), 67–100 (2000). https://doi.org/10.1023/A:1006314320276
12. Hamadi, Y., Jabbour, S., Sais, L.: Manysat: a parallel sat solver. J. Satisfiabil. Boolean Model. Comput. **6**(4), 245–262 (2010). https://doi.org/10.3233/SAT190070
13. Heule, M., Kullmann, O., Wieringa, S., Biere, A.: Cube and conquer: guiding CDCL sat solvers by lookaheads. In: Eder, K., Lourenço, J., Shehory, O. (eds.) Hardware and Software: Verification and Testing, pp. 50–65. Springer, Heidelberg (2012). https://doi.org/10.1007/978-3-642-34188-5_8
14. Katsirelos, G., Sabharwal, A., Samulowitz, H., Simon, L.: Resolution and parallelizability: barriers to the efficient parallelization of sat solvers. In: Proceedings of the AAAI Conference on Artificial Intelligence, vol. 27, no. 1, pp. 481–488 (2013). https://doi.org/10.1609/aaai.v27i1.8660
15. Michel, L., See, A., Van Hentenryck, P.: Transparent parallelization of constraint programming. INFORMS J. Comput. **21**(3), 363–382 (2009). https://doi.org/10.1287/ijoc.1080.0313
16. Perron, L.: Search procedures and parallelism in constraint programming. In: Jaffar, J. (ed.) CP 1999. LNCS, vol. 1713, pp. 346–360. Springer, Heidelberg (1999). https://doi.org/10.1007/978-3-540-48085-3_25

17. Perron, L., Didier, F., Gay, S.: The CP-SAT-LP solver. In: Yap, R. (ed.) 29th International Conference on Principles and Practice of Constraint Programming (CP). Dagstuhl, Germany (2023)
18. Régin, J.-C., Rezgui, M., Malapert, A.: Embarrassingly parallel search. In: Schulte, C. (ed.) CP 2013. LNCS, vol. 8124, pp. 596–610. Springer, Heidelberg (2013). https://doi.org/10.1007/978-3-642-40627-0_45
19. Schreiber, D., Sanders, P.: Mallobsat: scalable sat solving by clause sharing. J. Artif. Intell. Res. **80** (2024)
20. Schulte, C.: Parallel search made simple. In: Beldiceanu, N., Harvey, W., Henz, M., Laburthe, F., Monfroy, E., Müller, T., Perron, L., Schulte, C. (eds.) Proceedings of TRICS: Techniques foR Implementing Constraint Programming Systems, A Post-Conference Workshop of CP 2000, pp. 41–57. No. TRA9/00, 55 Science Drive 2, Singapore 117599 (2000)
21. Stuckey, P.J.: Lazy clause generation: combining the power of SAT and CP (and MIP?) solving. In: Lodi, A., Milano, M., Toth, P. (eds.) CPAIOR 2010. LNCS, vol. 6140, pp. 5–9. Springer, Heidelberg (2010). https://doi.org/10.1007/978-3-642-13520-0_3
22. Stuckey, P.J., Feydy, T., Schutt, A., Tack, G., Fischer, J.: The MiniZinc Challenge 2008–2013. AI Mag. **35**(2), 55–60 (2014)

Learning Primal Heuristics for 0–1 Knapsack Interdiction Problems

Luca Ferrarini[2(✉)], Stefano Gualandi[3], Letizia Moro[3,4], and Axel Parmentier[1]

[1] CERMICS, École des Ponts Paristech, Marne-la-Vallée, France
[2] LIPN, Université Sorbonne Paris Nord, Villetaneuse, France
ferrarini@lipn.univ-paris13.fr
[3] Università degli Studi di Pavia, Dipartimento di Matematica "F. Casorati", Pavia, Italy
[4] Agile Lab s.r.l., Paris, France

Abstract. In interdiction problems, two opposing decision-makers act sequentially: the leader plays first by selecting items to restrict the choices of the follower, while the follower selects those that maximize her profit from the remaining items. In knapsack interdiction, both decision-makers face different budget constraints. We propose a heuristic based on a single-level approximation of the leader-follower problem that we interpret as a combinatorial optimization layer in a machine learning pipeline. The ML pipeline includes a Generalized Linear Model as the first layer, which predicts the parameters of the single-level problem. Using a perturbation approach, we regularize the single-level problem, which enables to make it differentiable and provides a natural loss to train the model. Once trained, the pipeline provides effective ordering heuristics to solve Knapsack Interdiction problems. Extensive computational results on benchmarks from the literature show that the learned ML-based primal heuristics are extremely fast and compute solutions with a small optimality gap.

Keywords: Combinatorial Optimization · Knapsack Problem · Machine Learning · Fenchel-Young Loss

1 Introduction

The Knapsack Interdiction Problem dates back to 2011 when it was first introduced in DeNegre's Thesis [13], and since then has been deeply studied in the literature (e.g., see [9,12,16,30]). This interdiction problem can be considered bilevel version of the 0–1 Knapsack Problem (KP) [23] (see [7] for another variant). As described in [1], KPs have been objects of interest since 1896, when they were introduced by Mathews in [24]. The first dynamic programming algorithm for the KP was introduced by Bellman in 1954 [4], and in the following decade, Kolesar presented the first branch and bound algorithm [20]. Since then, several authors have contributed to the literature on this topic, and one of the most complete overviews of KPs can be found in [23]. Recent works on KPs are mainly focused on improving the solution algorithms, as in [27] and [22], on finding challenging instances, as in [28] and [18].

The Knapsack Interdiction Problem (KIP), that is known to be Σ_2^p-hard [8], belongs to a wider group of problems called bilevel interdiction problems [19]. These problems involve two interacting levels and have applications in various fields, such as engineering [2], economics [3], and military strategy [17]. Relevant works that focus on mixed-integer linear bilevel problems are presented in [13–15,25]. In [12], the authors present an exact approach based on mathematical programming lower bounds built around the leader's *critical item*, that is, the unique fractional item in the leader's fractional KP. In [16], the authors exploit a Benders decomposition and use interdiction cuts to tighten the LP relaxation. Another cutting plane approach is presented in [9], where the authors design a branch-and-cut algorithm using strengthened no-good cuts. A general approach for interdiction problems that use the KIP as a testbed is [29]. The state-of-the-art approach is the combinatorial branch-and-bound presented in [30], which exploits a new dynamic programming algorithm to compute strong lower bounds at every node of the branch-and-bound tree. Recently, in [21] the authors introduce a heuristic based on a deep learning architecture, utilizing a GNN framework.

Even the best algorithm can solve instances of only moderate size [30], and for this reason, in this work, we focus on efficient primal heuristics that trade optimality off with efficiency but allow to solve very large instances. We leverage the InferOpt framework introduced in [10] to design a Machine Learning pipeline with a Combinatorial Optimization (CO) layer for dealing with the two KP subproblems. Figure 1 presents the general pipeline that can be modeled using the InferOpt package: given a problem instance z, we design a machine learning layer φ_ω that maps the input into a weight vector θ, with ω as weight vector for the ML model. The vector θ is then fed into a CO oracle that solves a modified optimization problem over the set of feasible solutions $\mathcal{H}(z)$ with a linear objective with θ as a cost vector. The CO layer must be differentiable to permit the backpropagation of the Loss function to the weights of the ML model weights ω in the first layer. The complete details of the InferOpt package are in [10].

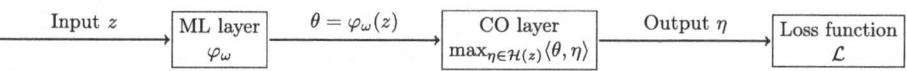

Fig. 1. Machine Learning pipeline with a Combinatorial Optimization layer [10].

Contributions. The main contributions of this paper are as follows:

1. We introduce a structured learning methodology to approximate interdiction problems by single-level problems. This methodology is generic and can

be straightforwardly applied to any bilevel optimization problem with full recourse.
2. We illustrate the methodology on the knapsack interdiction problem, for which we detail the full learning architecture.
3. The resulting single-level problem gives a primal heuristic for the interdiction problem. Through extensive numerical experiments on instances from the literature, we show the efficiency of the approach (see Table 4).

Outline. Section 2 reviews the bilevel formulation of the knapsack interdiction problem and fixes the notation. Section 3 presents the details of our learning architecture. Section 4 shows our computational results using benchmarks from the literature, and in Sect. 5, we discuss the benefits of our approach.

2 Backgrounds

In this section, we present the general backgrounds. First, we review interdiction problems formulated as bilevel programs. Second, we formally introduce the knapsack interdiction problem, and finally, we present a greedy algorithm from the literature.

2.1 Interdiction Problems

In interdiction problems, two opposing decision-makers act sequentially: the leader plays first by selecting items to restrict the choices of the follower, while the follower selects from the remaining items those that maximize her profit. This class of problems is naturally formulated as bilevel optimization problems by using objective functions $F, f : \mathbb{R}^n \times \mathbb{R}^m \to \mathbb{R}$, constraint functions $G : \mathbb{R}^n \times \mathbb{R}^m \to \mathbb{R}^l$, $g : \mathbb{R}^n \times \mathbb{R}^m \to \mathbb{R}^p$, and decision variables $x \in \mathcal{X} \subseteq \mathbb{R}^n$ and $y \in \mathcal{Y} \subseteq \mathbb{R}^m$. The general bilevel formulation is as follows.

$$\text{(Leader)} \quad \min_{x \in \mathcal{X}, y} \quad F(x,y)$$
$$\text{s.t.} \quad G(x,y) \geq 0, \quad (1)$$
$$y \in S(x),$$

where $S(x)$ is the set of optimal solutions of the x-parameterized follower's problem

$$\text{(Follower)} \quad \max_{y \in \mathcal{Y}} \quad f(x,y)$$
$$\text{s.t.} \quad g(x,y) \geq 0, \quad (2)$$

where $\mathcal{X}$ and $\mathcal{Y}$ can include integrality constraints. Problem (1) is the upper level that refers to the Leader while (2) is the lower level, representing the Follower's problem.

In the following subsection, we specialize this model to the case where the objective functions are linear, and the constraints are knapsack constraints.

2.2 The Knapsack Interdiction Problem

In the Knapsack Interdiction Problem, two opposing players interact with a set of items, where each item has a given profit and different weights for the leader and the follower. The leader plays first by selecting items to restrict the choices of the follower, who then selects among the remaining objects to maximize her profit. Both players face a budget constraint, but the follower wants to maximize her gain while the leader wants to prevent it.

Formally, given n items with a profit $p_i, \forall i \in [n]$, a weight w_i^L for the leader, and a weight w_i^F for the follower, we call C_L the leader capacity and C_F the follower budget. Note that herein, we denote everything related to the leader and the follower with superscripts L and F, respectively. We denote vectors $\mathbf{v} \in \mathbb{R}^N$ in bold and the scalar product of two vectors $\mathbf{a}, \mathbf{b} \in \mathbb{R}^n$ by $\langle \mathbf{a}, \mathbf{b} \rangle$. Using this notation, the Knapsack Interdiction Problem (KIP) is defined as follows.

$$\text{Leader} \begin{cases} \min\limits_{\mathbf{x} \in \{0,1\}^n} & \langle \mathbf{p}, \mathbf{y} \rangle \\ \text{s.t.} & \langle \mathbf{w}^L, \mathbf{x} \rangle \leq C_L, \end{cases}$$
$$\text{Follower} \begin{cases} \mathbf{y} \in \operatorname*{argmax}\limits_{\mathbf{y} \in \{0,1\}^n} & \langle \mathbf{p}, \mathbf{y} \rangle \\ \text{s.t.} & \langle \mathbf{w}^F, \mathbf{y} \rangle \leq C_F, \\ & \mathbf{y} \leq \mathbf{1} - \mathbf{x}. \end{cases} \qquad (3)$$

Interdiction constraints $\mathbf{y} \leq \mathbf{1} - \mathbf{x}$ represent the impossibility for the follower to choose any item selected by the leader. Herein, we denote the sets of feasible solutions for the leader by $\mathcal{X} = \{\mathbf{x} \in \{0,1\}^n : \langle \mathbf{w}^L, \mathbf{x} \rangle \leq C_L\}$, and for the follower by $\mathcal{Y} = \mathcal{Y}(\mathbf{x}) = \{\mathbf{y} \in \{0,1\}^n : \langle \mathbf{w}^F, \mathbf{y} \rangle \leq C_F, \mathbf{y} \leq \mathbf{1} - \mathbf{x}\}$. Once again, we remark that the feasible set of the follower is $\mathbf{x}$-parametric; that is, it depends on the leader's decision variables $\mathbf{x}$.

2.3 Greedy Algorithm from the Literature

The state-of-the-art exact algorithm to solve the Knapsack Interdiction Problem is introduced in [30], where the authors described a customized branch-and-bound algorithm. Their exact solver relies on a simple but effective heuristic to compute primal solutions. By exploiting the problem structure, the greedy heuristic, sketched in Algorithm 1, first solves a Knapsack problem for the leader and, second, solves the follower Knapsack problem by considering the leader decision variables as fixed. Overall, the greedy heuristic solves two 0–1 Knapsack Problems and, in practice, is very efficient and provides solutions with a small integrality gap. We refer the reader interested in the exact approach to the original paper [30]. In our computational experiments, we use the greedy algorithm presented in Algorithm 1 as a baseline. The next section introduces our approach to *learn* new primal heuristics that outperform this baseline.

Algorithm 1: State-of-the-art Greedy Heuristic [30]

1 **Function** GreedyHeuristic():
2 $\quad\bar{\mathbf{x}} \leftarrow \mathrm{argmax}\{\langle \mathbf{p}, \mathbf{x}\rangle : \mathbf{x} \in \mathcal{X}\}$;
3 $\quad\bar{\mathbf{y}} \leftarrow \mathrm{argmax}\{\langle \mathbf{p}, \mathbf{y}\rangle : \mathbf{y} \in \mathcal{Y}(\bar{\mathbf{x}})\}$;
4 $\quad$**return** $(\bar{x}, \bar{y})$;

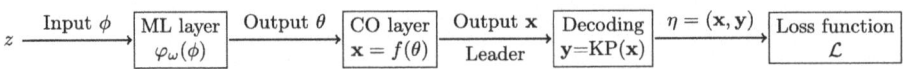

Fig. 2. Hybrid ML and CO pipeline architecture for KIP.

3 Hybrid Machine Learning Pipeline

Classical Machine Learning pipelines are composed of different *layers* tailored for the specific data to be treated. In our setting, we work on structured learning: we are interested in predictions over a large combinatorial set $\mathcal{H}$ and a large set of instance data $\mathcal{Z}$, where n represents the number of items of the instance z. In particular, we denote by $\mathcal{H}(z)$ the feasible set of the solutions associated with an instance $z \in \mathcal{Z}$. We focus on a machine learning pipeline, which embeds a combinatorial optimization problem as one of the pipeline layers. Hence, the two main ingredients for devising our pipeline are the design of the Combinatorial Optimization Layer and the Machine Learning Layer. Finally, designing the loss function that considers the optimization and learning parts is crucial. The reader can refer to [10] for more details.

In our settings, given an instance $z \in \mathcal{Z}$, the aim is to learn a parameter ω through a machine learning layer based on a generalized linear model φ_ω. The parameter ω is then used to define the cost vector θ, which is designed to lead to good predictions $\eta \in \mathcal{H}(z)$. Herein, we denote by $\mathcal{C}(z)$ the convex hull of the feasible solutions for the leader $\mathcal{X}$ for the instance z. At the end of the pipeline, a proper loss function is introduced to measure the quality of the output prediction. Figure 2 sketches our ML pipeline, and we describe the two layers in the following paragraphs. We first present the CO layer since it plays a crucial role.

3.1 Combinatorial Optimization Layer

Since the two levels of the bilevel formulation (3) deal with a Knapsack Problem each, a natural approach is to reduce the bilevel problem to the solution of two KPs, similarly to the approach of the greedy heuristics defined in [30]. However, instead of solving the first Knapsack exactly, we want to learn the weights θ that would lead to good solutions for the whole Knapsack Interdiction Problem. For this reason, we use the Knapsack Problem as a Combinatorial Optimization layer, which computes feasible solutions for the leader. Then, we retrieve the follower's solution by solving the follower's KP with an exact algorithm (KP(**x**) in 2). Given the weight vector θ defined as the output of the ML layer, we first solve the following problem:

$$\bar{\mathbf{x}} \in \operatorname{argmax} \langle \theta, \mathbf{x} \rangle$$
$$\text{s.t.} \langle \mathbf{w}^L, \mathbf{x} \rangle \leq C_L, \qquad (4)$$
$$\mathbf{x} \in \{0,1\}^n.$$

Notice that θ is unrestricted in sign. Problem (4) provides a *feasible solution* for the upper level: if we consider the KIP formulation in (3) we can notice that the only constraints on **x** are the integrality and budget constraints. Once we get $\bar{\mathbf{x}}$, we can then compute $\bar{\mathbf{y}}$ by solving:

$$\bar{\mathbf{y}} \in \operatorname{argmax} \langle \mathbf{p}, \mathbf{y} \rangle$$
$$\text{s.t} \langle \mathbf{w}^F, \mathbf{y} \rangle \leq C_F, \qquad (5)$$
$$\mathbf{y} \leq \mathbf{1} - \bar{\boldsymbol{x}}$$
$$\mathbf{y} \in \{0,1\}^n.$$

Indeed, the pair $(\bar{\mathbf{x}}, \bar{\mathbf{y}})$ is a feasible solution for the given KIP instance.

Remark 1. Given a feasible leader solution $\bar{\mathbf{x}}$, the corresponding follower solution $\bar{\mathbf{y}}$ is obtained by solving a KP with the items not selected by the leader. Thus, 4 finds the leader's optimal solution, while 5 determines the follower's optimal solution.

3.2 Machine Learning Layer

Given a KIP instance $z \in \mathcal{Z}$, we must first define a vector of features of z that might lead to good solutions of the combinatorial learning layer. The selection of features is based on the properties of the KP. We then employ a Generalized Linear Model (GLM) to *learn* a linear combination of feature values. We opt for a GLM since it offers easy interpretability with a reduced number of parameters. The GLM is formulated as a function $\varphi_w(z)$ that maps an instance z to a vector of weights $\theta \in \mathbb{R}^n$.

In practice, given an instance $z \in \mathcal{Z}$, we transform the instance z into a feature matrix ϕ, which has a fixed number of columns d (representing the number of features) and n rows, one for each item. We order the items by non increasing $\frac{p_i}{w_i^F}$ with ties broken by preferring instances with larger p_i. The feature matrix ϕ becomes the input of the machine learning layer, replacing instance z. Table 1 reports and describes all the $d = 30$ features used as input of our ML layer. Most features are defined as ratios of the given instance, while others depend on the *critical value* and the *critical index* of a KP instance. We recall that the *critical value* is defined as the value of the unique fractional variable in an optimal solution of a fractional KP [23,27]; the *critical index* is the index of the variable yielding the critical value. In the features numbered from 15 to 28, we indicate

Table 1. List of the $d = 30$ precomputed features of a KIP instance.

ID	Description	Definition
1	profit	p_i
2	leader's weight	w_i^L
3	follower's weight	w_i^F
4	ratio profit/weight	$\frac{p_i}{w_i^L}$
5	ratio profit/weight follower	$\frac{p_i}{w_i^F}$
6	ratio profit/product weight leader-follower	$\frac{p_i}{w_i^L w_i^F}$
7	ratio profit/weight follower squared	$\frac{p_i}{(w_i^F)^2}$
8	ratio weight leader/profit	$\frac{w_i^L}{p_i}$
9	ratio weight follower/profit	$\frac{w_i^F}{p_i}$
10	ratio weight follower/weight leader	$\frac{w_i^F}{w_i^L}$
11	ratio weight leader/weight follower	$\frac{w_i^L}{w_i^F}$
12	profit quantile	
13	leader's weight quantile	
14	follower's weight quantile	
15 → 28	distance of profit/weight follower to critical ratio	$\mathbb{1}((\frac{p_i}{w_i^F} - \frac{p_k}{w_k^F}) \in I)$
29	interaction with subsequent item	$\frac{p_i}{w_i^F} - \frac{p_{i+1}}{w_{i+1}^F}$
30	Boolean: if item i is in the solution of follower's KP	$\{0,1\}$

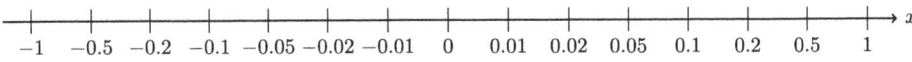

Fig. 3. Consecutive intervals for features 15–28. If $-1.0 \leq \left(\frac{p_i}{w_i^F} - \frac{p_k}{w_k^F}\right) \leq -0.5$, that is the first interval on the left, then the feature with ID = 15 is set to 1.

with $\frac{p_k}{w_k^F}$ the follower's critical value. In those 13 features, we check if $(\frac{p_i}{w_i^F} - \frac{p_k}{w_k^F})$ belongs to the I-th interval for all the intervals represented in Fig. 3.

Before passing the features matrix ϕ as input to the ML layer, we apply a regularization by dividing each column by its maximum value. Since we do not have negative values, every entry of ϕ is restricted to the interval $[0,1]$. Once the features matrix is preprocessed, we apply the GLM with parameters ω to an instance z to every i-th row ϕ_i by considering the following scalar product:

$$\varphi_\omega(\phi(z))_i = \langle \phi_i(z), \omega \rangle = \sum_{j=1}^{d} \phi_{ij}(z)\omega_j = \theta_i, \quad \forall i \in [n]. \tag{6}$$

Notice that, this model can be used with instances of any size n.

The vector $w \in \mathbb{R}^d$ is the vector of *weights* representing the linear combination of the features. Note that θ is indirectly defined in terms of an instance $z \in \mathcal{Z}$, and to make this dependence explicit, in the following, we write $\theta(z)$. In principle, finding the optimal w in the training process allows us to look for the most relevant leader's features for solving problem (4).

3.3 Learning by Imitation

The last component of our ML pipeline is the loss function. To define a proper loss function, we aim to integrate the optimization and learning steps in a unified framework by differentiating across the layers, see [5] and [10]. For this reason, the combinatorial optimization layer is replaced with a relaxation through a stochastic perturbation of the objective direction θ:

$$F_\varepsilon(\theta) := \mathbb{E}_Z\left[\max_{\mathbf{x} \in \mathcal{X}} \langle \theta(z) + \varepsilon Z, \mathbf{x}\rangle\right], \tag{7}$$

where Z is a random Gaussian variable $N(0, I)$ and ε controls the amplitude of the perturbation.

To implement the function (7), we use the flexible `InferOpt` package [10]. The `InferOpt` package turns the combinatorial layer into a differentiable layer compatible with the Julia programming language. The function F is convex with respect to θ, and, hence, we can compute the expectation taking the average approximation of m random samples $Z_1, \ldots, Z_m$. For this reason, $F_\varepsilon(\theta)$ is the sum of piece-wise linear functions.

The perturbation introduces a natural way to define a distribution over feasible solutions $\mathbf{x} \in \mathcal{X}$. Notice that it makes no difference to optimize over $\mathcal{C}(z)$ or $\mathcal{X}$ since the objective function is linear. Hence, the induced probability distribution is obtained as the expectation through the additive perturbation as

$$\hat{\mathbf{x}}(\theta) = \mathbb{E}_Z\left[\underset{\mathbf{x} \in \mathcal{C}(z)}{\operatorname{argmax}} \langle \theta + \varepsilon Z, \mathbf{x}\rangle\right] = \sum_{x \in \mathcal{X}} \mathbf{x} p(\mathbf{x}|\theta), \tag{8}$$

such that $p : \theta \to p(\cdot, \theta)$ is required to be differentiable. As in [5,6,26], let Ω_ε be the Fenchel-conjugate of $F_\varepsilon(\theta)$, that is

$$\Omega_\varepsilon(\mathbf{x}) := \sup_{\theta \in \mathbb{R}^n} \langle \theta, \mathbf{x}\rangle - F_\varepsilon(\theta). \tag{9}$$

Let us denote by $F_\varepsilon^*(\theta)$ the corresponding Fenchel-conjugate, and let $\Omega_\varepsilon^*(\mathbf{x})$ be the additional convex penalty introduced in the CO layer (4):

$$\Omega^*(\theta) := \underset{\mathbf{x} \in \mathcal{C}(z)}{\operatorname{argmax}} \langle \theta, \mathbf{x}\rangle - \Omega_\varepsilon(\mathbf{x}) \tag{10}$$

It is proved in [10] that $\Omega_\varepsilon(\mathbf{x})$ can be used as a regularization function, that is, $\Omega_\varepsilon(\mathbf{x})$ is a smooth and convex function $\Omega_\varepsilon : \mathbb{R}^n \longrightarrow \mathbb{R}$ taking finite values in $\mathcal{C}(z)$, which enforces that domain of Ω_ε is contained in $\mathcal{C}(z)$. The following proposition shows that perturbing the objective direction is equivalent to adding a regularization term to the optimization problem. In particular, we can derive its gradient through the associated probability distribution.

Proposition 1. *The function $F_\varepsilon(\theta)$ is convex and its gradient is computed as*

$$\nabla F_\varepsilon(\theta) = \hat{\mathbf{x}}(\theta) = \nabla \Omega^*(\theta), \tag{11}$$

where $\hat{\mathbf{x}}(\theta)$ is the expectation under the perturbed distribution, and the corresponding distribution for a given solution $\mathbf{x}$ is obtained as:

$$\nabla_\theta p(\mathbf{x}|\theta) = \frac{1}{\varepsilon}\mathbb{E}\left[\mathbf{1}\left\{\operatorname*{argmax}_{\mathbf{x}'\in\mathcal{H}(z)}\langle\theta + \varepsilon Z, \mathbf{x}'\rangle = \mathbf{x}\right\}Z\right]. \tag{12}$$

where $\mathbf{1}$ is the indicator function.

Notice that the expectation in Proposition 11 can be efficiently computed by Monte Carlo sampling approximation. Fenchel-Young Losses are grounded in the theory of the convex conjugates [6]. These losses integrate regularization into the learning process, where given a target solution $\hat{\mathbf{x}}$ and a regularization function Ω, the Fenchel-Young loss is defined as

$$\ell_\Omega(\theta, \mathbf{x}) := \Omega^*(\theta) + \Omega(\mathbf{x}) - \langle\theta, \mathbf{x}\rangle = \max_{\mathbf{x}\in\mathcal{X}}\langle\theta, \mathbf{x}\rangle - \Omega(\mathbf{x}) - \langle\theta, \bar{\mathbf{x}}\rangle + \Omega(\bar{\mathbf{x}}). \tag{13}$$

Differentiability of $\ell_\Omega(\theta, \mathbf{x})$ is guaranteed by considering $\Omega_\varepsilon(\mathbf{x})$ as regularization function, and the sub-gradient is a true gradient when the maximizer is unique, that is our case. The gradient computation leverages the structure of perturbation-based losses, allowing differentiation through maxima via Danskin's theorem [11]. This approach allows to easily compute the gradient by applying the definition in 9 at each step as follows

$$\nabla_\theta \ell(\theta, \bar{\mathbf{x}}) = \hat{\mathbf{x}}(\theta) - \bar{\mathbf{x}}$$

This permits us to employ the stochastic gradient descent method to find the best parameter θ, which minimizes $\min_\theta \ell(\theta, \bar{\mathbf{x}})$ due to the nature of the differentiable CO layer. These results provide an effective method to compute gradients in optimization layers integrated with machine learning, combining regularization and perturbation into a unified framework. Such techniques enable efficient gradient-based learning.

3.4 Hybrid ML-CO Greedy Heuristic

We conclude this section by presenting a simple but effective heuristic. The greedy heuristic [30] computes a feasible solution by solving two KPs instances to optimality: first, it solves the leader's problem, and later, it solves the follower's problem. Ultimately, our ML-CO heuristic also solves two KPs instances, but instead of using the original profit values, it relies on the learned weight vector $\theta(z)$. Hence, we can design a basic heuristic that runs the two heuristics in parallel and returns the solution yielding the best value for the objective function. As we will show in our computational results, this simple strategy is extremely effective.

4 Computational Results

This section presents our computational experiments to evaluate the primal heuristics learned by our ML-CO pipeline for the Knapsack Interdiction Problem. For the comparison, we use several instances from the literature rearranged into five datasets, as described in the following paragraphs. The objective of our experiments is to evaluate the following three primal heuristics in terms of percentage optimality gap and runtime:

1. The state-of-the-art greedy algorithm presented in [30].
2. The ML-CO algorithm obtained by applying our ML pipeline to two classes of datasets: the first considers instances from different sources but with the same size n, while the second considers the instances having the same structures regardless of their size.
3. The ML-CO greedy algorithm introduced in Subsect. 3.4, which exploits the most significant features learned by our ML-CO pipeline.

We can compute the optimality percentage gap of the three heuristics since we use only instances where, in the literature, the optimal solution is known.

Parameters. The parameters of our ML pipeline are ε defined in (12), the number of samples m used to approximate (7), the number of epochs for the learning phase, and the learning rate. We decided to fix $\varepsilon = 0.01$ while the number of training epochs and the number of samples are tuned such that the training phase terminates in a few hours on an entry-level desktop computer.

Datasets. We consider the following types of instances from the literature, as collected and described in [30]:

- **CCLW** [9] includes 50 instances.
- **DCS** [12] includes 500 instances.
- **DeNegre** [13] includes 160 instances.
- **FMS** [15] includes 450 instances.
- **TRS** [29] includes 180 instances.
- **New**, **None** and **Large Cap** [30] include 1 500 instances.

These instances mirror the structures of KP instances from the literature described in [28]. However, we build five different datasets where we mix the instances of different types. The first four datasets group all the instances with the same dimension n, with $n \in \{10, 15, 50, 100\}$. Table 2 reports the distribution of instances according to their source, and the last column gives the total number of instances for each dataset. The dataset D5 groups the instances from CCLW and DCS regardless of their dimension.

We observe that every instance has items with positive integer profits and weights and, hence, preprocess them: we normalize each instance by dividing the weight vectors with the instance capacity, which is then set to 1, and the profit vector by the highest profit of each instance. Preliminary computational tests

Table 2. Collection of the five datasets D1–D5 used for training.

Name	n	CCLW	DCS	DeNegre	FMS	TRS	New [30]	Tot.
D1	10	0	0	40	0	0	415	455
D2	15	0	0	0	0	30	500	530
D3	50	10	0	20	0	0	247	277
D4	100	0	100	0	0	0	222	322
D5	*	50	500	0	0	0	0	550

confirmed that this basic preprocessing is mandatory for obtaining satisfactory results.

In addition, we observed that whenever an instance has the leader's budget greater than or equal to the sum of all its weights, that is, $C_L \geq \sum_{i=1}^{n} w_i^L$, the leader can select all the items. In this situation, the follower selects none of the items, leading to a final profit equal to 0. Thus, we check whether the previous summation holds, mark the corresponding instance as trivial, and remove it from the dataset. All the instances available whose optimal profits are equal to zero belong to this group and are excluded from all the experiments.

4.1 Training over Fixed-Size Instances

Our first set of experiments involves training the ML pipeline to solve instances in the first four datasets, D1, D2, D3, and D4, which contain all instances of the same size. As a consequence of this choice, we train pipelines on problems with different difficulties, potentially leading to poor accuracy, even if we could overfit the instance size. To make the training more reliable, we selected those groups with a greater number of instances, reported in Table 2.

We randomly build two subsets for each dataset from D1 to D4: the training set with 85% of the instances and the testing set with the remaining 15%. The training set is repeatedly and randomly divided into training and validation sets to perform 10-fold cross-validation. We tune the learning rate using the training and validation percentage gaps, as reported in Table 3. We fix $\varepsilon = 0.01$, use 30 samples to approximate the expected value in the Fenchel-Young loss, and run the learning algorithm for 4 000 epochs. Lastly, we train a model on the training and validation set (*Train* in Table 3) and evaluate it with the testing set.

Table 3 presents our computational results. Here, we describe the results for the first four datasets and discuss the fifth dataset in the next subsection. The table reports the following details:

- **Lr**: the best learning rate found by cross-validation.
- **Train and Test gap**: Average percentage gap of the trained model on the training and testing sets.
- **Opt train and test**: Number of times the algorithm finds an optimal solution over the total number of instances in the training and testing sets.

Table 3. Results of the ML-CO heuristic vs. the Greedy algorithm. The last four columns refer to the testing instances (best results in bold).

		Runtime ML-CO		Train		Test ML-CO		Greedy [30]	
n	Lr	Avg.	Max	Gap	Opt	Gap	Opt	Gap	Opt
D1 10	$10^{-6.5}$	0.0026	0.0344	4.96%	236/379	**4.95%**	**44/68**	7.83%	42/68
D2 15	$10^{-6.5}$	0.0024	0.0327	6.15%	210/450	**7.67%**	28/80	10.05%	**33/80**
D3 50	$10^{-7.5}$	0.0050	0.0247	2.65%	113/234	**2.39%**	23/43	5.68%	**29/43**
D4 100	$10^{-5.0}$	0.0073	0.0330	4.39%	81/271	**3.71%**	18/49	9.44%	**24/49**
D5 *	$10^{-5.5}$	0.0241	0.1032	7.18%	11/437	**6.83%**	3/110	6.96%	**54/110**

- **Avg time**: Average runtime in seconds to solve an instance.
- **Max time**: Maximum runtime in seconds to solve an instance.

The first remark is that the primal heuristics, as expected, are very fast across all the datasets, running in just a few milliseconds. We do not distinguish the runtime across the heuristics since they all solve two KPs to optimality. However, the last four columns in Table 3 show that there is a striking difference between the ML-CO and the greedy heuristics since the first produces results that are more robust in terms of percentage gaps (smaller values across the different datasets), while the greedy heuristic is more reliable in finding the optimal solutions. We remark that the ML-CO pipeline does not overfit since the percentage gap between the training and testing sets is similar. More surprisingly, the average percentage gap in the testing sets is even better than the training sets, except for dataset D2.

The plots in Fig. 4 draw the values of the loss functions and the percentage gaps versus the number of epochs of the training phase. The ML pipeline reached convergence for all four datasets, even if the final percentage gaps range over different values. Only dataset D3 seems to require a larger number of epochs, but to show a fair comparison, we also fixed 4 000 epochs for this dataset.

4.2 Training over Instances of Different Size

For the second type of experiments, we used instances within similar groups without distinguishing the instance size n. We work with CCLW [9] and DCS [12] instances since they are generated with the same method and differ only for their sizes. We follow the same approach as the previous experiments, with a slightly different division of instances into the training-validation set (i.e., 80%) and testing set (i.e., 20%), ensuring the same percentage of instances from CCLW and DCS. We had to reduce the number of training epochs to 1 000 because of the large amount of data involved in these computations, which significantly increased the training time. The two plots in Fig. 5 show the values of the loss functions and percentage gap for dataset D5, which has the same trend as the results in the previous subsection. However, by looking at the last row of Table 3,

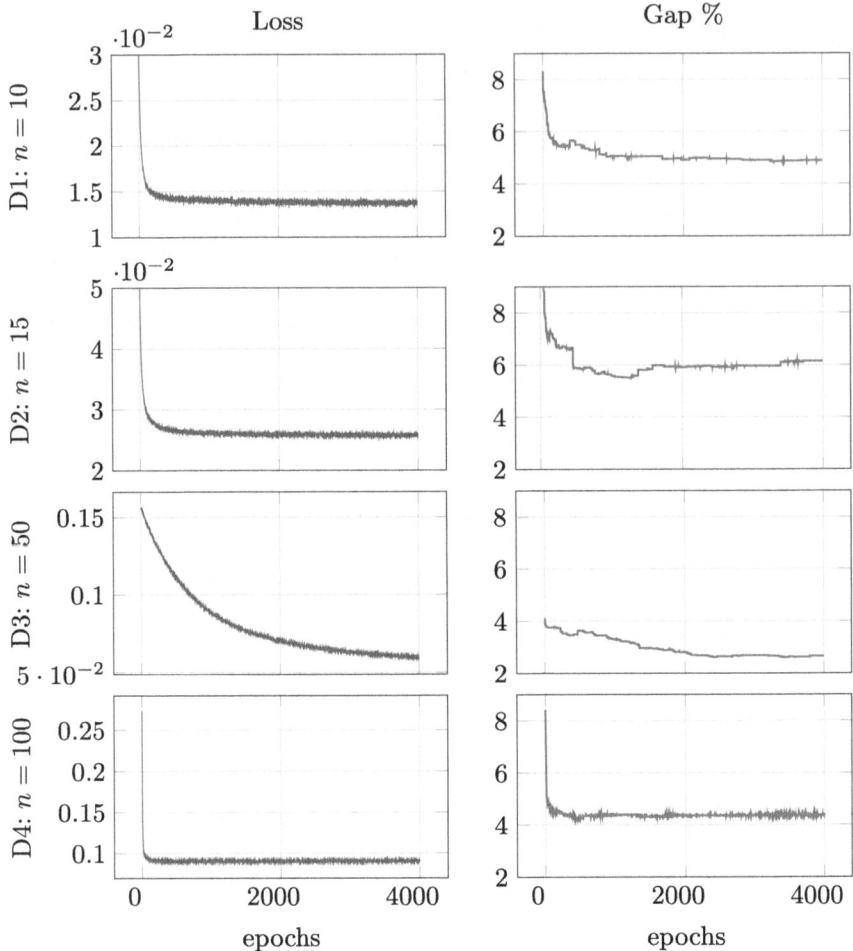

Fig. 4. Training losses and gaps w.r.t. learning epochs.

which reports the results for the dataset D5, we note a high percentage gap and a larger difference in the percentage gap between the training set and the testing set. We can partially explain this difference by the lower number of epochs used in these experiments: repeating the training phase with more powerful hardware would permit a higher number of epochs and, likely, a better final result.

4.3 Features Interpretation: What Did We Learn?

Overall, we trained our ML pipeline using five different datasets, obtaining solutions with different quality. In this subsection, we focus on *what* we learned with the Generalized Linear Model, trying to understand which features are useful to achieve good results. Recall that our ML layer consists of a linear combination of

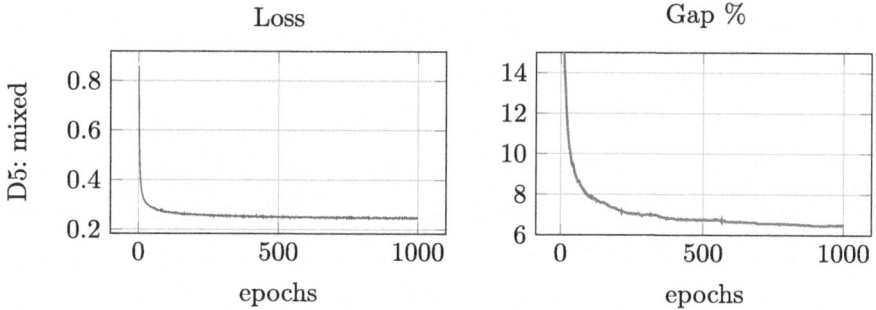

Fig. 5. Training losses and gaps w.r.t. learning epochs for dataset D5.

the feature matrix data described in Table 1. Since we performed an input regularization, all the values of the linear combination are in the interval $[0, 1]$ and are, hence, comparable. In particular, their coefficients, which are the weights of the model we optimize through the training process, can be directly compared, and they can give us a quantitative idea of which features have more importance while computing the final $\theta(z)$ vector.

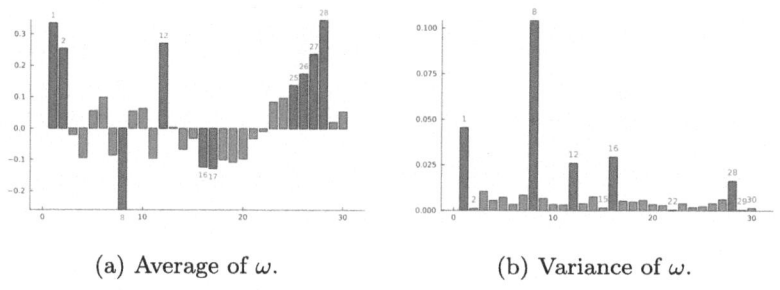

(a) Average of ω. (b) Variance of ω.

Fig. 6. Statistical analysis of the vector weight ω of the 30 input features.

Figure 6 shows the average and variance over our training phases for each of our 30 input features. We highlight the highest values of each feature in red (dark) and put the ID label over (or under) each bar. We remark that most of the features have a very low variance, and this suggests that for the five models trained, the ω vectors are very similar. The five most relevant features are: ID 1: the item's profit p_i; ID 8: the ratio leader's weight/profit w_i^L/p_i; ID 12: the profit quantile; ID 16 and 28: the second and last intervals, defined as in Fig. 3, for the distance to the critical ratio $\mathbb{1}((\frac{p_i}{w_i^F} - \frac{p_k}{w_k^F}) \in I)$. We tested this intuition by performing additional tests with all the available instances and considering the average of the ω vectors obtained in the experiments from the previous sections.

Table 4. Comparing the greedy from [30], the learned ML-CO heuristic, and the hybrid ML-CO greedy heuristic. We highlight the best results in bold.

Type	Num.	Greedy [30]		ML-CO		ML-CO greedy	
		Gap	Opt	Gap	Opt	Gap	Opt
CCLW	49	4.79%	27	7.36%	7	**1.47%**	**28**
DCS	498	6.11%	**271**	14.00%	1	**2.38%**	**271**
DeNegre	160	0.38%	150	7.05%	58	**0.14%**	**156**
FMS	371	2.52%	256	20.88%	26	**1.34%**	**260**
TRS	180	12.76%	48	**6.42%**	77	5.49%	**86**
New	1 141	8.54%	591	2.65%	579	**1.96%**	**690**
None	641	8.47%	414	7.40%	291	**3.54%**	**461**
Large Cap	154	11.93%	64	4.91%	83	**3.34%**	**100**
Total/Average:	*3 194*	*6.94%*	*1 821*	*8.83%*	*1 122*	*2.46%*	*2 052*

Table 4 reports our final results, where we compare the three considered primal heuristics. The table gives the average percentage gap and the number of instances solved to optimality. The second column gives the total number of instances in each class. The greedy heuristic is described in Sect. 2.3, while the ML-CO heuristic refers to the primal heuristics that we obtain by considering the average of the five learned vectors ω, considering only the five most relevant features. The last two columns refer to the hybrid ML-CO greedy heuristic, which consists of solving the same instance with both the Greedy and the learned ML-CO heuristics and then taking the best of the two solutions (see Sect. 3.4).

Combining the Greedy and ML-CO heuristics significantly pays off since this hybrid strategy consistently outperforms the other two. Indeed, we exploit the ability of the greedy algorithm to compute optimal solutions whenever possible, with more robust solutions whenever the greedy fails to find the optimum. We verified that the instances where the Greedy algorithm finds the optimal solutions differ from those where the ML-CO heuristic finds the optimum.

5 Conclusions

In this paper, we have presented a ML pipeline with a Combinatorial Optimization layer to learn primal heuristics for the Knapsack Interdiction Problem. Our approach exploits the InferOpt Julia package but carefully selects a set of instance features that lead to good performances. Using a large set of benchmark instances from the literature, we show that the ML-CO pipeline can produce reliable primal solutions with an average percentage gap better than the Greedy heuristic used in the state-of-the-art [30]. On the contrary, the Greedy heuristic either finds the optimal solution or computes a solution with a large optimality gap. Hence, we designed a hybrid ML-CO greedy heuristic that exploits the most significant features learned by our pipeline and returns the best solution between the greedy and our ML-CO heuristic. The hybrid ML-CO greedy heuristic is computationally extremely fast since it requires the solution of only four KPs and produces very small optimality percentage gaps. Future work includes refining the neural network architecture φ_w and integrating advanced combinatorial optimization techniques such as branch-and-bound.

Ackonwledgements. We kindly thank Noah Weininger for his precious help in collecting the instances and for useful discussions by email.

Stefano Gualandi acknowledges the contribution of the National Recovery and Resilience Plan, Mission 4 Component 2 - Investment 1.4 - National center for HPC, Big Data and Quantum Computing (project code: CN_00000013), funded by the European Union–NextGenerationEU.

References

1. Assi, M., Haraty, R.A.: A survey of the Knapsack Problem. In: 2018 International Arab Conference on Information Technology (ACIT), pp. 1–6. IEEE (2018)
2. Assimakopoulos, N.: A network interdiction model for hospital infection control. Comput. Biol. Med. **17**(6), 413–422 (1987)
3. Baggio, A., Carvalho, M., Lodi, A., Tramontani, A.: Multilevel approaches for the critical node problem. Oper. Res. **69**(2), 486–508 (2021)
4. Bellman, R.: The theory of dynamic programming. Bull. Am. Math. Soc. **60**(6), 503–515 (1954)
5. Berthet, Q., Blondel, M., Teboul, O., Cuturi, M., Vert, J.P., Bach, F.: Learning with differentiable perturbed optimizers. In: Larochelle, H., Ranzato, M., Hadsell, R., Balcan, M.F., Lin, H. (eds.) Advances in Neural Information Processing Systems, vol. 33, pp. 9508–9519 (2020)
6. Blondel, M., Martins, A., Niculae, V.: Learning with Fenchel-Young losses. J. Mach. Learn. Res. **21**(35), 1–69 (2020)
7. Brotcorne, L., Hanafi, S., Mansi, R.: A dynamic programming algorithm for the bilevel knapsack problem. Oper. Res. Lett. **37**(3), 215–218 (2009)
8. Caprara, A., Carvalho, M., Lodi, A., Woeginger, G.J.: A study on the computational complexity of the bilevel knapsack problem. SIAM J. Optim. **24**(2), 823–838 (2014)
9. Caprara, A., Carvalho, M., Lodi, A., Woeginger, G.J.: Bilevel knapsack with interdiction constraints. INFORMS J. Comput. **28**(2), 319–333 (2016)
10. Dalle, G., Baty, L., Bouvier, L., Parmentier, A.: Learning with combinatorial optimization layers: a probabilistic approach (2022)
11. Danskin, J.M.: The theory of max-min and its application to weapons allocation problems, vol. 5. Springer, Heidelberg (2012)
12. Croce, F.D., Scatamacchia, R.: An exact approach for the bilevel Knapsack Problem with interdiction constraints and extensions. Math. Program. **183**(1–2), 249–281 (2020)
13. Denegre, S.: Interdiction and discrete bilevel linear programming. PhD thesis, Lehigh University, USA (2011)
14. Fischetti, M., Ljubić, I., Monaci, M., Sinnl, M.: A new general-purpose algorithm for mixed-integer bilevel linear programs. Oper. Res. **65**(6), 1615–1637 (2017)

15. Fischetti, M., Monaci, M., Sinnl, M.: A dynamic reformulation heuristic for generalized interdiction problems. Eur. J. Oper. Res. **267**, 40–51 (2018)
16. Fischetti, M., Ljubić, I., Monaci, M., Sinnl, M.: Interdiction games and monotonicity, with application to Knapsack Problems. INFORMS J. Comput. **31**(2), 390–410 (2019)
17. Furini, F., Ljubić, I., Malaguti, E., Paronuzzi, P.: Casting light on the hidden bilevel combinatorial structure of the capacitated vertex separator problem. Oper. Res. **70**(4), 2399–2420 (2022)
18. Jooken, J., Leyman, P., De Causmaecker, P.: A new class of hard problem instances for the 0–1 Knapsack Problem. Eur. J. Oper. Res. **301**(3), 841–854 (2022). ISSN 0377-2217
19. Kleinert, T., Labbé, M., Ljubić, I., Schmidt, M.: A survey on mixed-integer programming techniques in bilevel optimization. EURO J. Comput. Optim. **9**, 100007 (2021). ISSN 2192-4406
20. Kolesar, P.: A branch and bound algorithm for the Knapsack Problem. Manag. Sci. **13**, 723–735 (1967)
21. Kwon, S., Choi, H., Park, S.: Deep learning based high accuracy heuristic approach for knapsack interdiction problem. Comput. Oper. Res. **176**, 106965 (2025). ISSN 0305-0548. https://doi.org/10.1016/j.cor.2024.106965. https://www.sciencedirect.com/science/article/pii/S0305054824004374
22. Li, D., et al.: A novel method to solve neural Knapsack Problems. In: Meila, M., Zhang, T. (eds.) Proceedings of the 38th International Conference on Machine Learning, vol. 139 of Proceedings of Machine Learning Research, pp. 6414–6424 (2021)
23. Martello, S., Toth, P.: Knapsack Problems: Algorithms and Computer Implementations. Wiley, Hoboken (1990). ISBN 9780471924203
24. Mathews, G.B.: On the of numbers. Proc. Lond. Math. Soc. **1**(1), 486–490 (1896)
25. Moore, J.T., Bard, J.F.: The mixed integer linear bilevel programming problem. Oper. Res. **38**(5), 911–921 (1990)
26. Niculae, V., Martins, A.F.T., Blondel, M., Cardie, C.: Sparsemap: differentiable sparse structured inference. In: International Conference on Machine Learning (2018)
27. Pisinger, D.: A minimal algorithm for the 0–1 Knapsack Problem. Oper. Res. **45**(5), 758–767 (1997)
28. Pisinger, D.: Where are the hard Knapsack Problems? Comput. Oper. Res. **32**(9), 2271–2284 (2005). ISSN 0305-0548
29. Tang, Y., Richard, J.-P.P., Smith, J.C.: A class of algorithms for mixed-integer bilevel min-max optimization. J. Glob. Optim. **66**, 225–262 (2016)
30. Weninger, N., Fukasawa, R.: A fast combinatorial algorithm for the bilevel knapsack problem with interdiction constraints. Math. Program., 1–33 (2024)

Bounded-Error Policy Optimization for Mixed Discrete-Continuous MDPs via Constraint Generation in Nonlinear Programming

Michael Gimelfarb[1](✉)[iD], Ayal Taitler[2][iD], and Scott Sanner[3][iD]

[1] Department of Computer Science, University of Toronto, Toronto, Canada
mike.gimelfarb@mail.utoronto.ca
[2] Department of Industrial Engineering and Management, Ben Gurion University of the Negev, Be'er Sheva, Israel
ataitler@bgu.ac.il
[3] Department of Mechanical and Industrial Engineering, University of Toronto, Toronto, Canada
ssanner@mie.utoronto.ca

Abstract. We propose the Constraint-Generation Policy Optimization (CGPO) framework to optimize policy parameters within compact and interpretable policy classes for mixed discrete-continuous Markov Decision Processes (DC-MDP). CGPO can not only provide bounded policy error guarantees over an infinite range of initial states for many DC-MDPs with expressive nonlinear dynamics, but it can also provably derive optimal policies in cases where it terminates with zero error. Furthermore, CGPO can generate worst-case state trajectories to diagnose policy deficiencies and provide counterfactual explanations of optimal actions. To achieve such results, CGPO proposes a bilevel mixed-integer nonlinear optimization framework for optimizing policies in defined expressivity classes (e.g. piecewise linear) and reduces it to an optimal constraint generation methodology that adversarially generates worst-case state trajectories. Furthermore, leveraging modern nonlinear optimizers, CGPO can obtain solutions with bounded optimality gap guarantees. We handle stochastic transitions through chance constraints, providing high-probability performance guarantees. We also present a roadmap for understanding the computational complexities of different expressivity classes of policy, reward, and transition dynamics. We experimentally demonstrate the applicability of CGPO across various domains, including inventory control, management of a water reservoir system, and physics control. In summary, CGPO provides structured, compact and explainable policies with bounded performance guarantees, enabling worst-case scenario generation and counterfactual policy diagnostics.

Keywords: Planning · Control · Mixed Discrete-Continuous MDP · Constraint Generation · Chance Constraints · Piecewise-Linear Policy · Sequential Decision Optimization · Policy Optimization

1 Introduction

An important aim of sequential decision optimization of challenging *Discrete-Continuous Markov Decision Processes* (DC-MDP) in the artificial intelligence, operations research, and control domains is to derive policies that achieve optimal control. A desirable property of such policies is *compactness* of representation, which provides efficient execution on resource-constrained systems such as mobile devices [27], as well as the potential for *introspection and explanation* [25]. Moreover, while the derived policy is expected to perform well in expectation, in many applications it is desirable to obtain *bounds on maximum policy error* and the scenarios that induce worst-case policy performance [9].

Popular policy optimization approaches used in model-free reinforcement learning [22] do not provide bounded error guarantees on policy performance, even if they use compact policy classes such as decision trees [25]. Model-based extensions that leverage gradient-based policy optimization [6] similarly do not provide bounded error guarantees due to the nonconvexity of the problem.

In contrast, there exists a rich tradition of work leveraging *mixed integer programming* (MIP) for bounded-error policy optimization [1,10,26], but these methods only apply to discrete state and action MDPs. More importantly, these previous policy optimization formulations cannot be directly extended to continuous state or action MDPs due to the infinite number of constraints arising from the infinite state and action space in these formulations. While some historical work has attempted to circumvent large or infinite constraint sets (from the continuous setting) via constraint generation [11,13,23], none of these works optimize for policies. Finally, while there exist bespoke bounded error policy solutions for niche MDP classes with continuous states or actions such as linear–quadratic controllers [4], ambulance dispatching [2], and the celebrated "(s, S)" threshold policies for [21]'s inventory management, these solution methods do not readily generalize beyond the specialized domains for which they are designed.

It remains an open question as to how to derive bounded error policies for the infinite state and action, discrete and continuous MDP (DC-MDP) setting. We address this open question through a novel bilevel MIP formulation. Unfortunately, unlike MIP solutions for discrete MDPs, DC-MDPs with continuous states and/or actions cannot be solved directly since the resulting MIP (a) requires a challenging bilevel MIP formulation, (b) has an uncountably infinite number of constraints, and (c) will often be nonlinear. Fortuitously, we show how a constraint generation form of this policy optimization – termed CGPO – is able to address (a), (b) and (c) to optimize policies with bounded error guarantees in many cases.

In summary, our aim is to provide a solution approach to optimize and provide strong performance bound guarantees on structured and compact DC-MDP policies under various expressivity classes of policy, reward, and nonlinear transition dynamics. Our specific contributions are:

1. *We propose a novel bilevel optimization framework for solving nonlinear DC-MDPs called Constraint-Generation Policy Optimization (CGPO), that*

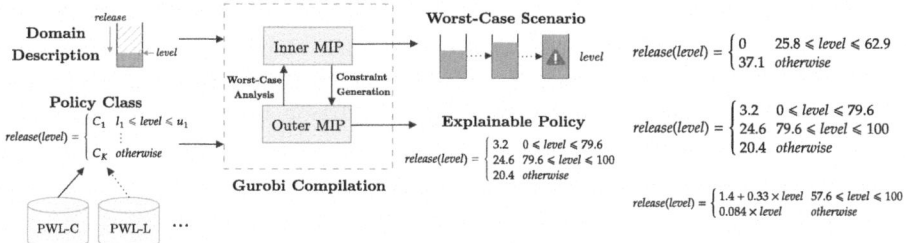

Fig. 1. Reservoir control is used as an illustrative example. **Left:** an overview of CGPO, which consists of a domain description and policy representation compiled to a bilevel mixed-integer program (MIP), in which the inner problem computes the worst-case trajectories for the current policy while the outer problem updates the policy via constraint-generation. The result is a worst-case scenario for the policy (facilitating policy failure analysis), a concrete policy within the expressivity class (for direct policy inspection), and a gap on its performance (error bound). **Right:** three optimal (i.e. zero-gap) policies produced upon termination across several piecewise policy classes. Crucially, our framework provides the ability to derive highly compact (e.g. memory and time-efficient to execute), intuitive and nonlinear policies, with strong bound guarantees on policy performance.

admits a clever reduction to an iterative constraint-generation algorithm that can be readily implemented using standard MIP solvers (Fig. 1). State-of-the-art MIP solvers often leverage spatial branch-and-bound techniques, which can provide not only optimal solutions for large mixed integer linear programs (MILP), but also bounded optimality guarantees for mixed integer nonlinear programs (MINLP) [7]. Chance constraints [12] are used for probabilistic guarantees.

2. If the constraint generation algorithm terminates, we guarantee that we have found the optimal policy within the given policy expressivity class (Theorem 1). Further, even when constraint generation does not terminate, our algorithm provides a tight optimality bound on the performance of the computed policy at each iteration and the scenario where the policy performs worst (and as a corollary, for any externally provided policy).
3. We provide a road map to characterize the optimization problems in (1) above – and their associated expressivity classes – for different expressivity classes of policies and state dynamics, ranging from MILP, to polynomial programming, to nonlinear programs (Table 1). This information is beneficial for reasoning about which optimization techniques are most effective for different combinations of DC-MDP and policy class expressivity.
4. Finally, we provide a variety of experiments demonstrating the range of rich applications of CGPO under linear and nonlinear dynamics and various policy classes. Critically, we derive bounded optimal solutions for a range of problems, including linear inventory management, reservoir control, and a highly nonlinear VTOL control problem. Furthermore, since our policy classes are compact by design, we can also directly inspect and analyze these policies

(Fig. 4). To facilitate policy interpretation and diagnostics, we can compute the state and exogenous noise trajectory scenario that attains the worst-case error bounds for a policy (Fig. 7) as well as a counterfactual explanation of what action should have been taken in comparison to what the policy prescribes.

2 Preliminaries

2.1 Function Classes

We first define some important classes of functions commonly referred to in the paper. Let $\mathcal{X}$ and $\mathcal{Y}$ be arbitrary sets. A function $f : \mathcal{X} \to \mathcal{Y}$ is called *piecewise* (PW) if there exist functions $f_1, \ldots f_K, f_{K+1} : \mathcal{X} \to \mathcal{Y}$ and disjoint Boolean case conditions $\mathcal{P}_1, \ldots \mathcal{P}_K : \mathcal{X} \to \{0, 1\}$, such that for any $x \in \mathcal{X}$, $\forall i, j \neq i$ we have $\mathcal{P}_i(x) = 1 \Rightarrow \mathcal{P}_j(x) = 0$, and for case functions $f(x) = f_i(x)$ if $\mathcal{P}_i(x) = 1$ and $f(x) = f_{K+1}(x)$ if $\mathcal{P}_i(x) = 0$, $\forall i$.

Both case conditions $\mathcal{P}_i$ and case functions f_i can be linear or nonlinear. In this paper, we consider the following classes:

C *constant*, i.e. $f_i(\mathbf{x}) = C$
D *discrete*, same as restricted values of C
S *simple (axis-aligned) linear functions*, i.e. $f_i(\mathbf{x}) = b_i + w_i \times x_j$, where $j \in \{1, 2 \ldots n\}$ and $b_i, w_i \in \mathbb{R}$
L *linear*, i.e. $f_i(\mathbf{x}) = b_i + \mathbf{w}_i^T \mathbf{x}$
B *bilinear*, i.e. $f_i(\mathbf{x}) = x_1 \times x_2 + x_2 \times x_3$
Q *quadratic*, i.e. $f_i(\mathbf{x}) = b_i + \mathbf{w}_i^T \mathbf{x} + \mathbf{x}^T M_i \mathbf{x}$, where $M_i \in \mathbb{R}^{n \times n}$
P *polynomial of order* m, i.e. $f_i(\mathbf{x}) = \sum_{j_1=1}^{m} \cdots \sum_{j_n=1}^{m} w_{i, j_1, \ldots j_n} \prod_{k=1}^{n} x_k^{j_k}$
N *general nonlinear*, i.e. $f(\mathbf{x}) = e^{b_i + \mathbf{w}_i^T \mathbf{x}}$.

Analogously, we consider analogous functions over integer (I) domains $\mathcal{X}$ or $\mathcal{Y}$, or mixed discrete-continuous (M) domains. $\mathcal{P}_i$ can be characterized similarly based on whether the constraint set defining it is constant, linear, bilinear, etc.

We also introduce a convenient notation to describe general PW functions by concatenating the expressivity classes of their constraints $\mathcal{P}_i$ and case functions f_i. For example, PWS-C describes a piecewise function on $\mathbb{R}^n$ with simple (axis-aligned) linear constraints and constant values, while PWS-L describes a function with similar constraints but linear values. We can also append the number of cases K to the notation, i.e. PWS1-L describes a piecewise linear function with $K = 1$ case conditions. More generally, for PWL, PWP or PWN, constraints could also be logical conjunctions of other simpler constraints.

2.2 Mathematical Programming

Given a function $f : \mathcal{X} \to \mathcal{Y}$ and Boolean case condition $\mathcal{P} : \mathcal{X} \to \{0, 1\}$, the *mathematical program* is defined as:

$$\min_{x \in \mathcal{X}} \quad f(x) \quad \text{s.t.} \quad \mathcal{P}(x) = 1.$$

Important expressivity classes of *mixed-integer program* (MIP) optimization problems (i.e. those with discrete and continuous decision variables) include:

MILP *mixed-integer linear*: f and $\mathcal{P}$ both in L
MIBCP *mixed-integer bilinearly constrained*: f is in L and $\mathcal{P}$ is in B
MIQP *mixed-integer quadratic*: f is in Q and $\mathcal{P}$ is in L
QCQP *quadratically constrained quadratic*: f and $\mathcal{P}$ both in Q
PP *polynomial*: f and $\mathcal{P}$ both in P
MINLP *mixed-integer nonlinear*: f and $\mathcal{P}$ both in N.

Branch-and-bound [15] solvers are commonly used to maintain upper and lower bounds on the minimal objective value of any linear or nonlinear MIP, whose difference is the so-called *optimality gap*; when this gap is zero, an optimal solution has been found. Moreover, some packages such as Gurobi, which we use to perform the necessary compilations in CGPO in our experiments, also support a variety of nonlinear mathematical operations via piecewise-linear approximation [7].

2.3 Discrete-Continuous Markov Decision Processes

A *Discrete-Continuous Markov decision process* (DC-MDP) is a tuple of the form $\langle \mathcal{S}, \mathcal{A}, P, r \rangle$, where $\mathcal{S}$ is a set of states, and $\mathcal{A}$ is a set of actions or controls. $\mathcal{S}$ and $\mathcal{A}$ may be discrete, continuous, or mixed. $P(\mathbf{s}, \mathbf{a}, \mathbf{s}')$ is the probability of transitioning to state $\mathbf{s}'$ immediately upon choosing action $\mathbf{a}$ in state $\mathbf{s}$, and $r(\mathbf{s}, \mathbf{a}, \mathbf{s}')$ is the corresponding reward received. We assume that r is uniformly bounded, i.e. there exists $B < \infty$ such that $|r(\mathbf{s}, \mathbf{a}, \mathbf{s}')| \leq B$ holds for all $\mathbf{s}, \mathbf{a}, \mathbf{s}'$.

Given a fixed planning horizon of length T, the *value* of a (open-loop) *plan* $\alpha = [\mathbf{a}_1, \ldots \mathbf{a}_T] \in \mathcal{A}^T$ starting in state $\mathbf{s}_1$ is

$$V(\alpha, \mathbf{s}_1) = \mathbb{E}_{\mathbf{s}_{t+1} \sim P(\mathbf{s}_t, \mathbf{a}_t, \cdot)} \left[\sum_{t=1}^{T} r(\mathbf{s}_t, \mathbf{a}_t, \mathbf{s}_{t+1}) \right].$$

A (closed-loop) *policy* $\pi = [\pi_1, \ldots \pi_T]$ is a sequence of mappings $\pi_t : \mathcal{S} \to \mathcal{A}$, whose value is defined as[1]

$$V(\pi, \mathbf{s}_1) = \mathbb{E}_{\mathbf{s}_{t+1} \sim P(\mathbf{s}_t, \pi_t(\mathbf{s}_t), \cdot)} \left[\sum_{t=1}^{T} r(\mathbf{s}_t, \pi_t(\mathbf{s}_t), \mathbf{s}_{t+1}) \right].$$

Dynamic programming approaches such as value and policy iteration can compute an optimal horizon-dependent policy π_H^* [18], but do not directly apply to DC-MDPs with infinite or continuous state or action spaces.

Our goal is to compute an optimal stationary[2] policy π^* that minimizes the error in value relative to π_H^* over all initial states of interest $\mathcal{S}_1 \subseteq \mathcal{S}$, and all stationary policies Π

$$\pi^* \in \arg\min_{\pi \in \Pi} \max_{\mathbf{s}_1 \in \mathcal{S}_1} \left[V(\pi_H^*, \mathbf{s}_1) - V(\pi, \mathbf{s}_1) \right]. \tag{1}$$

---
[1] Deterministic (time-dependent) policies are sufficient for finite-horizon control [18].
[2] We focus on stationary policies for notational convenience; the extension is trivial.

In practical applications, the policy class is often restricted to function approximations $\tilde{\Pi} \subset \Pi$. A variety of planning approaches can compute plans $\tilde{\pi}^* \in \tilde{\Pi}$ for this problem, including straight-line planning that scales well in practice but does not learn policies [19,28]. Reactive policy optimization [6,14] and variants of MCTS [24] that learn neural network policies cannot provide concrete error bounds on policy performance nor do they facilitate worst-case analysis, or ease of interpretation of learned compact policy classes as we provide.

3 Methodology

We begin with a derivation of CGPO for general DC-MDPs in the deterministic setting. We then derive CGPO in the more general stochastic setting using chance constrained optimization. We end with a discussion of problem class complexity and convergence. A worked example illustrating the overall execution of CGPO is provided in the appendix.

3.1 Constraint Generation for Deterministic DC-MDPs

We assume π can be compactly identified by a vector $\mathbf{w} \in \mathcal{W}$ of decision variables, and we use the shorthand $V(\mathbf{w}, \mathbf{s}_1)$ to denote the value $V(\pi_\mathbf{w}, \mathbf{s}_1)$ of policy $\pi_\mathbf{w}$ parameterized by $\mathbf{w}$. We focus our attention on approximate policy sets $\tilde{\Pi}_\mathcal{W} = \{\pi_\mathbf{w} : \mathbf{w} \in \mathcal{W}\}$ enumerated by a compact set of parameters $\mathcal{W}$.

First, observe that for every possible initial state $\mathbf{s}_1$, there exists a fixed optimal plan $\alpha^* = [\mathbf{a}_1^*, \ldots \mathbf{a}_T^*]$ such that $V(\alpha^*, \mathbf{s}_1) = V(\pi_\Pi^*, \mathbf{s}_1)$. On the other hand, since we only have access to an expressivity class of approximate policies $\tilde{\Pi}_\mathcal{W}$, the error $\varepsilon(\mathbf{w}, \mathbf{s}_1)$ of $\pi_\mathbf{w}$ in $\mathbf{s}_1$ relative to $V(\pi_H^*, \mathbf{s}_1)$ according to (1) is

$$\varepsilon(\mathbf{w}, \mathbf{s}_1) \geq \max_{\alpha \in \mathcal{A}^T} V(\alpha, \mathbf{s}_1) - V(\mathbf{w}, \mathbf{s}_1),$$

and thus the worst-case error $\varepsilon(\mathbf{w})$ of $\mathbf{w}$ is

$$\varepsilon(\mathbf{w}) \geq \max_{\mathbf{s}_1 \in \mathcal{S}_1} \max_{\alpha \in \mathcal{A}^T} \left[V(\alpha, \mathbf{s}_1) - V(\mathbf{w}, \mathbf{s}_1) \right]. \tag{2}$$

However, since we seek the approximate optimal policy $\tilde{\pi}^* \in \tilde{\Pi}_\mathcal{W}$, we can directly minimize (2) over $\mathcal{W}$, obtaining the *infinitely-constrained mixed-integer program*:

$$\begin{aligned}\min_{\mathbf{w} \in \mathcal{W},\, \varepsilon \in [0,\infty)} & \quad \varepsilon \\ \text{s.t.} & \quad \varepsilon \geq \max_{\mathbf{s}_1 \in \mathcal{S}_1} \max_{\alpha \in \mathcal{A}^T} \left[V(\alpha, \mathbf{s}_1) - V(\mathbf{w}, \mathbf{s}_1) \right].\end{aligned} \tag{3}$$

However, the constraint is highly nonlinear, turning (3) into a bilevel program and making analysis of this particular formulation difficult. Instead, since the max's must hold for all states and actions, we can rewrite the problem (3) as:

$$\begin{aligned}\min_{\mathbf{w} \in \mathcal{W},\, \varepsilon \in [0,\infty)} & \quad \varepsilon \\ \text{s.t.} & \quad \varepsilon \geq V(\alpha, \mathbf{s}_1) - V(\pi, \mathbf{s}_1), \quad \forall \alpha \in \mathcal{A}^T, \mathbf{s}_1 \in \mathcal{S}_1.\end{aligned} \tag{4}$$

The goal is therefore to solve (4), which is still infinitely-constrained when $\mathcal{S}$ or $\mathcal{A}$ are infinite spaces. Instead, we solve it by splitting the min over $\mathcal{W}$ and the max over $(\alpha, \mathbf{s}_1)$ into two problems and apply *constraint generation* [5,8].

Specifically, starting with a fixed arbitrary scenario $(\mathbf{s}_1, \alpha)$, we form the constraint set $\mathcal{C} = \{(\mathbf{s}_1, \alpha)\}$, and then solve the following two problems:

Outer Solve (4) within the set of finite constraints $\mathcal{C}$ to obtain $\mathbf{w}^* \in \mathcal{W}$.
Inner Solve (2) $\operatorname{argmax}_{\mathbf{s}_1 \in \mathcal{S}_1, \alpha \in \mathcal{A}^T}[V(\alpha, \mathbf{s}_1) - V(\mathbf{w}^*, \mathbf{s}_1)]$, for the highest-error scenario for $\mathbf{w}^*$, and append it to $\mathcal{C}$.

These two steps are repeated until the constraint added to the outer problem is no longer binding, i.e. $V(\alpha^*, \mathbf{s}_1^*) - V(\mathbf{w}^*, \mathbf{s}_1^*) \leq 0$, since the solution of the outer problem will not change with the addition of the new constraint. We name our approach *Constraint-Generation Policy Optimization* (CGPO). *In retrospect, we can view CGPO as a policy iteration algorithm where the inner problem adversarially critiques the policy with a worst-case trajectory and the outer problem improves the policy w.r.t. all critiques.*

Remarks. Upon termination, CGPO guarantees an optimal (lowest error) policy within the specified policy class (Theorem 1). While we cannot provide a general finite-time guarantee of termination with an optimal policy, CGPO is an anytime algorithm that provides a best policy and a bound on performance at each iteration. At each iteration, the solution to the inner problem allows analysis of the worst-case trajectory with respect to π (Fig. 7), and generates a counterfactual explanation of what actions should have been made.

3.2 Chance Constraints for Stochastic DC-MDPs

Chance-Constrained Optimization. [3,12,16] allows us to derive high probability intervals on stochastic MDP transitions and reduce our solution to a robust optimization problem with probabilistic performance guarantees. To achieve this, we will require that the state transition function $P(\mathbf{s}, \mathbf{a}, \mathbf{s}')$ has a natural reparameterization as a deterministic function g of $(\mathbf{s}, \mathbf{a})$ and some exogenous independent and identically distributed noise variable ξ with density $q(\cdot)$ on support Ξ, e.g. $\mathbf{s}' = g(\mathbf{s}, \mathbf{a}, \xi)$ [6,17]. Given a threshold $p \in (0,1)$ close to 1, we also assume the existence of a computable interval $\Xi_p = [\xi_l, \xi_u]$ such that $\mathbb{P}(\xi_l \leq \xi \leq \xi_u) \geq p$.

Thus, we can repeat the derivation of (3) by considering the worst case not only over $\mathbf{s}_1$, but also over possible noise variables $\xi_{1:T} = [\xi_1, \dots \xi_T] \in \Xi_p^T$. The final problem is:

$$\min_{\mathbf{w} \in \mathcal{W}, \varepsilon \in [0, \infty)} \varepsilon \qquad \qquad (5)$$
$$\text{s.t.} \quad \varepsilon \geq \max_{\mathbf{s}_1 \in \mathcal{S}_1} \max_{\xi_{1:T} \in \Xi_p^T} \max_{\alpha \in \mathcal{A}^T} [V(\alpha, \mathbf{s}_1, \xi_{1:T}) - V(\mathbf{w}, \mathbf{s}_1, \xi_{1:T})],$$

where $V(\cdot, \mathbf{s}_1, \xi_{1:T})$ corresponds to the total reward of the policy or plan accumulated over trajectory $\xi_{1:T}$ starting in $\mathbf{s}_1$. In the stochastic setting, ε only holds with probability p^T for a horizon T problem. Experimentally, we found that choosing p such that $p^T = P$, where P is a desired probability bound on the full planning trajectory, was sufficient[3].

Once again, (5) can be reformulated as:

$$\min_{\mathbf{w} \in \mathcal{W}, \varepsilon \in [0,\infty)} \varepsilon \tag{6}$$
$$\text{s.t. } \varepsilon \geq V(\alpha, \mathbf{s}_1, \xi_{1:T}) - V(\mathbf{w}, \mathbf{s}_1, \xi_{1:T}), \quad \forall \alpha \in \mathcal{A}^T, \mathbf{s}_1 \in \mathcal{S}_1, \xi_{1:T} \in \Xi_p^T.$$

Therefore, we can again apply constraint generation to solve this problem in two stages, in which the inner optimization produces not only a worst-case initial state and action sequence, but also the disturbances $\xi_{1:T}^*$ that reproduce all future worst-case state realizations $\mathbf{s}_2, \ldots \mathbf{s}_T$. The overall workflow of the computations is summarized as Algorithm 1.

Algorithm 1. Constraint-Generation Policy Optimization (CGPO)

Initialize $p \in (0,1)$, Ξ_p, $\mathbf{s}_1 \in \mathcal{S}_1$, $\xi_{1:T} \in \Xi_p^T$, $\alpha \in \mathcal{A}^T$
Set $t = 0$ and $\mathcal{C}_0 = \{(\mathbf{s}_1, \xi_{1:T}, \alpha)\}$
Step 1: Solve the **outer problem**:

$$(\mathbf{w}_t^*, \varepsilon_t^*) \in \text{argmin}_{\mathbf{w} \in \mathcal{W}, \varepsilon \in [0,\infty)} \varepsilon$$
$$\text{s.t. } \varepsilon \geq V(\alpha, \mathbf{s}_1, \xi_{1:T}) - V(\mathbf{w}, \mathbf{s}_1), \quad \forall (\mathbf{s}_1, \xi_{1:T}, \alpha) \in \mathcal{C}_t$$

Step 2: Solve the **inner problem**:

$$(\mathbf{s}_1^*, \xi_{1:T}^*, \alpha^*) \in \text{argmax}_{\mathbf{s}_1 \in \mathcal{S}_1, \xi_{1:T} \in \Xi_p^T, \alpha \in \mathcal{A}^T} [V(\alpha, \mathbf{s}_1, \xi_{1:T}) - V(\mathbf{w}_t^*, \mathbf{s}_1, \xi_{1:T})]$$

Step 3: Check convergence:
if $V(\mathbf{w}_t^*, \mathbf{s}_1^*, \xi_{1:T}^*) \geq V(\alpha^*, \mathbf{s}_1^*, \xi_{1:T}^*)$ (or $\|\mathbf{w}_t^* - \mathbf{w}_{t-1}^*\| < \delta$) **then**
    Terminate with policy $\pi_{\mathbf{w}_t^*}$ and error ε_t^*
else
    Set $t = t+1$, $\mathcal{C}_{t+1} = \mathcal{C}_t \cup \{(\mathbf{s}_1^*, \xi_{1:T}^*, \alpha^*)\}$ and **go to Step 1**
end if

Remark. Algorithm 1 may not terminate in general, unless convexity assumptions are placed on $V(\alpha, \mathbf{s})$ or $V(\mathbf{w}, \mathbf{s})$. Thus, in typical applications, one would terminate when $\|\mathbf{w}_t^* - \mathbf{w}_{t-1}^*\| < \delta$ for chosen hyper-parameter δ. However, our experiments have shown empirically that Algorithm 1 terminates with an exact optimal solution for most policy classes considered, at which point both the return and error curves "plateau" (cf. Fig. 3).

3.3 Optimality of CGPO

Under mild conditions, if CGPO terminates at some iteration t, then $\mathbf{w}_t^*$ is optimal in $\mathcal{W}$ (with probability at least p^T in the stochastic case).

---

[3] Hence, we arbitrarily chose $p = 0.995$ for our empirical evaluations.

Theorem 1. *If $\mathcal{S}_1$, $\mathcal{A}$, Ξ_p and $\mathcal{W}$ are non-empty compact subsets of Euclidean space, $V(\alpha, \mathbf{s}, \xi_{1:T})$ and $V(\mathbf{w}, \mathbf{s}, \xi_{1:T})$ are continuous, and CGPO terminates at iteration t, then $\mathbf{w}_t^*$ is optimal for problem (6).*

Proof. In the notation of [5], we define $X^0 = [0, 2BT]$, $X = \mathcal{W}$, $Y = \mathcal{A}^T \times \mathcal{S}_1 \times \Xi_p^T$. Making the change of variable $\varepsilon = x_0 \in X^0$, $x = \mathbf{w}$, and $y = (\alpha, \mathbf{s}_1, \xi_{1:T}) \in Y$, and defining the continuous function $f(x, y) = \varepsilon(\alpha, \mathbf{s}_1, \xi_{1:T}) = V(\alpha, \mathbf{s}_1, \xi_{1:T}) - V(\mathbf{w}, \mathbf{s}_1, \xi_{1:T})$, the problem (6) is equivalent to the problem:

$$\min_{(x_0, x) \in X^0 \times X} x_0 \quad \text{s.t.} \quad f(x, y) - x_0 \leq 0, \quad \forall y \in Y$$

as discussed on p. 262 of [5]. Similarly, Algorithm 1 can be reparameterized as Algorithm 2.1 in the aforementioned paper, with $x_t = \mathbf{w}_t^*$. Thus, the assumption of Theorem 2.1 in [5] holds and $\mathbf{w}_t^*$ is a solution of (6) as claimed. □

Remark. Termination of CGPO guarantees that $\mathbf{w}_t^*$ is optimal in $\tilde{\Pi}_\mathcal{W}$, and ε_t^* is the corresponding optimality gap. Moreover, if $\varepsilon_t^* = 0$ and the MDP is deterministic, then $\pi_{\mathbf{w}_t^*}$ is the optimal policy in Π. This implies that the gap estimate can be used as a principled way to assess the suitability of different policy classes.

3.4 Problem Expressivity Class Analysis

Table 1. Problem classes of the inner/outer optimization problems in CGPO for different expressivity classes of dynamics/reward (columns) and policies (rows).

Policy	Dynamics/Reward		
	L	P	N
PWS-{C, D}	MILP	PP	MINLP
PWL-{C, D}	MILP/MIBCP	PP	MINLP
S, L	MILP/PP	PP	MINLP
PW{S,L}-{S,L}	MILP/PP	PP	MINLP
PW{S,L,P}-P	PP	PP	MINLP
Q	PP	PP	MINLP
PWN-N	MINLP	MINLP	MINLP

In Table 1, we present the relationship between the expressivity classes of policies and state transition dynamics and the corresponding classes of the inner and outer optimization problems. The policy classes of interest include PWS-C, PWL-C, PWS-L, PWL-L, PWN-N and the different variants of piecewise polynomial policies. Meanwhile, expressivity classes for state dynamics and reward include linear, polynomial and general nonlinear functions (their piecewise counterparts generally fall under the same expressivity classes, and are excluded for brevity). Interestingly, as shown in the appendix, when the policy and state dynamics are both linear, the outer problem is a PP. In a similar vein, a PWL-C policy and linear dynamics result in a MIBCP outer problem due to the bilinear interaction between successor state decision variables and policy weights in the linear conditions.

Our experiments in the next section empirically evaluate PWS-C and PWS-L policies with linear dynamics and Q policies with nonlinear dynamics. This requires solving mixed-integer problems with large numbers of decision variables ranging from MILP to PP to MINLP.

4 Empirical Evaluation

We empirically validate the performance of CGPO on several MDPs, aiming to answer the following questions:

Q1 Does CGPO recover exact solutions when the ideal policy class is known? How does it perform if the optimal policy class is not known?
Q2 How do different policy expressivity classes perform for each problem?
Q3 Does the worst-case analysis provide further insight about a policy?

4.1 Domain Descriptions

To answer these questions, we evaluate on linear Inventory, linear Reservoir, and nonlinear VTOL (vertical take-off and landing) domains as summarized below. Inventory has provably optimal PWS-S policies, whereas no optimal policy class is known explicitly for Reservoir. A public GitHub repository[4] allows reproduction of all results and application of CGPO to arbitrary domains, and also contains an appendix with experimental details and additional results.

Linear Inventory. State s_t describes the discrete level of stock available for a single good, action $a_t \in [0, B]$ is the discrete reorder quantity, and demand ξ_t is stochastic and distributed as discrete uniform:

$$s_{t+1} = s_t + a_t - \lfloor \xi_t \rfloor, \quad \xi_t \sim \text{Uniform}(L, U).$$

We permit backlogging of inventories, represented as negative s_t. We define costs C, P and S which represent, respectively, the purchase cost, excess inventory cost and shortage cost, and the reward function can be written as

$$r(\mathbf{s}_{t+1}) = -C \times a_t - P \times (s_{t+1})_+ - S \times (-s_{t+1})_+,$$

---

[4] https://github.com/pyrddlgym-project/pyRDDLGym/tree/GurobiCompilerBranch/.

where $(\cdot)_+ = \max[0,\cdot]$. We set $B = 10, C = 0.5, P = S = 2, L = 2, U = 6$. If $P > C$ and $T = \infty$, then a PWS-S policy is provably optimal (otherwise if $T < \infty$, then it may be non-stationary). A planning horizon of $T = 8$ is used. For this domain and reservoir below, we focus on learning factorized piecewise policies, i.e. C, S, PWS-C and PWS-S.

Linear Reservoir. The goal is to manage the water level in a system of interconnected reservoirs. State $s_{t,r}$ represents the continuous water level of reservoir r with capacity M_r, action $a_{t,r} \in [0, s_{t,r}]$ is the continuous amount of water released, and rainfall $\xi_{t,r}$ is a truncated normally-distributed random variable. Each reservoir r is connected to a set $U(r)$ of upstream reservoirs, thus:

$$s_{t+1,r} = \min\left[M_r, \left(s_{t,r} + (\xi_{t,r})_+ - a_{t,r} + \sum_{d \in U(r)} a_{t,d}\right)_+\right], \quad \xi_{t,r} \sim \mathcal{N}(m_r, v_r).$$

Reward linearly penalizes any excess water level above H_r and below L_r

$$r(\mathbf{s}_{t+1}) = \sum_{r=1}^{N} l_r (L_r - s_{t+1,r})_+ + h_r (s_{t+1,r} - H_r)_+.$$

Our experiment uses a two-reservoir system with the following values for reservoir 1: $L_1 = 20, H_1 = 80, M_1 = 100, m_1 = 5, v_1 = 5, U(1) = \emptyset$, and the following values for reservoir 2: $L_2 = 30, H_2 = 180, M_2 = 200, m_2 = 10, v_2 = 10, U(2) = \{1\}$. Costs are identical for all reservoirs and are set to $l_r = -10, h_r = -100$. We use $T = 10$.

Nonlinear VTOL. The goal is to balance two masses on opposing ends of a long pole. The state consists of the angle θ_t and angular velocity ω_t of the pole, and the action is the force $F_t \in [0, 1]$ applied to the mass (Fig. 2):

$$\theta_{t+1} = \max\left[-\sin(h/l_1), \min\left[\sin(h/l_2), \theta_t + \tau \omega_t\right]\right]$$
$$\omega_{t+1} = \omega_t + \frac{\tau}{J}(9.8(m_2 l_2 - m_1 l_1)\cos\theta_t + 150 l_1 F_t)$$
$$J = m_1 l_1^2 + m_2 l_2^2$$
s.t. $l_1 m_1 \geq l_2 m_2, \quad l_1 > l_2 > h, \quad 0 \leq F_t \leq 1.$

Time is discretized into intervals of $\tau = 0.1$ seconds, and we set $T = 6$. The reward penalizes the difference between the pole angle at epoch t and a target angle θ_{target}

$$r(F_t, \theta_{t+1}, \omega_{t+1}) = -|\theta_{t+1} - \theta_{target}|.$$

We use $l_1 = 1, l_2 = 0.5, m_1 = 10, m_2 = 1, h = 0.4, g = 9.8$. We optimize for nonlinear quadratic policies.

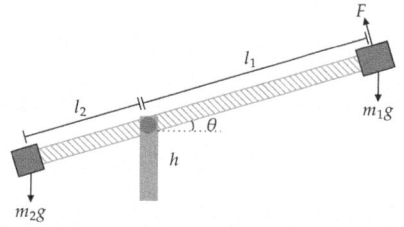

Fig. 2. Nonlinear VTOL system.

Nonlinear Intercept. To demonstrate an example of discrete (Boolean) action policy optimization over nonlinear continuous state dynamics with independently moving but interacting projectiles, we turn to the Intercept problem [20]. Space restrictions require us to relegate details and results to the appendix, but we remark here that CGPO is able to derive an optimal policy.

Initial State Bounds $\mathcal{S}_1$. For Reservoir, the initial state bounds are set to $[50, 100]$ for reservoir 1 and $[100, 200]$ for reservoir 2. For Inventory, the bounds are $[0, 2]$, and for VTOL and Intercept they are fixed to the initial state of the system.

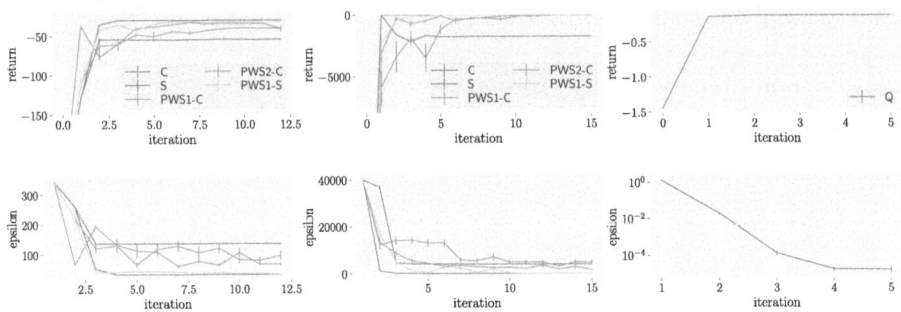

Fig. 3. Reservoir (left), Inventory (middle), VTOL (right): Simulated return (top row) and worst-case error ε^* (bottom row) over 100 roll-outs as a function of the number of iterations of constraint generation. Bars represent 95% confidence intervals calculated using 10 independent runs of CGPO.

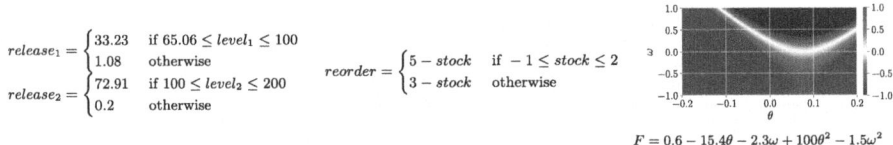

Fig. 4. Reservoir (left), Inventory (middle), VTOL (right): Examples of optimal policies computed at the end of CGPO.

4.2 Empirical Results

Figure 3 compares the empirical (simulated) returns and error bounds ε^* (optimal objective value of the inner problem) across different policy classes. Figure 4 illustrates examples of policies learned in a typical run of CGPO. As hypothesized, CGPO can recover exact (s, S) control policies for Inventory in the PWS-S class, which together with S policies achieve the lowest error and best return across all policy classes. Interestingly, as Fig. 3 shows, C and PWS-C policies perform relatively poorly even for $K = 2$, but as expected, the error decreases as the expressivity of the policy class, and in particular the number of cases, increases. This is intuitive, as Fig. 4 shows, the reorder quantity is strongly linearly dependent on the current stock, which is not easily represented as PWS-C. In constrast, on Reservoir the class of PWS-C policies perform comparatively well for $K \geq 1$, and achieve (near)-optimality despite requiring a larger number of iterations. As expected, policies typically release more water as the water level approaches the upper target bound. Finally, the optimal policy for VTOL applies a large upward force when the angle or angular velocity are negative, with relative importance being placed on angle. As Fig. 4 (right) shows, the force is generally greater when either the angle is below the target angle or the angular velocity is negative, and equilibrium is achieved by applying a modest upward force between the initial angle (0.1) and the target angle (0).

Figure 5 and 6 provide examples of policies learned in other expressivity classes for Inventory and Reservoir problems, respectively. In all cases, policies remain intuitive and easy to explain even as the complexity increases. For instance, PWS2-C policies for Inventory order more inventory as the stock level decreases, while for Reservoir, the amount of water released is increasing in the water level.

$$order = 3 \qquad order = 4 - stock \qquad order = \begin{cases} 8 & \text{if } -13 \leq stock \leq -2 \\ 2 & \text{otherwise} \end{cases} \qquad order = \begin{cases} 6 & \text{if } -10 \leq stock \leq -1 \\ 1 & \text{if } 2 \leq stock \leq 2 \\ 3 & \text{otherwise} \end{cases}$$

Fig. 5. Inventory: Examples of C, S, PWS1-C and PWS2-C (left to right) policies computed by CGPO.

$$release_1 = 19.73$$
$$release_2 = 55.87$$

$$release_1 = \begin{cases} 1.39 & \text{if } 30.62 \leq level_1 \leq 68.66 \\ 40.71 & \text{if } 70.93 \leq level_1 \leq 100 \\ 10.622 & \text{otherwise} \end{cases}$$

$$release_2 = \begin{cases} 80.78 & \text{if } 159.93 \leq level_2 \leq 200 \\ 38.69 & \text{if } 144.21 \leq level_2 \leq 159.11 \\ 18.25 & \text{otherwise} \end{cases}$$

$$release_1 = level_1 - 20$$
$$release_2 = 0.93 \times level_2 - 29.74$$

$$release_1 = \begin{cases} level_1 - 50 & \text{if } 20 \leq level_1 \leq 100 \\ 2 \times level_1 - 100 & \text{otherwise} \end{cases}$$

$$release_2 = \begin{cases} level - 2 - 37.8 & \text{if } 38.9 \leq level_2 \leq 200 \\ 1.17 \times level_2 - 235.5 & \text{otherwise} \end{cases}$$

Fig. 6. Reservoir: Examples of C, S, PWS2-C and PS1-S (left to right, top to bottom) policies computed by CGPO.

4.3 Ablations

Worst-Case Analysis. Figure 7 plots the state trajectory, actions and rainfall that lead to worst-case performance of C and S policies at the last iteration of CGPO on Reservoir. Here, we observe that the worst-case performance for the optimal C policy occurs when the rainfall is low and both reservoirs become empty. Given that the cost of overflow exceeds underflow, this is expected as the C policy must release enough water to prevent high-cost overflow events during high rainfall at the risk of water shortages during droughts. In contrast, the optimal S policy maintains safe water levels even in the worst-case scenario, since it avoids underflow and overflow events by releasing water (linearly) proportional to the water level. Similar to Reservoir, a non-trivial worst-case scenario is identified by CGPO for Inventory (Fig. 8) for optimal C policies but not S policies. This scenario corresponds to very high demand that exceeds the constant reorder quantity, causing stock-outs. For VTOL (Fig. 9), there is no worst-case scenario with non-zero cost upon convergence at the final iteration of CGPO (bottom row). We have also shown the worst-case scenario computed at the first iteration prior to achieving policy optimality, where a worst-case scenario corresponds to no force being applied causing the system to eventually lose balance.

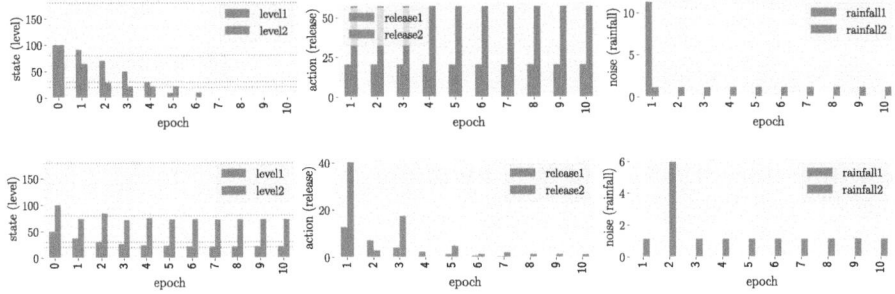

Fig. 7. Reservoir: Worst-case state trajectory (left), actions (middle), and noise (right) for C (top) and S policies (bottom) computed by CGPO. The more expressive S policy provides much more stable water levels.

Analysis of Problem Size. To understand how the problem sizes in CGPO scale across differing expressivity classes of policies and dynamics, Table 2 shows the number of decision variables and constraints in the MIPs at the last iteration of constraint generation. For VTOL with a quadratic (Q) policy, the number of variables in the inner and outer problems are $0/0/212$ and $0/0/512$, respectively, while the number of constraints are $114/18/72$ and $270/95/155$. The number of decision variables/constraints are fixed in the inner problem and grow linearly with iterations in the outer problem. The number of variables and constraints do not grow significantly with the policy class expressiveness. Please note that we do not report runtimes due to the highly stochastic and unpredictable nature of the runs.

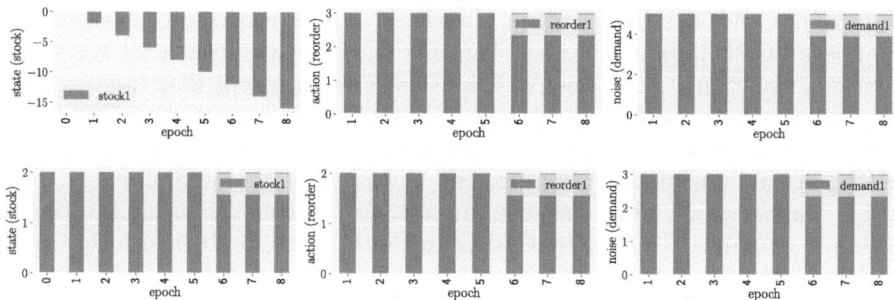

Fig. 8. Inventory: Worst-case state trajectory (left), actions (middle), and disturbances/noise (right) for C (top) and S policies (bottom) computed by CGPO.

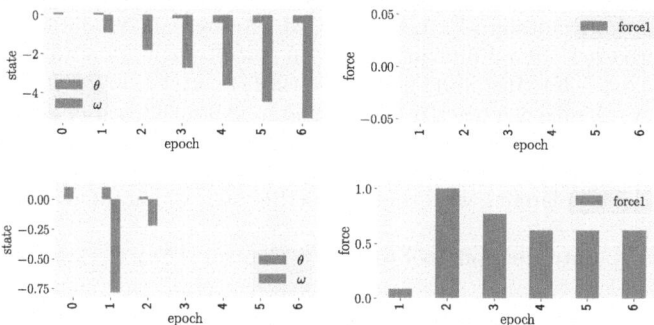

Fig. 9. VTOL: Worst-case state trajectory (top), and actions (bottom) for Q policies. First (last) column corresponds to the first (last) iterations of CGPO.

Table 2. Number of binary/integer/real-valued decision variables in the inner and outer optimization problems at the last iteration of CGPO (top half), and the number of linear/quadratic/general constraints (bottom half).

	Inventory		Reservoir	
Policy	Inner	Outer	Inner	Outer
C	8/121/72	96/673/385	20/0/822	300/0/5823
PW-S	32/129/72	384/772/387	80/0/842	1200/0/7133
C	160/0/64	780/0/480	540/0/320	4035/0/2400
PWS-S	168/0/112	709/0/168/1056	560/0/440	3797/540/4200

5 Conclusion

We presented CGPO, a novel bilevel mixed-integer programming formulation for DC-MDP policy optimization in predefined expressivity classes with bounded optimality guarantees. **To the best of our knowledge, CGPO is the first algorithm capable of providing bounded error guarantees for a broad**

range of DC-MDPs and policy classes. We used constraint generation to decompose CGPO into an inner problem that produces an adversarial worst-case constraint violation (i.e., trajectory), and an outer problem that improves the policy w.r.t. these trajectories. Across diverse domains and policy classes, we showed that the learned policies (some provably optimal) and their worst-case performance/failure modes were easy to interpret, while maintaining a manageable number of variables and constraints. To overcome the significant computational demands required by CGPO to tackle larger problems, future work should leverage tools from reinforcement learning to approximately solve the inner/outer problems.

References

1. Ahmed, A., Varakantham, P., Lowalekar, M., Adulyasak, Y., Jaillet, P.: Sampling based approaches for minimizing regret in uncertain markov decision processes (mdps). J. Artif. Int. Res. **59**(1), 229–264 (2017)
2. Albert, L.A.: A mixed-integer programming model for identifying intuitive ambulance dispatching policies. J. Oper. Res. Soc., 1–12 (2022)
3. Ariu, K., Fang, C., da Silva Arantes, M., Toledo, C., Williams, B.C.: Chance-constrained path planning with continuous time safety guarantees. In: AAAI Workshop (2017)
4. Åström, K.J.: Introduction to Stochastic Control Theory. Courier Corporation (2012)
5. Blankenship, J.W., Falk, J.E.: Infinitely constrained optimization problems. J. Optim. Theory Appl. **19**, 261–281 (1976)
6. Bueno, T.P., de Barros, L.N., Mauá, D.D., Sanner, S.: Deep reactive policies for planning in stochastic nonlinear domains. In: AAAI, vol. 33, pp. 7530–7537 (2019)
7. Castro, P.M.: Tightening piecewise mccormick relaxations for bilinear problems. Comput. Chem. Eng. **72**, 300–311 (2015)
8. Chembu, A., Sanner, S., Khurram, H., Kumar, A.: Scalable and globally optimal generalized l1 k-center clustering via constraint generation in mixed integer linear programming. In: AAAI, vol. 37, pp. 7015–7023 (2023)
9. Corso, A., Moss, R., Koren, M., Lee, R., Kochenderfer, M.: A survey of algorithms for black-box safety validation of cyber-physical systems. J. Artif. Intell. Res. **72**, 377–428 (2021)
10. Dolgov, D., Durfee, E.: Stationary deterministic policies for constrained mdps with multiple rewards, costs, and discount factors. In: Proceedings of the 19th International Joint Conference on Artificial Intelligence. IJCAI'05, pp. 1326–1331. Morgan Kaufmann Publishers Inc., San Francisco (2005)
11. Farias, V.F., Van Roy, B.: Tetris: a study of randomized constraint sampling, pp. 189–201. Springer, London (2006). https://doi.org/10.1007/1-84628-095-8_6
12. Farina, M., Giulioni, L., Scattolini, R.: Stochastic linear model predictive control with chance constraints-a review. J. Process Control **44**, 53–67 (2016)
13. Hauskrecht, M.: Approximate linear programming for solving hybrid factored mdps. In: International Symposium on Artificial Intelligence and Mathematics, AI&Math 2006, Fort Lauderdale, Florida, USA, 4–6 January 2006 (2006)
14. Low, S.M., Kumar, A., Sanner, S.: Sample-efficient iterative lower bound optimization of deep reactive policies for planning in continuous mdps. In: AAAI, vol. 36, pp. 9840–9848 (2022)

15. Morrison, D.R., Jacobson, S.H., Sauppe, J.J., Sewell, E.C.: Branch-and-bound algorithms: a survey of recent advances in searching, branching, and pruning. Disc. Optim. **19**, 79–102 (2016)
16. Ono, M., Pavone, M., Kuwata, Y., Balaram, J.: Chance-constrained dynamic programming with application to risk-aware robotic space exploration. Auton. Robot. **39**, 555–571 (2015)
17. Patton, N., Jeong, J., Gimelfarb, M., Sanner, S.: A distributional framework for risk-sensitive end-to-end planning in continuous mdps. In: AAAI, vol. 36, pp. 9894–9901 (2022)
18. Puterman, M.L.: Markov Decision Processes: Discrete Stochastic Dynamic Programming. John Wiley & Sons, Hoboken (2014)
19. Raghavan, A., Sanner, S., Khardon, R., Tadepalli, P., Fern, A.: Hindsight optimization for hybrid state and action mdps. In: AAAI, vol. 31 (2017)
20. Scala, E., Haslum, P., Thiébaux, S., Ramirez, M.: Interval-based relaxation for general numeric planning. In: ECAI, pp. 655–663. IOS Press (2016)
21. Scarf, H., Arrow, K., Karlin, S., Suppes, P.: The optimality of (s, s) policies in the dynamic inventory problem. In: Optimal Pricing, Inflation, and the Cost of Price Adjustment, pp. 49–56 (1960)
22. Schulman, J., Wolski, F., Dhariwal, P., Radford, A., Klimov, O.: Proximal policy optimization algorithms. ArXiv arxiv:1707.06347 (2017)
23. Schuurmans, D., Patrascu, R.: Direct value-approximation for factored MDPs. In: Dietterich, T., Becker, S., Ghahramani, Z. (eds.) Advances in Neural Information Processing Systems, vol. 14. MIT Press (2001)
24. Świechowski, M., Godlewski, K., Sawicki, B., Mańdziuk, J.: Monte carlo tree search: a review of recent modifications and applications. Artif. Intell. Rev. **56**(3), 2497–2562 (2023)
25. Topin, N., Milani, S., Fang, F., Veloso, M.: Iterative bounding mdps: learning interpretable policies via non-interpretable methods. In: AAAI, vol. 35, pp. 9923–9931 (2021)
26. Vos, D., Verwer, S.: Optimal decision tree policies for markov decision processes. In: IJCAI, pp. 5457–5465 (2023)
27. Wang, Y., et al.: A survey on deploying mobile deep learning applications: a systemic and technical perspective. Digital Commun. Netw. **8**(1), 1–17 (2022)
28. Wu, G., Say, B., Sanner, S.: Scalable planning with tensorflow for hybrid nonlinear domains. In: NeurIPS, vol. 30 (2017)

Minimising Source-Plate Swaps for Robotised Compound Dispensing in Microplates

Ramiz Gindullin[1(✉)], María Andreína Francisco Rodríguez[1], Brinton Seashore-Ludlow[2], and Ola Spjuth[3,4]

[1] Department of Information Technology, Uppsala University, Uppsala, Sweden
{Ramiz.Gindullin,Maria.Andreina.Francisco}@it.uu.se
[2] Department of Oncology-Pathology and Science for Life Laboratory, Karolinska Institutet, Stockholm, Sweden
brinton.seashore-ludlow@ki.se
[3] Department of Pharmaceutical Biosciences and Science for Life Laboratory, Uppsala University, Uppsala, Sweden
ola.spjuth@uu.se
[4] Phenaros Pharmaceuticals AB, Uppsala, Sweden

Abstract. Liquid-handling instruments are indispensable tools in modern biomedical laboratories, streamlining compound and sample management tasks with precision and efficiency. Compound dispensing from large chemical libraries divided over hundreds of microwell plates can require substantial swapping of plates, particularly if compounds from multiple source plates are to be dispensed in each destination plate. Despite robotisation, plate swapping is a time-consuming necessity for high-throughput experiments, posing a significant bottleneck. In this paper, we explore the application of constraint programming (CP) to the planning of liquid-handling tasks to minimise plate swaps in automated dispensing. We formulate the problem as a combination of a set partitioning problem and the construction of a bipartite network. We present six CP models implemented in MiniZinc and evaluate their performance on synthetic benchmarks using three state-of-the-art constraint solvers. This work highlights the potential of CP to enhance the scalability and efficiency of automated compound transfer systems.

Keywords: Constraint Programming · Experiment Design · Planning

1 Introduction

High-throughput technologies have transformed biomedical research and enabled large-scale exploration of chemical and biological space, facilitating the discovery of new drugs, the identification of therapeutic targets, and the detailed characterisation of complex biological systems [1]. Central to these workflows are liquid-handling systems, which ensure the precise and efficient transfer of compounds

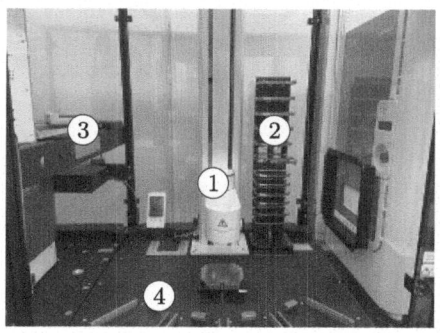 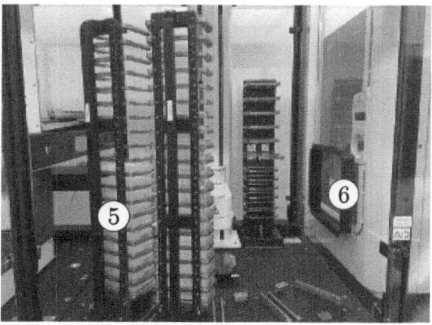

Fig. 1. Photos of the physical automated dispensing solution without (left) and with (right) plate hotels in place, where 1) the robotic arm, 2) the plate deliddler, 3) the sealer, 4) base for plate hotels, 5) plate hotels and 6) the dispenser.

and samples to individual wells in microplates [24,25]. Microplates, also known as microwell plates or simply 'plates', are flat, multi-well plates widely used in biological and biochemical assays, providing a standardised format for high-throughput experimentation by enabling parallel processing of samples or reactions. Microplates are available in various standardised sizes, typically featuring 6, 24, 96, 384, or 1536 wells. Each well holds a separate experiment where, commonly, multiple reagents and samples are intermixed. For large-scale biomedical experiments, robotised equipment is commonly used to manage the processing of multiple plates. This typically involves a robot arm to move plates between different positions, such as plate hotels holding multiple plates, plate sealers, plate delidders, incubators, microscopes, etc. (see Fig. 1).

Compound dispensing from large chemical libraries constitutes an early and important step in drug discovery [27]. The objective is to set up experiments on multiple microplates which involve taking selected compounds from a relevant *source plate* and dispensing them in the right quantities to a *destination plate* (see Fig. 2). This process can require extensive swapping of source and destination plates in and out of the dispensing instrument when the compounds are located in different source plates or even in different chemical libraries, or when dispensing multiple compounds in the same well for combination studies [21].

Because of a large number of swaps, the dispensing process requires running time measured in days or even weeks. It renders overly complex experimental designs infeasible and necessitates simplifications. Adding further constraints, humans in the loop are not uncommon, depending on the automation system, e.g., they can compensate for limited intermediate plate storage. Another practical consideration is that plates holding compound stocks require a lid or seal to avoid evaporation. The lid needs to be removed for the dispensing and added back afterwards. The time without a lid should be as short as possible, especially for the source plates with compound stocks in higher volumes which are meant

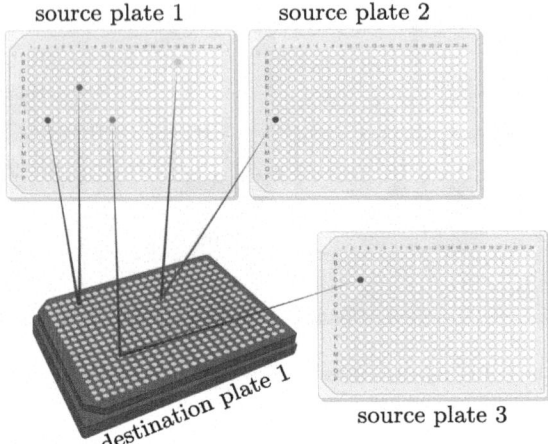

Fig. 2. Dispensing drugs from source plates to the wells of a destination plate for conducting experiments.

to generate 100 s to 1000 s of wells in destination plates. This is important for expensive or difficult-to-synthesise or difficult-to-acquire compounds.

As a practical example, let us consider a realistic research with 10 destination plates and cherry-picked compounds from 50 source plates. In the best-case scenario, where each source plate needs to be accessed only once, the plate-swapping part of the whole dispensing process would take a bit over 1 h. However, if every destination plate requires access to every source plate, the plate swapping would already take more than 8.5 h. This is problematic not only because of the people who need to be present to start the process and store all the plates in the end, but also because the time that microplates are exposed without a lid is significantly larger. Combinatorial optimisation can be used to address this problem. It has been used to model and solve problems in scientific discovery and experimental design [9,14].

In this paper, we propose a CP approach to tackle the problem of minimising the total number of plate swaps given a list of *experiments*, defined as compounds or combinations thereof, that need to be tested, and the source plate where each compound is located. In Sect. 2, we describe the problem in detail and present its formalisation. Section 3 provides several CP models based on the proposed formalisation. These CP models are evaluated on different solvers in Sect. 4. Finally, we present our conclusions and future work in Sect. 5.

2 The Microplate Swapping Problem

This section first describes the dispensing procedure and then describes and formalises all aspects of the problem: allocation of the experiments between destination plates and the construction of the sequence with the minimal number of swaps and input data with a running example.

2.1 Problem Description

Given a list of experiments to be tested, the problem consists of assigning every experiment to a destination plate to perform the following procedure:

1. An empty destination plate is placed in the dispensing robot.
2. An appropriate source plate is placed in the dispensing robot.
3. The robot dispenses compounds from wells of the source plate to the assigned wells of the destination plate.
4. (a) If the destination plate has all its wells filled with the planned experiments (i.e. it is serviced), then:
      i. if all destination plates are serviced, terminate the process,
      ii. else, return to step 1 to swap to the next empty destination plate.
   (b) Otherwise, return to step 2 (swap the source plate).

We need to plan (i) the allocation of experiments across destination plates in sequence, i.e. destination plate 1, destination plate 2, etc., and (ii) the sequence of placements of source plates in the dispensing robot. The goal is to minimise the number of plate swaps, i.e. the number of times steps 1 and 2 are performed. Note that we do not have control over the content of the source plates.

An alternative strategy is to switch steps 1 and 2, i.e. service source plates in sequence instead. In this paper, we assume that the swap time between destination plates and the swap time between source plates is equal.

2.2 Allocation of Experiments Between destination plates

The allocation of experiments between destination plates is a set partitioning problem (SPP) [17] with the number of swaps used as the minimisation criteria. SPP is often a crucial aspect of scheduling problems widely reported in the literature since 1973 [8]. A recent paper [19] applies several integer and CP models of SPPs to flight-gate schedules. Assignment problems, e.g. a team assignment problem [6], are also formulated as SPPs. CP has been successful in solving SPPs either on its own [15] or in combination with linear programming [7]. Some works use a solution to a different problem to calculate the criteria for the SPP, e.g. a vehicle routing problem [2]. CP can be applied to clusterings [16], specifically constraint clustering [5,12,29], which can be seen as a generalisation of SPPs.

2.3 Calculating the Number of Microplate Swaps

The problem's objective is to find the shortest feasible plate-swap sequence. Each swap in the dispensing machine is done with the goal of transferring compounds from a given source plate to a given destination plate, i.e. these plates are connected. Thus, we can imagine the problem as a bipartite graph: all the source plates on one side, and all the destination plates on the other. An arc denotes the transfer of materials from a source plate to a destination plate (see

Fig. 3). Bipartite graphs are widely used in various applications to describe e.g. social, ecological, biomedical, biomolecular, epidemiological networks [20]. Bipartite graphs are also used in scheduling problems [3,26].

If we consider a small example with only 2 source plates, S_1 and S_2, and 1 destination plate D_1 (see Fig. 3A) with 2 arcs connecting each source plate to D_1. A swap sequence $\mathcal{L}$ then contains only 1 swap between 2 source plates, either $\mathcal{L} = [\langle S_1, D_1 \rangle, \langle S_2, D_1 \rangle]$ or $\mathcal{L} = [\langle S_1, D_1 \rangle, \langle S_2, D_1 \rangle]$.

Consider another example with 4 source plates, 2 destination plates and 4 arcs (see Fig. 3B). An optimal swap sequence $\mathcal{L}$ has 3 swaps, e.g. $\mathcal{L} = [\langle S_1, D_1 \rangle, \langle S_2, D_1 \rangle, \langle S_3, D_2 \rangle, \langle S_4, D_2 \rangle]$. The length of sequence $\mathcal{L}$ is equal to the number of arcs. The number of swaps $\mathcal{K}$ within the sequence $\mathcal{L}$ is calculated as:

$$\mathcal{K} = \sum_{i=1}^{|\mathcal{L}|-1} (S_{\mathcal{L}_i} \neq S_{\mathcal{L}_{i+1}}) + \sum_{i=1}^{|\mathcal{L}|-1} (D_{\mathcal{L}_i} \neq D_{\mathcal{L}_{i+1}}) \qquad (1)$$

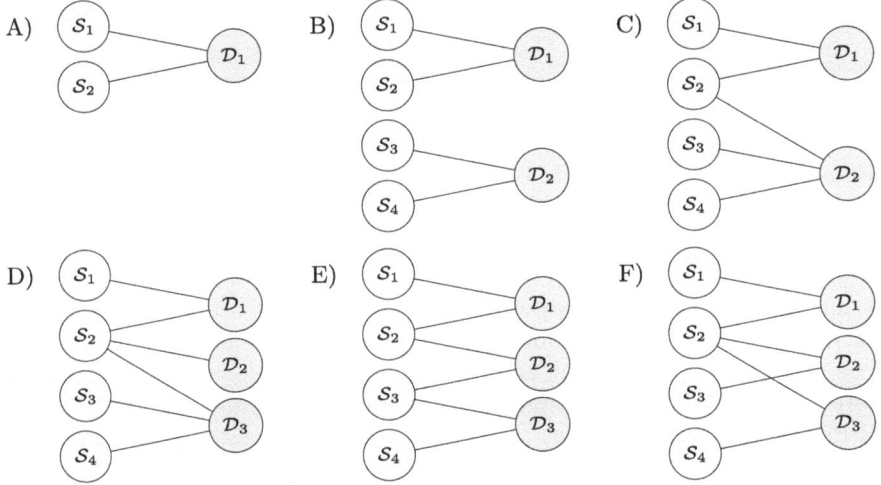

Fig. 3. Examples of bipartite graphs in relation to plate swaps

From the examples in Fig. 3A and Fig. 3B, we deduce that for graphs where the destination plates have no source plate in common, the number of swaps is equal to the number of arcs plus the number of destination plates minus 2, e.g. if we have 1 destination plate, 1 source plate and 1 arc connecting them then, by this formula, there are 0 swaps. Otherwise, it is possible to reduce the number of swaps by reordering the arcs of some bipartite graph: if 2 subsequent destination plates are connected to the same source plate, then this source plate can be placed last when the preceding destination plate is serviced by the pipetting machine. We swap the destination plate but not the source plate – it is already installed. See the graph of Fig. 3C with 5 arcs and 3 destination plates, i.e. the number of swaps would be equal to $\mathcal{K} = 5 + 2 - 2 = 5$. Destination plates $\mathcal{D}_1$ and $\mathcal{D}_2$ are connected to the same source plate $\mathcal{S}_3$, meaning that we can construct sequence $\mathcal{L} = [\langle\mathcal{S}_1, \mathcal{D}_1\rangle, \langle\mathcal{S}_2, \mathcal{D}_1\rangle, \langle\mathcal{S}_2, \mathcal{D}_2\rangle, \langle\mathcal{S}_3, \mathcal{D}_2\rangle, \langle\mathcal{S}_4, \mathcal{D}_2\rangle]$ with $\mathcal{K} = 4$ swaps instead.

In this paper, we define a *valid transition* as a source plate that connects, as a between node, two subsequent destination plates. The bipartite network in Fig. 3B has no valid transitions. The bipartite network in Fig. 3C uses source plate $\mathcal{S}_1$ as a valid transition between the destination plates $\mathcal{D}_1$ and $\mathcal{D}_2$. A valid transition can reduce the number of swaps by 1. We consider two scenarios: a destination plate connected to only 1 or to multiple source plates. If a destination plate has only 1 source plate connected to it, e.g. destination plate $\mathcal{D}_2$ in Fig. 3D, then the number of swaps is minimal when both preceding and succeeding destination plates have a connection to the same source plate. e.g. there is no need to swap the source plate $\mathcal{S}_2$ while switching between destination plates $\mathcal{D}_1$, $\mathcal{D}_2$ and $\mathcal{D}_3$. This results in sequence $\mathcal{L} = [\langle\mathcal{S}_1, \mathcal{D}_1\rangle, \langle\mathcal{S}_2, \mathcal{D}_1\rangle, \langle\mathcal{S}_2, \mathcal{D}_2\rangle, \langle\mathcal{S}_2, \mathcal{D}_3\rangle, \langle\mathcal{S}_3, \mathcal{D}_3\rangle, \langle\mathcal{S}_4, \mathcal{D}_3\rangle]$ with $\mathcal{K} = 5$ swaps. If a preceding and/or succeeding destination plates are not connected to the same source plate then the number of valid transitions is reduced by 1, if there is only 1 valid transition, or not reduced, if there are no valid transitions.

If a destination plate is connected to 2 or more source plates, in the optimal case, it has different source plates used as valid transitions with the preceding and the succeeding destination plates. Destination plate $\mathcal{D}_2$ of the bipartite graph in the Fig. 3E has valid transitions both from $\mathcal{D}_1$ and to $\mathcal{D}_3$ using source plates $\mathcal{S}_2$ and $\mathcal{S}_3$ resp., and thus we can construct the sequence $\mathcal{L} = [\langle\mathcal{S}_1, \mathcal{D}_1\rangle, \langle\mathcal{S}_2, \mathcal{D}_1\rangle, \langle\mathcal{S}_2, \mathcal{D}_2\rangle, \langle\mathcal{S}_3, \mathcal{D}_2\rangle, \langle\mathcal{S}_3, \mathcal{D}_3\rangle, \langle\mathcal{S}_4, \mathcal{D}_3\rangle]$ with $\mathcal{K} = 5$ swaps.

Otherwise, only 1 valid transition is possible. e.g. for the bipartite graph in Fig. 3F destination plate $\mathcal{D}_2$ uses source plate $\mathcal{S}_2$ as valid transitions both from $\mathcal{D}_1$ and to $\mathcal{D}_3$. The optimal sequence is then $\mathcal{L} = [\langle\mathcal{S}_1, \mathcal{D}_1\rangle, \langle\mathcal{S}_2, \mathcal{D}_1\rangle, \langle\mathcal{S}_2, \mathcal{D}_2\rangle, \langle\mathcal{S}_3, \mathcal{D}_2\rangle, \langle\mathcal{S}_2, \mathcal{D}_3\rangle, \langle\mathcal{S}_4, \mathcal{D}_3\rangle]$ with $\mathcal{K} = 6$ swaps.

As discussed in Sect. 2.1 we can build swap sequences either by fully servicing destination plates one by one in sequence or by fully servicing source plates one by one. The description above explains the construction of a swap sequence in accordance with the former. To achieve the latter, the sequence construction is the same, with a possible difference that the order of source plates can differ from the default $\mathcal{S}_1, \mathcal{S}_2, \ldots, \mathcal{S}_s$. The question, then, is whether or not one strategy is better than another. Currently, we have not been able to find an example, either manually or by enumerating, where an optimal swap sequence produced by a different strategy has a different number of swaps. Thus, we concluded that, in most cases, if we construct an optimal swap sequence with one approach, we can construct a sequence with the same number of swaps with another strategy.

The choice between the strategies depends on additional criteria, e.g. swap times between destination and source plates. In this paper, we assume that sequences with the same number of swaps incur same costs. This assumption allows us to model only one strategy – servicing destination plates in sequence.

In conclusion, the initial problem of minimising microplate swaps can be reformulated as a combination of three interlinked sub-problems:

- SPP to allocate experiments between destination plates,
- building a bipartite network corresponding to the allocated experiments,
- constructing a sequence of bipartite arcs to minimise the number of swaps.

All of these problems must be solved together simultaneously to get a sequence with the minimal number of swaps. Note that this structure allows for the development of heuristics which can solve each sub-problem separately, step by step. In this paper we construct CP models to find exact solutions – many constraints for building bipartite networks and search for valid transitions are more naturally expressed in CP rather than MILP or SAT.

2.4 The Input Data

According to Sects. 1 and 2.1 we know:

- m, the number of drug compounds,
- n, the number of drug concentrations,
- s, the number of source plates,
- d, the number of destination plates,
- w, the number of wells in a single destination plate,
- c, the number of experiments, i.e. compound combinations, that must fill the destination plates, $c \leq d \cdot w$,
- tensor $\mathcal{S}$, where $\mathcal{S}_{i,j,k} \geq 0, i \in [1,s], j \in [1,m], k \in [1,n]$ denotes the amount of wells in source plate i that contain compound j in concentration k.
- tensor $\mathcal{C}$, where $\mathcal{C}_{i,j,k} \geq 0, i \in [1,c], j \in [1,m], k \in [1,n]$ denotes the amount of of compound j in concentration k which is present in experiment i. This allows to use not just compound pairs, but also single compounds, compound triplets, quadruplets, etc. For purposes of this paper we assume that we only need compound pairs, i.e. $\mathcal{C}_{i,j,k} \in \{0,1\}$ and $\sum_{j=1}^{m} \sum_{k=1}^{n} \mathcal{C}_{i,j,k} = 2, i \in [1,c]$.

If only one source plate contains any given compound at any given concentration we transform tensors $\mathcal{S}$ and $\mathcal{C}$ into matrix $\mathcal{C}^1$, where $\mathcal{C}^1_{i,j} \in \{\texttt{false}, \texttt{true}\}$, $i \in [1,c], j \in [1,s]$ denotes if experiment i uses materials of source plate j or not.

2.5 The Running Example

We first introduce a small example of a problem represented in accordance with Sect. 2.4. In it we have $s = 4$ source plates which contain $m = 4$ compounds, $\{A, B, C, D\}$, in $n = 2$ concentrations each, $\{c_1, c_2\}$. The placement of compounds on source plates is stored in the tensor $\mathcal{S}$ in Table 1.

To denote combinations of compounds to test we fill the tensor $\mathcal{C}$ in Table 2 where, e.g., experiment $i = 1$ denotes that we want to combine compound $j = 1$ in concentration $k = 1$ with compound $j = 2$ in concentration $k = 1$.

Table 1. Tensor $\mathcal{S}$ to denote the distribution of compounds in various concentrations between source plates for the running example.

	Compounds:	$j=1$		$j=2$		$j=3$		$j=4$	
	Concentrations:	$k=1$	$k=2$	$k=1$	$k=2$	$k=1$	$k=2$	$k=1$	$k=2$
	$i=1$	6	0	6	0	0	0	0	0
Source	$i=2$	0	0	0	0	6	0	6	0
plates:	$i=3$	0	6	0	6	0	0	0	0
	$i=4$	0	0	0	0	0	6	0	6

Table 2. Tensor $\mathcal{C}$ to denote experiments of compounds in various concentrations that are required to be synthesised for the running example, where 0 denotes `false` and 1 denotes `true`.

	Compounds:	$j=1$		$j=2$		$j=3$		$j=4$	
	Concentrations:	$k=1$	$k=2$	$k=1$	$k=2$	$k=1$	$k=2$	$k=1$	$k=2$
	$i=1$	1	0	1	0	0	0	0	0
	$i=2$	1	0	0	1	0	0	0	0
	$i=3$	1	0	0	0	1	0	0	0
	$i=4$	1	0	0	0	0	1	0	0
	$i=5$	1	0	0	0	0	0	1	0
	$i=6$	1	0	0	0	0	0	0	1
	$i=7$	0	1	1	0	0	0	0	0
	$i=8$	0	1	0	1	0	0	0	0
	$i=9$	0	1	0	0	1	0	0	0
	$i=10$	0	1	0	0	0	1	0	0
	$i=11$	0	1	0	0	0	0	1	0
Experiments	$i=12$	0	1	0	0	0	0	0	1
	$i=13$	0	0	1	0	1	0	0	0
	$i=14$	0	0	1	0	0	1	0	0
	$i=15$	0	0	1	0	0	0	1	0
	$i=16$	0	0	1	0	0	0	0	1
	$i=17$	0	0	0	1	1	0	0	0
	$i=18$	0	0	0	1	0	1	0	0
	$i=19$	0	0	0	1	0	0	1	0
	$i=20$	0	0	0	1	0	0	0	1
	$i=21$	0	0	0	0	1	0	1	0
	$i=22$	0	0	0	0	1	0	0	1
	$i=23$	0	0	0	0	0	1	1	0
	$i=24$	0	0	0	0	0	1	0	1

Tensor $\mathcal{S}$ in Table 1 indicates that each concentration of each compound is provided only by a single source plate, i.e. we can transform tensors $\mathcal{S}$ and $\mathcal{C}$ into matrix $\mathcal{C}^1$ (see Table 3).

For the purposes of the example we assume that we need to fill $d = 6$ destination plates $w = 4$ wells each.

Table 3. Matrix $\mathcal{C}^1$ to denote which source plates are required for any given experiment, where 0 denotes `false` and 1 denotes `true`.

Source plates	destination plates, $i \in [1, 24]$
$j = 1$	1 1 1 1 1 1 1 0 0 0 0 0 1 1 1 1 0 0 0 0 0 0 0 0
$j = 2$	0 0 1 0 1 0 0 0 1 0 1 0 1 0 1 0 1 0 0 0 1 1 1 0
$j = 3$	0 1 0 0 0 0 1 1 1 1 1 1 0 0 0 0 1 1 1 1 0 0 0 0
$j = 4$	0 0 0 1 0 1 0 0 0 1 0 1 0 1 0 1 0 1 0 1 0 1 1 1

3 Constraint Models for the Simplified Problem

A model that finds the shortest swap sequence must cover the following aspects:

1. allocation of experiments between destination plates, stored in *destination plates placement variables*,
2. calculate *material transfers variables*, i.e. transfers of materials between source and destination plates which (i) satisfy requirements of experiments on a given destination plate and (ii) within capacities of source plates,
3. record an arc in *bipartite network variables* between a source and a destination plates if there is at least 1 transfer of materials between them,
4. find *valid transitions variables* and a number of swaps in a bipartite network.

In practice, the most common case is that for any given compound in any given concentration there is only one source plate that contains it. In this case, we omit transfer variables as they would be calculated automatically: if an experiment placed on a given destination plate uses a material that comes from only 1 source plate then there is an arc in the bipartite graph between these destination and source plates. We can consider a model that covers all 4 aspects, i.e. a *full model*, or a model that omits transfer variables, i.e. a *simplified model*.

In this section, we provide a number of simplified models which are evaluated in Sect. 4. We focus on simplified models because they are very common in real-life applications. Additionally, simplified models can be used to quickly assess the performance of an available solver on a given hardware.

3.1 Destination plates placement variables

Destination plates placement variables are used for solving the SPP. There are several ways to represent partitioned sets in CP [13].

Representation I – Boolean matrix: variables $P_{i,j}^I \in \{\texttt{false}, \texttt{true}\}, i \in [1,d], j \in [1,c]$, where $P_{i,j}^I = 1$ iff experiment j is placed on destination plate i. Each experiment is placed only on one destination plate:

$$\sum_{i=1}^{d} P_{i,j}^I = 1, \ j \in [1,c] \tag{2}$$

The number of experiments placed on a single destination plate must not exceed its maximum size w:

$$\sum_{j=1}^{c} P_{i,j}^I \leq w, \ i \in [1,d] \tag{3}$$

Representation II – integer array: variables $P_i^{II} \in \{1,d\}, i \in [1,c]$. Variable P_i^{II} denotes that experiment i if placed on destination plate P_i^{II}. The number of experiments placed on a single destination plate must not exceed its maximum size w. This is expressed by the following constraint:

$$\text{GLOBAL\_CARDINALITY} \left(\begin{array}{c} \langle P_1^{II}, P_2^{II}, \ldots, P_c^{II} \rangle, \\ \langle 1:w, \ 2:w, \ \ldots, \ d:w \rangle \end{array} \right) \tag{4}$$

Representation III – array of sets of integers: sets $P_i^{III} = \{j \mid \exists j \in [1,c]\}$, $i \in [1,d]$. Each set P_i^{III} contains all the experiments j placed on destination plate i. The number of experiments placed on a single destination plate must not exceed its maximum size w:

$$|P_i^{III}| \leq w, i \in [1,d] \tag{5}$$

Each experiment must be placed on one and only one destination plate. This is expressed by the next PARTITION_SET [22] constraint:

$$\text{PARTITION\_SET} \left(\langle P_1^{III}, P_2^{III}, \ldots, P_d^{III} \rangle, \ \langle 1, 2, \ \ldots, \ c \rangle \right) \tag{6}$$

3.2 Bipartite Network Variables

The Bipartite network variables are derived from the allocation of experiments between destination plates.

Representation A – Boolean matrix (adjacency matrix): the variables $\mathcal{A}_{i,j}^A \in \{\texttt{false}, \texttt{true}\}, i \in [1,s], j \in [1,d]$ track the arcs of the bipartite graph, where $\mathcal{A}_{i,j} = \texttt{true}$ iff there is at least one experiment on destination plate j which uses a material from source plate i.

Constraints 7, 8 and 9 connect variables $\mathcal{A}^A$ with placement variables P^I, P^{II} and P^{III}, respectively:

$$\mathcal{A}^A_{i,j} \Leftrightarrow \bigvee_{k=1}^{c} \left(\mathcal{C}^1_{k,i} \wedge P^I_{j,k}\right), \ i \in [1,s], j \in [1,d] \tag{7}$$

$$\mathcal{A}^A_{i,j} \Leftrightarrow \bigvee_{k=1}^{c} \left(\mathcal{C}^1_{k,i} \wedge (P^{II}_k = j)\right), \ i \in [1,s], j \in [1,d] \tag{8}$$

$$\mathcal{A}^A_{i,j} \Leftrightarrow \bigvee_{k \in P^{III}_j} \mathcal{C}^1_{k,i}, \ i \in [1,s], j \in [1,d] \tag{9}$$

Representation B – array of sets of integers (adjacency list): sets $\mathcal{A}^B_i = \{j \mid \exists j \in [1,s]\}, i \in [1,d]$. Each set $\mathcal{A}^B_i$ contains all source plates j from which at least one experiment placed on destination plate i uses a material.

Constraints 10, 11 and 12 connect variables $\mathcal{A}^B$ with placement variables P^I, P^{II} and P^{III}, respectively:

$$\mathcal{A}^B_i = \{j \mid j \in [1,s], \bigvee_{k=1}^{c} \left(\mathcal{C}^1_{k,i} \wedge P^I_{j,k}\right)\}, \ i \in [1,d] \tag{10}$$

$$\mathcal{A}^B_i = \{j \mid j \in [1,s], \bigvee_{k=1}^{c} \left(\mathcal{C}^1_{k,i} \wedge (P^{II}_k = j)\right)\}, \ i \in [1,d] \tag{11}$$

$$\mathcal{A}^B_i = \{j \mid j \in [1,s], \bigvee_{k \in P^{III}_j} \mathcal{C}^1_{k,i}\}, \ i \in [1,d] \tag{12}$$

3.3 Valid Transitions Variables

The variables $\mathcal{N}_i \in [0,s]$, for $i \in [1,d]$ track the number of arcs connected to a destination plate i. To track valid transitions we introduce array of potential valid transitions $\mathcal{T}$, where $\mathcal{T}_i \in [0,s]$, for $i \in [1,d-1]$. If $\mathcal{T}_i = 0$ then there is no valid transition between destination plates i and $i+1$. Otherwise, source plate $\mathcal{T}_i > 0$ becomes a valid transition between destination plates i and $i+1$.

Constraints 13 and 14 connect variables $\mathcal{N}$ and $\mathcal{T}$ with variables $\mathcal{A}^A$:

$$\mathcal{N}_i = \sum_{j=1}^{s} \mathcal{A}^A_{j,i}, \ i \in [1,d] \tag{13}$$

$$\mathcal{T}_i > 0 \Rightarrow \left(\mathcal{A}^A_{\mathcal{T}_i,i} = 1 \wedge \mathcal{A}^A_{\mathcal{T}_i,i+1} = 1\right), \ i \in [1,d-1] \tag{14}$$

Constraints 15 and 16 connect variables $\mathcal{N}$ and $\mathcal{T}$ with variables $\mathcal{A}^B$:

$$\mathcal{N}_i = |\mathcal{A}^B_i|, \ i \in [1,d] \tag{15}$$

$$\mathcal{T}_i > 0 \Rightarrow \left(\mathcal{T}_i \in \mathcal{A}^B_i \wedge \mathcal{T}_i \in \mathcal{A}^B_{i+1}\right), \ i \in [1,d-1] \tag{16}$$

In accordance with Sect. 2.3, subsequent destination plates i, $i+1$ and $i+2$ can share same valid transition, i.e. $\mathcal{T}_i = \mathcal{T}_{i+1} > 0$, iif destination plate $i+1$ connected to only 1 source plate:

$$(\mathcal{T}_i > 0 \wedge \mathcal{N}_{i+1} > 1) \Rightarrow \mathcal{T}_i \neq \mathcal{T}_{i+1}, \quad i \in [1, d-2] \tag{17}$$

Finally, to calculate the number of swaps $\mathcal{K}$ in the sequence $\mathcal{L}$:

$$\mathcal{K} = d - 2 + \sum_{i=1}^{d} \mathcal{N}_i - \sum_{i=1}^{d-1} (\mathcal{T}_i > 0), \quad \text{minimize } \mathcal{K} \tag{18}$$

3.4 Interpreting the Results of Constraint Model

The optimal solution for the running example from Sect. 2.5 is a swap sequence with $\mathcal{K} = 12$ swaps. For simplicity, we omit the target placement variables and only provide the bipartite network $\mathcal{A}^A$ with valid-transition variables $\mathcal{T}$:

$$\mathcal{A}^A = \begin{matrix} i=1 \\ i=2 \\ i=3 \\ i=4 \end{matrix} \begin{pmatrix} \begin{matrix} j=1 & j=2 & j=3 & j=4 & j=5 & j=6 \end{matrix} \\ \textbf{true} & \text{false} & \text{false} & \textbf{true} & \textbf{true} & \text{false} \\ \text{false} & \text{false} & \textbf{true} & \text{true} & \textbf{true} & \textbf{true} \\ \text{false} & \textbf{true} & \text{true} & \textbf{true} & \text{false} & \text{false} \\ \textbf{true} & \text{true} & \text{false} & \text{false} & \text{false} & \textbf{true} \end{pmatrix} \tag{19}$$

$$\mathcal{T} = [\,4,\quad 3,\quad 2,\quad 1,\quad 2\,] \tag{20}$$

Matrix $\mathcal{A}^A$ corresponds to the bipartite graph in Fig. 4 with 13 arcs. For clarity, all valid transitions of $\mathcal{T}$ use the same corresponding colours in the graph.

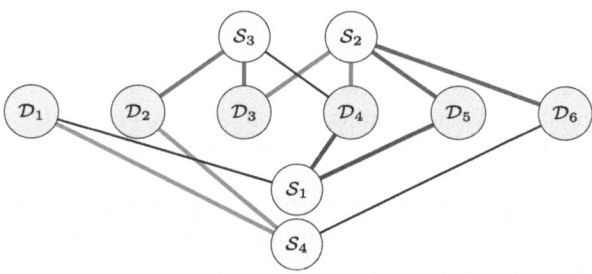

Fig. 4. Bipartite graph corresponding to the solution in Eqs. 19 and 20 for the running example in Sect. 2.5.

We start with an empty swap sequence $\mathcal{L}$. We take the destination plate 1, i.e. column $j = 1$ in matrix $\mathcal{A}^A$. We note that destination plate $\mathcal{D}_1$ is connected to $\mathcal{S}_1$ and $\mathcal{S}_4$. One of them, $\mathcal{S}_4$, is used as a valid transition to destination plate $\mathcal{D}_2$.

Thus, we first put all the source plates which are connected to the destination plate $\mathcal{D}_1$, in a default order, except the plate that is a valid connection into the swap sequence $\mathcal{L}$. In this case, we add plate $\mathcal{S}_1$ to the sequence. Only after that, we add the source plate acting as the valid transition to the next destination plate, i.e. $\mathcal{S}_4$. Thus, we get the sequence $\mathcal{L} = [\langle \mathcal{S}_1, \mathcal{D}_1 \rangle, \langle \mathcal{S}_4, \mathcal{D}_1 \rangle]$.

On destination plate $\mathcal{D}_2$, i.e. column $j = 2$ in matrix $\mathcal{A}^A$, we have two connected source plates, one of which, $\mathcal{S}_4$ is currently the last plate in the sequence, thus it already services destination plate $\mathcal{D}_2$. Lastly, we add the remaining source plate $\mathcal{S}_3$, which is also a valid transition to the next destination plate $\mathcal{D}_3$.

We apply same logic for destination plate $\mathcal{D}_3$ after which we get the sequence $\mathcal{L} = [\langle \mathcal{S}_1, \mathcal{D}_1 \rangle, \langle \mathcal{S}_4, \mathcal{D}_1 \rangle, \langle \mathcal{S}_4, \mathcal{D}_2 \rangle, \langle \mathcal{S}_3, \mathcal{D}_2 \rangle, \langle \mathcal{S}_3, \mathcal{D}_3 \rangle, \langle \mathcal{S}_2, \mathcal{D}_3 \rangle]$.

For the destination plate $\mathcal{D}_4$, i.e. column $j = 4$ in matrix $\mathcal{A}^A$, we note that it is connected to source plate $\mathcal{S}_1$, which is a valid transition to plate $\mathcal{D}_5$, source plate $\mathcal{S}_3$ which has no transition to either destination plates $\mathcal{D}_3$ and $\mathcal{D}_5$, and source plate $\mathcal{S}_2$, which is both a valid transition from destination plate $\mathcal{D}_3$ and the last source plate in the sequence $\mathcal{L}$. Thus, we first add to the sequence $\mathcal{L}$ all remaining source plates connected to destination plate $\mathcal{D}_4$ that are not a valid transitions to the subsequent destination plate $\mathcal{D}_5$, i.e. we add source plate $\mathcal{S}_3$. Lastly, we add the valid transition $\mathcal{S}_1$ from $\mathcal{D}_4$ to $\mathcal{D}_5$.

The same logic is applied to the remaining destination plates, and we finish with the swap sequence $\mathcal{L} = [\langle \mathcal{S}_1, \mathcal{D}_1 \rangle, \langle \mathcal{S}_4, \mathcal{D}_1 \rangle, \langle \mathcal{S}_4, \mathcal{D}_2 \rangle, \langle \mathcal{S}_3, \mathcal{D}_2 \rangle, \langle \mathcal{S}_3, \mathcal{D}_3 \rangle, \langle \mathcal{S}_2, \mathcal{D}_3 \rangle, \langle \mathcal{S}_2, \mathcal{D}_4 \rangle, \langle \mathcal{S}_3, \mathcal{D}_4 \rangle, \langle \mathcal{S}_1, \mathcal{D}_4 \rangle, \langle \mathcal{S}_1, \mathcal{D}_5 \rangle, \langle \mathcal{S}_2, \mathcal{D}_5 \rangle, \langle \mathcal{S}_2, \mathcal{D}_6 \rangle, \langle \mathcal{S}_4, \mathcal{D}_6 \rangle]$ with $\mathcal{K} = 6 - 2 + 13 - 5 = 12$ swaps.

3.5 Six Constraint Programming Models

With 3 representations for destination plate placement variables and 2 representations for bipartite network variables, we can construct 6 candidate CP models:

- constraint model I-A: constraints 2, 3, 7, 13, 14, 17 and 18;
- constraint model I-B: constraints 2, 3, 10, 15, 16, 17 and 18;
- constraint model II-A: constraints 4, 8, 13, 14, 17 and 18;
- constraint model II-B: constraints 4, 11, 15, 16, 17 and 18;
- constraint model III-A: constraints 5, 6, 9, 13, 14, 17 and 18;
- constraint model III-B: constraints 5, 6, 12, 15, 16, 17 and 18.

We explored the inclusion of some symmetry-breaking constraints. Because preliminary tests showed no consistent improvement in the performance of any of the models we did not present them in the paper.

4 Evaluation

In this section, we compare the performance of the 6 models in various environments. For this reason, we are not interested in comparing the performance of the solvers with each other, as some solvers can be configured to utilise all available computational resources while others cannot.

4.1 Test-Dataset Generation

In most cases, experiments are allocated and sorted by hand or modified to lower the maximum number of swaps.

The example that inspired this work had 3748 combinations from 30 available source plates to be distributed between 12 destination plates. A greedy search provided a solution with approximately 270 swaps. While a proof of optimality was not found, we obtained a solution with 114 swaps after running the CP-SAT solver on a model on 8 threads for 10 h. We considered using it as a benchmark, e.g. by executing models multiple times and comparing the average number of swaps. Ultimately, we decided that a dedicated dataset with multiple smaller, diverse examples would measure model performance better and allow comparison with future models.

We generated a synthetic dataset for the simplified problem consisting of 10 random instances written in MiniZinc [23] for each number of destination plates $d \in \{2, 5, 10, 15\}$, number of wells in a single destination plate $w \in \{2, 4, 6, 8\}$ and number of source plates $s \in \{2, 5, 10\}$. The total number of experiments is $t = d \cdot w$, where each experiment uses materials from either 1 or 2 source plates. The total number of instances is 480 divided into the following 3 categories – 180 small instances ($d \cdot w \cdot s \leq 80$), 160 medium ($81 < d \cdot w \cdot s \leq 280$) and 140 large instances ($d \cdot w \cdot s > 280$).

4.2 Computational Experiments

We compare the performance of 6 models of Sect. 3.5 written in MiniZinc on 3 solvers – Chuffed 0.13.2 [4], CP-SAT 9.11.4210 [11] and Gurobi 11.0.3 [18]. We configured the Chuffed solver to run on 1 CPU thread, CP-SAT and Gurobi solvers – on 4 CPU threads. We set a 300-second timeout for every test. In practice, letting real instances run for longer if needed is possible.

The evaluation is performed on a 2024 MacBook Air with 8 cores Apple M3 and 16Gb. The dataset, model files, solver configurations, evaluation results and instructions on how to perform the evaluation are available on [10].

4.3 Results and Discussion

For each test we record the following statistics (see Table 4):

- N_{to}, number of instances where the timeout is reached;
- N_{ns}, number of instances with no solution found within the time limit;
- N_b, number of instances when the model produced the best result, i.e. it found the optimal solution with the fastest proof or found a solution with the least number of swaps within the time limit. If two or more models are tied, we count all of them as best;
- T_p, average execution time per instance, including timeouts, in seconds.

Table 4. Performance of 6 CP models on the dataset with different solvers

Solver	Model	small				medium				large			
		N_{to}	N_{ns}	N_b	T_p	N_{to}	N_{ns}	N_b	T_p	N_{to}	N_{ns}	N_b	T_p
Chuffed	I-A	0	0	**78**	0.5	7	0	**100**	16.9	93	0	**93**	210.1
	I-B	0	0	24	0.5	7	0	19	16.3	**92**	0	70	**205.0**
	II-A	0	0	77	0.5	6	0	45	17.9	98	0	36	220.1
	II-B	0	0	17	0.5	5	0	20	**15.8**	98	0	47	216.7
	III-A	0	0	22	0.5	11	0	13	28.0	111	0	30	243.6
	III-B	0	0	14	0.6	12	0	6	28.1	111	0	28	244.8
CP-SAT	I-A	0	0	**76**	0.4	6	6	**75**	14.4	83	83	**19**	189.4
	I-B	0	0	44	0.4	6	6	28	15.2	84	84	7	188.4
	II-A	0	0	36	0.4	6	6	30	14.5	83	83	15	**187.4**
	II-B	0	0	16	0.4	5	5	11	14.0	**82**	**82**	7	187.7
	III-A	0	0	43	0.4	5	5	20	**13.9**	83	83	8	188.4
	III-B	0	0	18	0.5	6	6	2	15.1	83	83	2	189.5
Gurobi	I-A	0	0	**75**	3.2	17	0	**73**	42.5	106	0	73	244.6
	I-B	0	0	23	3.2	18	0	37	43.9	106	0	74	244.8
	II-A	0	0	60	**3.1**	17	0	48	44.8	105	0	70	244.3
	II-B	0	0	18	3.3	**16**	0	23	**42.5**	105	0	81	243.3
	III-A	0	0	46	3.2	17	0	40	43.4	105	0	80	**241.5**
	III-B	0	0	20	3.4	17	0	25	44.8	106	0	73	244.4

For Chuffed, model I-A is often faster and, when the timeout is reached, produces solutions with significantly fewer swaps than other models, e.g. on one instance, 33 swaps are reached by model I-A and 54 swaps – by model II-B.

CP-SAT tends to perform worse with the increased complexity of the instances. On 60 large instances, it does not produce an initial solution within a time limit with any of the models. Models I-A and II-A perform marginally better on large instances. Model I-A performs the best across small and medium instances. Execution time does not differ between models.

For Gurobi, model I-A dominates other models on small and medium instances. On large instances, models II-B and III-A demonstrate a slightly better performance. Execution time does not differ between models.

When the timeout is reached, for CP-SAT and Gurobi we observed no significant spread between a worst solution and a best solution, i.e. the choice of a model does not significantly improve or degrade the solution quality.

Representation III of placement variables leads to a disproportional number of created variables and constraints on all solvers. e.g. for an instance with $d = 15$, $w = 8$, and $s = 10$, Chuffed creates 3480 variables and 28560 constraints related to the PARTITION_SET constraint, CP-SAT – 3480 and 28560, and Gurobi – 1560 and 17520. Despite it being the heaviest representation in terms of constraints, on Gurobi, model III-A performs almost as well as model II-B, which creates only 15 constraints related to GLOBAL_CARDINALITY constraint.

5 Conclusion

We investigated the application of CP for minimising plate swaps when setting up high-throughput experiments in biomedical laboratories. We formulated the problem as a combination of an SPP and the construction of a bipartite network between source and destination plates. In the most common scenario, a compound is stored in only 1 source plate. We constructed 6 CP models to solve this use case. The models use bipartite graphs to construct swapping sequences which is, to the knowledge of the authors, a novel technique. It could be applied in other areas, e.g. packing and scheduling when multiple containers arrive at a facility, and their contents must be mixed and repackaged for multiple destinations.

We evaluated 6 CP models on a synthetic dataset. The achieved performance is adequate in practice. The dispensing procedure is time-consuming and expensive; thus, the goal of the problem is to get an optimal or near-optimal solution in a reasonable time, e.g. within a day. In the example with 3748 combinations discussed in Sect. 4.1, a 270-swap solution means the dispensing robot works for about 2 weeks. The 114-swap solution reduces the dispensing process to less than a week, i.e. acquiring a near-optimal solution in 10 h is a net gain in total time. Access to a dispensing robot is a more expensive and scarce resource than a computer for running the models. CP models can be run unsupervised, while the dispensing process requires human intervention and must happen during work hours with specific personnel available.

Several promising research opportunities arise from this work: 1. designing and evaluating models for the full problem, 2. exploration of symmetry-breaking constraints for existing models, 3. implementing a system for an automatic selection of the best model and solver, similar to that of [28].

Acknowledgements. We thank Aleksandr Karakulev and Prashant Singh for the discussions.

References

1. Blay, V., Tolani, B., Ho, S.P., Arkin, M.R.: High-throughput screening: today's biochemical and cell-based approaches. Drug Discov. Today **25**(10), 1807–1821 (2020)
2. Bourreau, E., Garaix, T., Gondran, M., Lacomme, P., Tchernev, N.: A constraint-programming based decomposition method for the generalised workforce scheduling and routing problem (GWSRP). Int. J. Prod. Res. **60**(4), 1265–1283 (2022)
3. Cardinaels, E., Borst, S., van Leeuwaarden, J.S.: Redundancy scheduling with locally stable compatibility graphs. arXiv preprint arXiv:2005.14566 (2020)
4. Chu, G., Stuckey, P.J., Schutt, A., Ehlers, T., Gange, G., Francis, K.: Chuffed, a lazy clause generation solver. https://github.com/chuffed/chuffed. Accessed 04 Dec 2024
5. Dao, T., Duong, K.C., Vrain, C.: Constrained clustering by constraint programming. Artif. Intell. **244**, 70–94 (2017)

6. Daş, G.S., Altınkaynak, B., Göçken, T., Türker, A.K.: A set partitioning based goal programming model for the team formation problem. Int. Trans. Oper. Res. **29**(1), 301–322 (2022)
7. De Silva, A.: Combining constraint programming and linear programming on an example of bus driver scheduling. Ann. Oper. Res. **108**(1), 277–291 (2001)
8. Ernst, E.A., Lasdon, L.S., Ostrander, L.E., Divell, S.S.: Anesthesiologist scheduling using a set partitioning algorithm. Comput. Biomed. Res. **6**(6), 561–569 (1973)
9. Francisco Rodríguez, M.A., Puigvert, J.C., Spjuth, O.: Designing microplate layouts using artificial intelligence. Artif. Intell. Life Sci. **3**, 100073 (2023)
10. Gindullin, R., Francisco Rodríguez, M.A.: Minimising source-plate swaps. https://github.com/astra-uu-se/multiplates. Accessed 04 Dec 2024
11. Google: CP-sat solver. https://developers.google.com/optimization/cp/cp_solver. Accessed 04 Dec 2024
12. Guilbert, M., Vrain, C., Dao, T.B.H.: Towards explainable clustering: a constrained declarative based approach. arXiv preprint arXiv:2403.18101 (2024)
13. Jefferson, C., Frisch, A.M.: Representations of sets and multisets in constraint programming. In: Proceedings of the 4th International Workshop on Modelling and Reformulating Constraint Satisfaction Problems, pp. 102–116. Citeseer (2005)
14. Kell, D.B.: Scientific discovery as a combinatorial optimisation problem: how best to navigate the landscape of possible experiments? BioEssays **34**(3), 236–244 (2012)
15. Laborie, P., Rogerie, J., Shaw, P., Vilím, P.: IBM ILOG CP optimizer for scheduling: 20+ years of scheduling with constraints at IBM/ILOG. Constraints **23**, 210–250 (2018)
16. Mariz, J.L.V., Peroni, R.D.L., Silva, R.M.D.A.: A constraint programming approach to solve the clustering problem in open-pit mine planning. REM-Int. Eng. J. **77**(2), e230060 (2024)
17. Müller, T.: Solving set partitioning problems with constraint programming. In: Proceedings of PAPPACT 1998, pp. 313–332 (1998)
18. Gurobi optimizer. https://www.gurobi.com/solutions/gurobi-optimizer/. Accessed 04 Dec 2024
19. Ornek, M.A., Ozturk, C., Sugut, I.: Integer and constraint programming model formulations for flight-gate assignment problem. Oper. Res. Int. J. **22**(1), 135–163 (2022)
20. Pavlopoulos, G.A., Kontou, P.I., Pavlopoulou, A., Bouyioukos, C., Markou, E., Bagos, P.G.: Bipartite graphs in systems biology and medicine: a survey of methods and applications. GigaScience **7**(4), giy014 (2018)
21. Rietdijk, J., Aggarwal, T., Georgieva, P., Lapins, M., Carreras-Puigvert, J., Spjuth, O.: Morphological profiling of environmental chemicals enables efficient and untargeted exploration of combination effects. Sci. Total Environ. **832**, 155058 (2022)
22. Stuckey, P.J., Marriott, K., Tack, G.: Set-related constraints (2020). https://docs.minizinc.dev/en/2.8.7/lib-globals-set.html. Accessed 13 Nov 2024
23. Minizinc. https://www.minizinc.org/. Accessed 04 Dec 2024
24. Tegally, H., San, J.E., Giandhari, J., de Oliveira, T.: Unlocking the efficiency of genomics laboratories with robotic liquid-handling. BMC Genomics **21**, 1–15 (2020)
25. Torres-Acosta, M.A., Lye, G.J., Dikicioglu, D.: Automated liquid-handling operations for robust, resilient, and efficient bio-based laboratory practices. Biochem. Eng. J. **188**, 108713 (2022)
26. Weng, W., Zhou, X., Srikant, R.: Optimal load balancing in bipartite graphs. arXiv preprint arXiv:2008.08830 (2020)

27. Yasi, E.A., Kruyer, N.S., Peralta-Yahya, P.: Advances in g protein-coupled receptor high-throughput screening. Curr. Opin. Biotechnol. **64**, 210–217 (2020)
28. Zhang, Z., Xiao, C., Wang, S., Yu, W., Bai, Y.: Automatic constraint programming solver selection method based on machine learning for the cable tree wiring problem. J. Eng. **2024**(3), e12368 (2024)
29. Zuenko, A., Fridman, O., Zuenko, O., Zhuravleva, O.: An approach to solution of constrained clustering problems using the constraint programming paradigm and the multiset theory. In: Journal of Physics: Conference Series, vol. 1801, pp. 012–041. IOP Publishing (2021)

Author Index

A
Akgün, Özgür I-152
Artigues, Christian II-173
Assaf, George I-1
Avgerinos, Ioannis I-17

B
Baier, Hendrik I-86
Balogh, Andrea I-34
Barrass, Rosemary I-52
Barrault, Romain I-68
Beck, J. Christopher II-137
Begnardi, Luca I-86
Beldiceanu, Nicloas II-155
Benslimane, Wyame I-103
Besis, Apostolos I-17
Bierlee, Hendrik I-113, II-1
Boyer, Romain II-173

C
Caceres, Rajmonda S. II-191
Cai, Junyang I-134
Camps, Frédéric II-173
Cappart, Quentin II-103, II-252
Catanzaro, Daniele II-70
Chang, Mun See I-152
Chmiela, Antonia II-218
Codish, Michael I-169
Coffrin, Carleton I-52
Coppé, Vianney II-70

D
Danzinger, Philipp I-188
Davies, Toby O. I-205
Dekker, Jip J. I-113
Desaulniers, Guy II-103
Didier, Frédéric I-205
Dilkina, Bistra I-134
Douence, Rémi II-155

E
Eliassi-Rad, Tina II-191
Escamocher, Guillaume I-34

F
Farges, Jean-Loup II-235
Ferrarini, Luca I-222
Francisco Rodríguez, María Andreína I-256
Frank, Jeremy II-86

G
Garnier, Philippe II-173
Geibinger, Tobias I-188
Gent, Ian P. I-152
Gimelfarb, Michael I-239
Gindullin, Ramiz I-256, II-155
Gjergji, Ida II-1
Gleixner, Ambros II-35
Gomes, Carla P. II-18
Greenstreet, Laura II-18
Grigas, Paul I-103
Grimson, Marc II-18
Gualandi, Stefano I-222

H
Hebrard, Emmanuel II-173
Hoen, Alexander II-35
Hofstedt, Petra I-1
Huang, Taoan I-134

J
Jacquet, Thomas II-103
Janota, Mikoláš I-169
Jefferson, Christopher I-152

K
Kalyanakrishnan, Shivaram II-119
Kletzander, Lucas II-1
Koch, Thorsten II-218
Kuroiwa, Ryo II-137

L

Lawless, Connor II-51
Legrand, Emma II-70
Levinson, Richard II-86
Li, Yingxi II-51
Lodi, Andrea II-18
Löffler, Sven I-1
Lombardi, Michele II-209
Lopez, Pierre II-173

M

Miller, Benjamin A. II-191
Mischek, Florian I-188
Montanaro, Yoann Sabatier II-103
Moro, Letizia I-222
Mourtos, Ioannis I-17
Mukherjee, Dibyangshu II-119
Musliu, Nysret I-188, II-1

N

Nagarajan, Harsha I-52
Narita, Minori II-137
Ngouonou, Jovial Cheukam II-155

O

O'Sullivan, Barry I-34

P

Parmentier, Axel I-222
Pavero, Philippe II-235
Pereira, Mickaël II-173
Perron, Laurent I-205
Pesant, Gilles II-252
Picard, Gauthier I-68, II-235
Pralet, Cédric I-68
Psarros, Athanasios I-17

Q

Quimper, Claude-Guy II-155

R

Ravindra, Vinay II-86

Rouzot, Julien II-173

S

Saephan, Meghan II-86
Sanner, Scott I-239
Sawyer, Eric I-68
Schaus, Pierre II-70
Seashore-Ludlow, Brinton I-256
Sethi, Suresh A. II-18
Shafi, Zohair II-191
Shi, Qinru II-18
Shmoys, David B. II-18
Signorelli, Gaetano II-209
Simon, Franz W. II-18
Spjuth, Ola I-256
Stuckey, Peter J. I-113, I-205, II-1

T

Taitler, Ayal I-239
Turner, Mark II-218

U

Udell, Madeleine II-51

V

van Jaarsveld, Willem I-86
Vatikiotis, Stavros I-17
Vitercik, Ellen II-51
von Meijenfeldt, Bart I-86

W

Wikum, Anders II-51
Willot, Henoïk II-235
Winkler, Michael II-218

Y

Yin, Chao II-252

Z

Zhang, Yingqian I-86
Zois, Georgios I-17

The manufacturer's authorised representative in the EU is Springer Nature Customer Service Centre GmbH, Europaplatz 3, 69115 Heidelberg, Germany. If you have any concerns regarding our products, please contact ProductSafety@springernature.com

Printed and bound by CPI Group (UK) Ltd, Croydon, CR0 4YY

26/03/2026

02078976-0003